Naomi Kato, Shinji Kamimura (Eds.)

Bio-mechanisms of Swimming and Flying

Fluid Dynamics, Biomimetic Robots, and Sports S[illegible]

Naomi Kato, Shinji Kamimura (Eds.)

Bio-mechanisms of Swimming and Flying

Fluid Dynamics, Biomimetic Robots, and Sports Science

With 303 Figures, Including 18 in Color

Springer

Naomi Kato, Ph.D.
Department of Naval Architecture and Ocean Engineering
Graduate School of Engineering, Osaka University
2-1 Yamadaoka, Suita, Osaka 565-0871, Japan

Shinji Kamimura, Ph.D.
Division of Biology, Department of Life Sciences
Graduate School of Arts and Sciences, The University of Tokyo
3-8-1 Komaba, Meguro, Tokyo 153-8902, Japan

ISBN978-4-431-99829-7 e-ISBN 978-4-431-73380-5

Springer is a part of Springer Science+Business Media
springer.com

Printed in Japan

Printed on acid-free paper

Preface

Tens of thousands of animals are living on this planet. They have survived for hundreds of millions of years through adaptation and evolution and are living with a variety of structures and functions in various circumstances in this world. Biomechanics studies of animals' swimming and flying motions can play an increasingly larger role in understanding the mechanisms that enable animals to move effectively and efficiently in fluid, as well as analyzing the characteristics of their various forms of behavior in fluid. Based on such understanding and analysis, we can develop environment-friendly devices that simulate the natural movements and responses of biological creatures in fluid.

This book is a collection of peer-reviewed papers originally submitted for the Proceedings of the 3rd International Symposium on Aero Aqua Bio-Mechanisms (ISABMEC 2006) held July 3–7, 2006, in Okinawa, Japan. ISABMEC 2006 was organized by The Study Group of Aero Aqua Bio-Mechanisms (ABMECH) in Japan, seeking to provide a common forum for interdisciplinary discussions between biologists and engineers on natural autonomous systems and locomotion mechanisms, particularly swimming and flying, in animals ranging from flagellate microorganisms to marine mammals. This book follows *Bio-mechanisms of Animals in Swimming and Flying* that was published by Springer in 2004 as a collection of papers originally submitted for the Proceedings of the 2nd International Symposium on Aqua Bio-Mechanisms (ISABMEC 2003).

In 1997, ABMECH was established in Japan. The goals of the group are, first, to understand the mechanisms that enable flying and swimming animals to move effectively and efficiently in their environments, and second, to reveal the characteristics of their structures, functions, and behaviors. Based on this information we plan to develop novel machines and vehicles that mimic or simulate natural motions in environmentally benign ways. To encourage broad participation in these interdisciplinary fields of basic science and applied engineering, ABMECH organized the 1st and the 2nd International Symposia on Aqua Bio-Mechanisms in Honolulu in 2000 and 2003, respectively, and the 3rd International Symposium on Aero Aqua Bio-Mechanisms in Okinawa in 2006. "Aqua Bio-Mechanisms" was renamed "Aero Aqua Bio-Mechanism" in the hopes of further expanding our group to include those studying various types of organisms and mechanical devices under aerodynamic circumstances.

This book, with 31 chapters, covers the following four major topics: 1) the biological aspects of locomotive mechanisms and behaviors of animals while swimming and flying, ranging from microorganisms to dolphins; 2) the hydrodynamics of swimming and flying; 3) biomimetic swimming or flying robots; and 4) sports science.

The first topic includes 11 chapters. Chapters 1 to 4 deal with microorganisms, Chapters 5 to 10 with fishes, and Chapter 11 with flying insects. The book begins with observations and numerical analysis of asymmetric swimming motion of singly flagellated bacteria near a rigid surface. Chapter 2 deals with rheological and diffusive properties of a suspension of microorganisms both experimentally and numerically following interactions between two swimming microorganisms. Chapter 3 discusses dynamics modeling and real-time observation of galvanotaxis in microorganisms. Chapter 4 describes a phototaxis-based motion control system for *Euglena* by using a blue laser and galvano scanners. Motion of a huge group of *Euglena* was controlled to manipulate objects. Chapter 5 focuses on passive mechanisms of posture and trajectory control in swimming fishes, showing comparative studies of swimming biomechanics and hydrodynamics, based on observations of tetraodontiform fishes. Chapter 6 investigates the effect of caudal fin behavior resulting from a caudal skeletal structure on propulsive performance, considering the interaction between the mechanical properties on the caudal fin and the hydrodynamic forces acting on the caudal fin. Chapter 7 describes the analysis of artificial tail flukes for a bottlenose dolphin that lost most or part of its tail flukes by disease, and the field tests that showed the dolphin was able to swim as before with the help of the artificial tail flukes. Chapter 8 investigates drag reduction of an angled wavy silicon-rubber in turbulent flow by experiments showing the time-averaged wall shear stress is lower than that in the case of the flat plate. Chapter 9 reviews the neuronal activities of the nucleus of the medial longitudinal fasciculus (Nflm) during swimming, the connections between Nflm neurons and neurons in spinal CPGs, and the cytoarchitecture of Nflm neurons, including their shapes, numbers, and distributions. It also attempts to identify the cerebellar efferent neurons and their transmitters and relationships with Purkinje cells using retrograde labeling and immunohistochemical techniques. Chapter 10 analyzes the biosonar of porpoises by a miniature stereo acoustic data logger and demonstrates that these animals inspect the area ahead of them before swimming silently into it. Chapter 11 investigates the microstructure of honeybees to clarify the indirect flight mechanism from the viewpoint of the resonance model.

The second topic deals with hydrodynamics of swimming and flying, including three chapters on fishes and three on insects. Chapter 12 reviews some typical work, including measurement on kinematics of free-swimming fish and prediction of dynamics acting on an arbitrarily deformable body, numerical and experimental simulations of flow over flapping and traveling wavy bodies, and theoretical modeling and analysis on flow around a waving plate and a flapping plate. Chapter 13 deals with optimization of fish shape and swim mode in fully resolved two-dimensional (2-D) flow fields around the moving 2-D fish swimming by undulatory motion propagating from head to tail tip by genetic algorithm facilitated by an efficient prediction method of target functions. Chapter 14 presents a computational framework for the modeling, simulation, and evolutionary optimization of anguilliform swimmers. The hydrodynamics of anguilliform swimming was investigated by solving the 3-D Navier-Stokes equations of incompressible flow past a self-propelled body. An inverse design procedure was employed in order to identify optimal swimming behavior. Chapter 15 deals with Reynolds number dependence of unsteady aerodynamics in a flapping-flying insect by using a multi-block- and overset-grid-based CFD solver with special emphasis on the structure of leading-edge vortex and spanwise flow. Chapter 16 investigates the stabilization of flapping-of-wings flight of a butterfly through derivation of a dynamics model and simulator construction, experimental measurement of butterfly motions and aerodynamic forces, and discussions for the stable flapping-of-wings flight through simulations. Chapter 17 describes computational investigations on the unsteady aerodynamic performance of a swimming tuna with an oscillating caudal fin, the flapping flight of the fruit fly and a pectoral fin swimmer, the bird wrasse, to incorporate those into the design of unmanned underwater and unmanned air vehicles.

The third topic deals with biomimetic swimming robots and flying ones, and includes seven chapters on biomimetic swimming robots and two on biomimetic flying robots. Chapter 18 describes the design and kinematic simulations of a virtual fishlike robot actuated by a fish's superficial red muscle model. The physical and kinematic parameters of the specimen of Chinese sturgeon were examined to design the virtual fishlike robot. Chapter 19 shows the investigation of the propulsion mechanism of a fish both by computational fluid dynamics analysis and measurement of flow field by particle image velocimetry, and the development of fish robots powered by polymer electrolyte fuel cells (PEFCs) with the help of a genetic algorithm for better swimming performance. Chapter 20 describes the design for control of a biomimetic robot fish equipped with a caudal fin with one degree of freedom (DOF) as a main propulsor, a pair of pectoral fins with one DOF for each for auxiliary locomotion, and an artificial swim bladder for ascending and descending. Chapter 21 describes the development of a propulsion mechanism in fluid using a fin with a variable-bending-stiffness fin for which stiffness can be changed dynamically, and the thrust force characteristics. Chapter 22 deals with development of two types of elastic pectoral fins for biomimetic underwater vehicles, and the numerical and experimental analyses of those dynamic behaviors in air by FEM and in water, respectively. One is composed of silicone with three built-in FMAs (flexible micro actuators) and is actively controlled. The other is composed of silicone rubber with grooves and is passively controlled. Chapter 23 describes the design, construction, and testing of a biomimetic pectoral fin with actively controlled curvature for underwater vehicles. It also describes the design and development of a vehicle that incorporates the fins using computational fluid dynamics and the design of a control system for vertical plane motion. Chapter 24 deals with electronic nervous systems composed of analog and computed neurons and synapses for biomimetic robots based on neurobiological model systems, the lobster and the lamprey, instead of finite state systems based on the organizational principles of the animal model nervous systems. Chapters 25 and 26 deal with biomimetic flying robots. Chapter 25 describes the development of a micro wind energy converter using insect wings with comparison of the performance using artificial wings. Chapter 26 describes the development and flight of an insect-size clapping micro air vehicle (MAV), making use of the "Weis-Fogh clap and fling mechanism" that increases the efficiencies for generating the axial thrust and the lift of MAV.

The book moves on to the final topic to discuss the locomotive mechanisms in sports, with five chapters. Chapter 27 describes applications of a vortex element method providing a Lagrangian simulation of unsteady and vertical flows around a 100-m runner and a ski jumper. Chapter 28 shows computational fluid dynamics simulation of the rowing motion of a single scull comparing with in-situ experiments. Chapter 29 analyzes breast, back, and butterfly strokes using a swimming human simulation model from the viewpoint of propulsive efficiency. Chapter 30 experimentally investigates the effect of the palm of a hand on the characteristics of lift and drag forces in freestyle swimming. The results show that the characteristics of hand palms in swimming are similar to those of delta wings. The last chapter describes the development of a measurement system of flexural vibration of a ski jumper in flight, and the measurement of the vibration that is not subject to fluid dynamics simulation using rigid bodies.

We hope that this book will promote exchanges of ideas between biologists and engineers for further collaboration on natural autonomous systems and locomotion mechanisms of animals used in swimming and flying.

Naomi Kato
Shinji Kamimura
May 2007

Acknowledgments

The editors thank the following researchers for reviewing the chapters of this book. Each chapter was assessed by two reviewers.

Y. Doi, F. Fish, Y. Hagiwara, K. Isogai, S. Ito, A. Itoh, K.D. Jones, J.D. Eldredge, A.J. Ijspeert, K. Kamemoto, T. Kajishima, R. Kanzaki, Y. Kawamura, T. Kinoshita, S. Kobayashi, A. Konno, S. Kudo, G.V. Lauder, H. Liu, X.-Y. Lu, M.S. Gordon, T. Goto, Y. Magariyama, H. Miyake, H. Morikawa, T. Murayama, M. Nakashima, T. Nomura, Y. Oda, Y. Ohgi, H. Rikimaru, W. Sandberg, K. Senda, K. Seo, W. Shyy, S. Sudo, H. Suzuki, H. Takagi, Y. Toda, K. Uematsu, I. Yamamoto, and M. Watanabe.

Acknowledgments

[illegible]

[illegible]

Contents

Part 3: Biomimetic Swimming or Flying Robots

Part 4: Sports Science

Part I

Biological Aspects of Locomotive Mechanisms and Behaviors of Animals While Swimming and Flying

Chapter 1

Asymmetric Swimming Motion of Singly Flagellated Bacteria near a Rigid Surface

Tomonobu Goto[1], Seishi Kudo[2], and Yukio Magariyama[3]

[1]Department of Applied Mathematics and Physics, Faculty of Engineering, Tottori University, Koyama, Tottori 680-8552, Japan
[2]Department of Robotics and Bio-mechatronics, Faculty of Engineering, Toin University of Yokohama, Yokohama 225-8502, Japan
[3]Food Engineering Division, National Food Research Institute, Tsukuba 305-8642, Japan

Summary. This paper gives an overview of consecutive studies on the asymmetrical motion of *Vibrio alginolyticus* cells, which possess a single polar flagellum. Inertial forces are negligible because of the cell size and the motion is expected to be symmetrical. However, asymmetrical characteristics between forward and backward motions were observed. The asymmetry observed in trajectory, swimming speed, and residence time appears only when a cell swims close to a surface. In backward motion, a cell traces circular path, while in forward motion the cell moves in a straight line. The backward swimming speed is faster than the forward speed. Backward swimming cells tend to stay close to a surface longer than forward swimming cells do. An explanation for these asymmetrical characteristics is given based on the results of boundary element analyses of creeping flow around a cell model that consists of a cell body and a rotating flagellum. According to the explanation, the attitude of a cell relative to a surface produces the asymmetry. The studies presented here indicate that the fluid-dynamic interaction between bacterial cells and a surface produces the unexpected asymmetrical motion. This asymmetry may help cells search for preferable states on a surface or to attach to the surface.

Key words. Bacterial motion, Swimming speed, Trajectory, Asymmetry, Fluid-dynamic interaction, Boundary element analysis, Creeping flow

1 Introduction

Most bacteria possess the ability to swim in order to seek suitable circumstances for reproduction and to evade adversarial conditions. Bacterial cells are so small (on the order of 1 μm) that it might be assumed that the swimming motion of these cells is insignificant compared to the motion caused by the surrounding convective flow. Nevertheless, most bacteria possess organs for locomotion called flagella. Many other eukaryotic microorganisms also have locomotion organelles, e.g. cilia, flagella. This suggests that the ability to swim is essential, even for the microorganisms, because swimming ability was selected and developed over a long evolutionary process.

Figure 1 illustrates a schematic bacterial cell, which consists of two noticeable elements. The cell body is a capsule that contains genetic information and the chemical systems for reproduction, adaptation, locomotion, sensing, etc. The cell body is the core of the small creature. The other element is the flagellar filament, which is a helical thin filament made of protein. Unlike the flagella of eukaryotic cells that contain microtubules and bend by themselves, the flagella of bacterial cells do not deform actively. Instead, each filament is driven by a rotary motor (flagellar motor) embedded in the cell body. The rotation of the helical flagellum produces a helical wave within the neighboring fluid that exerts a reaction force on the filament in the direction opposite to the wave motion. Thus, the cell is propelled by the reaction force due to the rotation of the flagellum.

A flagellar motor can drive a flagellar filament in either a clockwise (CW) or counterclockwise (CCW) direction. Some bacterial cells such as *Vibrio alginolyticus*, which this paper specifically addresses, possess a single left-handed helix flagellum. When the motor rotates CCW, as seen from the distal end of the flagellum, the helical wave propagates from the proximal end to the distal end. As a result, the cell swims with the cell body preceding the flagellum, hereafter called "forward" motion. When the motor rotates CW, the cell swims in an orientation in which the flagellum pulls the cell body, which is hereafter called "backward" motion. Some bacterial cells like *Salmonella typhimurium* and *Escherichia coli* possess several flagella. Their motion differs from the singly flagellated bacterial cells and is usually referred to as "run" and "tumble." When the motors rotate CCW, the filaments form a helical-shaped bundle and the cell swims forward. However, these bacteria are incapable of backward motion. When the motors switch their rotational direction from CCW to CW, the inverted torque induces polymorphic transformations of the filaments, which result in the dissociation of the bundle. During the process, the cell changes its orientation due to fluid forces exerted on each filament. When the motors rotate CCW again, the cell swims in a different direction.

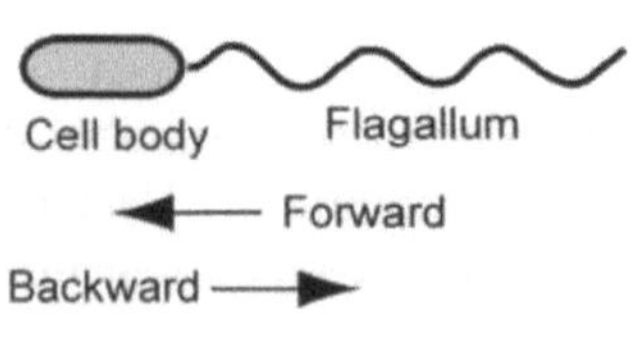

Fig. 1. Schematic drawing of a bacterial cell.

The size of a bacterial cell is so small that its swimming motion is governed by viscous forces rather than inertial forces. The Reynolds number, the ratio of the inertial forces to the viscous forces, is estimated to be 10^{-5}. If the motor of a singly flagellated bacterial cell changes its rotational direction from CCW to CW, the cell instantaneously switches its swimming direction. Furthermore, the forward and backward swimming is expected to be symmetrical. In a highly viscous fluid, the flow is reversible; the direction of flow cannot be distinguished by only observing the stream lines. If an object moves in one direction and is then moved back along the same path in the opposite direction to the previous starting position, the total momentum imparted to the neighboring fluid will be zero and the object will not gain any net momentum. The actively bending flagella of eukaryotic cells change their shapes between effective strokes and recovery strokes to generate a net momentum. By contrast, the filament of a singly flagellated bacterial cell does not change its shape between forward and backward motions. Therefore, the cell is expected to follow the same path as long as there are no disturbances like Brownian motion.

Despite this assumption, previous experimental works have revealed that there are asymmetrical characteristics between forward and backward motions of singly flagellated bacterial cells (Magariyama et al. 2001, Kudo et al. 2005, Magariyama et al. 2005). These differing characteristics have been observed in swimming speed and trajectory. However, the asymmetrical motion appears only when the bacterial cell swims close to a surface. In the following sections of this paper, the experimental results will be described, and an explanation based on fluid dynamics for the cause of this peculiar motion near a surface will be detailed.

2 Observed asymmetric trajectory

The motion of singly flagellated *Vibrio alginolyticus* bacterial cells was observed (Kudo et al. 2005). Figure 2 shows a dark-field microscope image of a cell. The flagellum is attached to one end (pole) of the cell body. The motions of three strains of *V. alginolyticus* were compared. Wild type YM4 cells swim forward and backward according to the rotational direction of the flagellum.

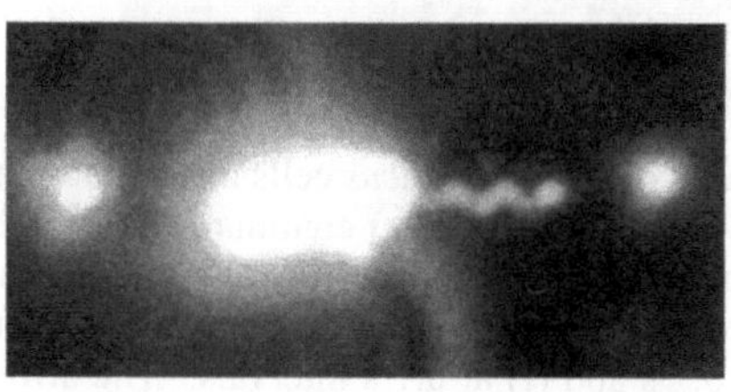

Fig. 2. Dark-field microscope image of a *V. alginolyticus* cell.

YM42 cells almost always swim forward, and NMB102 cells almost always swim backward (Table 1).

A drop of each cell suspension was sealed between a glass slide and a cover slip, the distance between which is estimated to be about 7 μm. The samples were observed with a high-intensity dark-field microscope, enabling the flagellum to be observed.

Figure 3(a) shows a typical example of swimming traces for a YM4 cell. Initially, the cell moves clockwise (the lower part of the trace) and travels straight in the eleven o'clock direction, then turns counterclockwise. Figure 3(b) is the same trace with a time interval of 0.1 s. In this figure, it is apparent whether the cell swims forward or backward since the flagellum is clearly

Table 1. Bacterial strains used in this study.

Strain	Phenotype
YM4	Wild type
YM42	Forward swimming
NMB102	Backward swimming

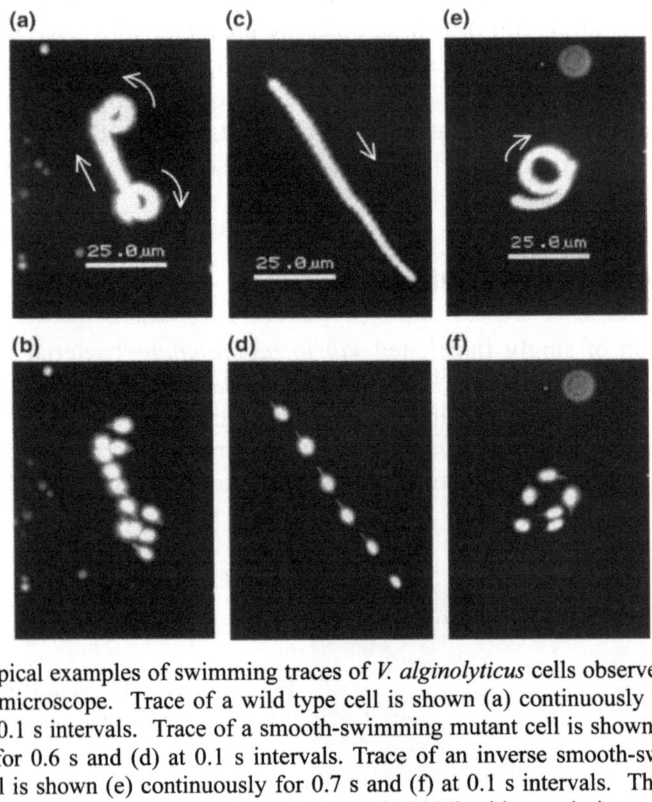

Fig. 3. Typical examples of swimming traces of *V. alginolyticus* cells observed with a dark-field microscope. Trace of a wild type cell is shown (a) continuously for 1.2 s and (b) at 0.1 s intervals. Trace of a smooth-swimming mutant cell is shown (c) continuously for 0.6 s and (d) at 0.1 s intervals. Trace of an inverse smooth-swimming mutant cell is shown (e) continuously for 0.7 s and (f) at 0.1 s intervals. The arrows indicate the swimming direction. From Kudo et al. (2005) with permission.

shown. During the first and last circular motions, the flagellum precedes the cell body; hence, the cell moves backward. In straight motion, on the other hand, the cell swims forward. Figures 3(c) and (d) are the swimming traces of a YM42 cell, and Figs. 3(e) and (f) are for a NMB102 cell. The differences in trajectory between forward and backward motions are plainly demonstrated: a forward swimming cell moves straight and a backward swimming cell moves circularly.

As mentioned above, the thickness of the liquid layer in this experiment was about 7 μm. Because the size of the cell was around 1 μm, which is comparable to its thickness, the cells were likely to be affected by the glass surfaces. The influence of a surface on cell motion was investigated by comparing the trajectories of cells swimming near a glass slide with the trajectories of cells swimming at a distance from a surface. In order to do so, a simple chamber was made. Spacers were inserted between the glass slide and the cover slip so as to make the distance between them about 170 μm. High-intensity dark-field microscopy cannot be applied to this thick sample, so inverted phase-contrast microscopy was used for observation.

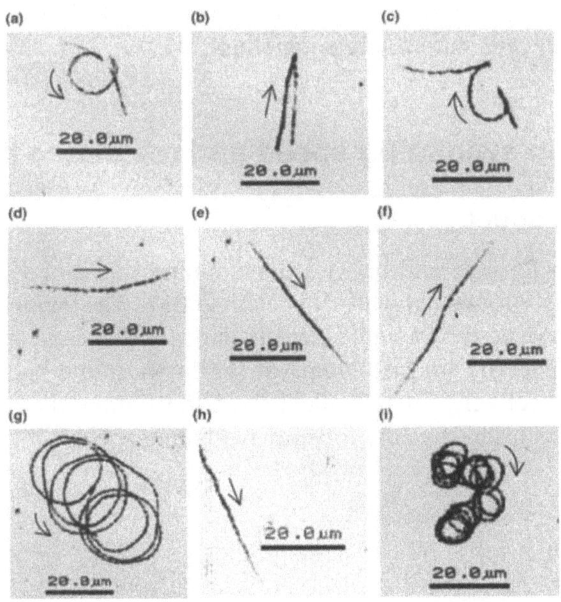

Fig. 4. Typical examples of swimming traces of *V. alginolyticus* cells observed with an inverted phase-contrast microscope. (a)-(c) Wild type cells. (d)-(f) Smooth-swimming mutant cells. (g)-(i) Inverse smooth-swimming mutant cells. (a), (d) and (g) Cells swimming near slide glasses. (b), (e) and (h) Cells swimming several tens of micrometers from glass surfaces. (c), (f) and (i) Cells swimming near cover glasses. The arrows indicate the swimming direction. From Kudo et al. (2005) with permission.

Typical examples of the trajectories for YM4 cells are shown in Figs. 4(a)-(c). The trajectory contains circular motion and straight lines when the cells swim close to a top or bottom boundary, as shown in Figs. 4(a) and (c), while the trajectory is a series of straight lines when the cell swims at a distance from both of the glass surfaces, as shown in Fig. 4(b). Figures 4(d)-(f) indicate that YM42 cells always move in straight lines regardless of their distance from the surfaces. The NMB102 cells draw circular traces when the cells swim close to a surface, as depicted in Figs. 4(g) and (i). When there is not surface influence, the cells move in straight lines (Fig. 4(h)).

These results suggest that asymmetric trajectory appears only when the cell swims close to a surface. A forward swimming cell moves in straight lines, while a backward swimming cell curves.

Circular motions have also been reported on bacterial cells with several flagella. Berg and Turner (1990) described the circular motion of *E. coli* cells. The three-dimensional tracking of the bacterial cells showed that circular motion appears when the cells swim very close to a surface (Frymier et al. 1995). DiLuzi et al. (2005) demonstrated sorting of cells using branched narrow passages. The motion of bacterial cells with several flagella is referred to as "run" and "tumble", or in other words, "forward" and "change directions." The asymmetrical characteristics between forward and backward motions are unique to bacterial cells with a single flagellum.

3 Difference in swimming speed and residence time

3.1 Swimming speed

The difference in forward and backward swimming speed of *V. alginolyticus* cells was previously found (Magariyama et al. 2001). The swimming speed of YM4 cells was measured by a high-intensity dark-field microscopy.

Figures 5(a) and (b) are histograms of the forward and backward swimming speeds, respectively. It is evident from the figures that the backward swimming speed is faster than the forward swimming speed.

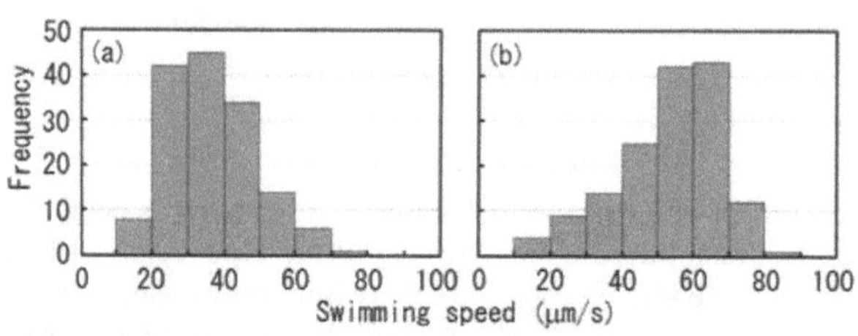

Fig. 5. Swimming characteristics of a *V. alginolyticus* strain, YM4. There were 150 measurements. (a) Histogram of forward swimming speeds. (b) Histogram of backward swimming speeds. From Magariyama et al. (2001) with permission.

3.2 Residence time

Since the focal depth of a microscope is limited, a bacterial cell appears when the cell enters a certain thickness layer that is in focus, and disappears when the cell leaves that layer. Here, the residence time of a cell is defined as the period in which the cell is in focus.

A suspension of cells was sealed in the simple chamber constructed from a glass slide, a cover slip and spacers. This sample was observed with an inverted phase-contrast microscope. The residence time of three strains of *V. alginolyticus*, YM4, YM42, NMB102 were measured (Magariyama 2005).

Figure 6 shows the frequency distributions of the residence time of the cells swimming the upper (a, b, c), middle (d, e ,f) and lower (g, h, i) layers. The YM4 cells tend to stay longer within the layer of focus when they are near either surface (a, g) than when they are in the middle layer. The NMB102 cells display the same tendency (c, f, i). On the other hand, the residence time of YM42 cells is irrespective of the distance from a surface. The residence time of YM4 and NMB102 swimming close to a surface (a, g, c, i) is longer than the residence time of YM42 (b, e, h). The residence time of YM4 and NMB102 swimming in the middle layer (d, f) is as same as the residence time of YM42 (b, e, h).

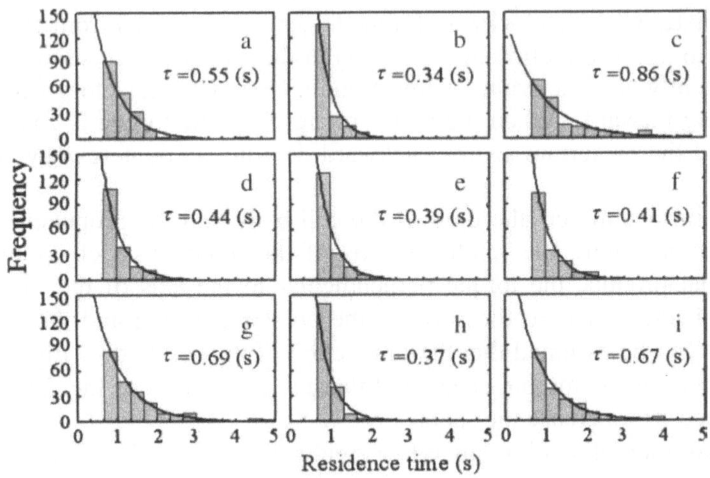

Fig. 6. Frequency distribution of residence time. The motions were recorded near the upper surface (a, b, and c), in the middle (d, e, and f), and near the base (g, h, and i). YM4 (a, d, and g), YM42 (b, e, and h), and NMB102 (c, f, and i) were used in this experiment. The bars refer to the measured data. The number of data points is 200 for each case. Data are not shown in the two left-most columns of each graph, since trajectories that had 20 frames (2/3 s) of residence time were not analyzed. The line represents an exponential function used to fit the measured data . From Magariyama et al. (2005) with permission.

In summary, backward swimming cells tend to stay close to a surface longer than forward swimming cells, and this asymmetry between the forward and backward motions appears only when the cells swim under the influence of a surface.

4 A fluid-dynamic interpretation for the asymmetrical motion

4.1 Fluid dynamic interactions between a surface and a cell

When the motor rotates in one direction, the cell body rotates in the opposite direction, as shown in Fig. 7. If there are no interfering surfaces, these rotations of the flagellum and the cell body do not induce any lateral motion. If there is a surface, however, the rotational motions of the cell body and the flagellar filament are both affected by the surface. This explains the difference between bacterial motion close to a rigid surface and its motion in free space.

The resistive force exerted on the side facing the surface and the force on the opposite side are different. As a result, a lateral force perpendicular to the rotational axis and parallel to the surface is exerted on the filament. Another lateral force in the opposite direction is exerted on the cell body. These two forces produce a torque around an axis perpendicular to the surface. When the motor rotates CCW, a clockwise torque is generated as seen from above (from the plus to minus direction in the z-axis) in the configuration shown in Fig. 7. Consequently, the bacterium swims along a clockwise trajectory because it is swimming forward. When the motor rotates CW, the torque is counterclockwise, and the bacterium swims backward drawing a counterclockwise trajectory.

The surface affects the drag of the cell body and the propulsive force of the flagellum along the rotational axis of the motor, as well as the lateral forces that produce the torque perpendicular to the axis of the motor. The closer a bacterium is to the surface, the greater the drag and the propulsive force. It can be assumed that the interaction of the cell body with the surface will reduce the swimming speed and that the interaction of the filament with the surface will increase the speed.

The attitude of a cell relative to a rigid surface is another factor that influ-

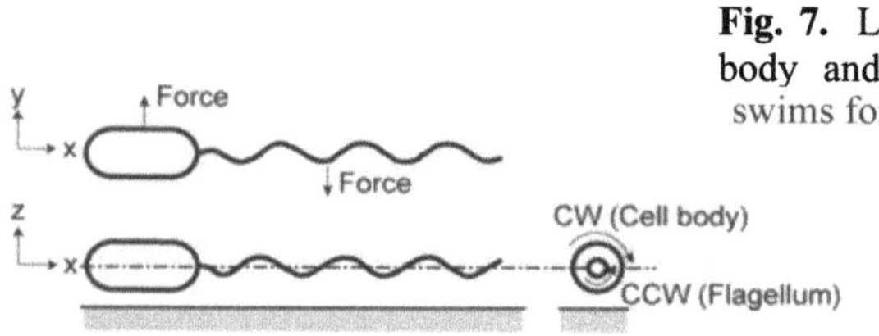

Fig. 7. Lateral forces exerted on the cell body and the flagellum when the cell swims forward.

ences the motion of the cell. If the cell body swims backward toward the surface, the interaction between the flagellar filament and the surface is stronger than that between the cell body and the surface. Thus, both the swimming speed and the trajectory depend upon the attitude of the cell.

4.2 Numerical simulation

Boundary element analyses (Ramia et al. 1993) were carried out to investigate the effect of the fluid-dynamic interaction between a bacterial cell and a rigid surface on the asymmetrical motion illustrated in previous sections by changing the cell's distance from a rigid boundary and the attitude of the cell (Goto et al. 2005). Since the Reynolds number of the flow associated with the motion of a bacterial cell is very small, it was assumed that the fluid motion is steady at any instant and is a creeping flow governed by the steady Stokes equation. The velocities and the angular velocities of the cell body were calculated from the condition that the net force on the whole cell is zero and that the net torque is also zero. In order to take into account a rigid surface that produces a no-slip boundary condition, a basic solution derived by Blake (1971) was adopted.

The dimension of the cell model was determined according to the measured values of YM4 cells (Goto et al. 2001). The calculations were performed by changing the pitch angle θ and the distance d shown in Fig. 8. The velocities and the angular velocities were obtained for each set of θ and d. The swimming speed was defined as the velocity in the x direction Ux. The pitch rate Ωy and the yaw rate Ωz" were also examined. Here, the yaw rate was defined as the projection of Ωz' to the *z* direction, as shown in Fig. 8.

The swimming speed Ux, the pitch rate Ωy and yaw rate Ωz" of the cell model swimming forward are shown as functions of the pitch angle θ in Fig. 9(a)-(c), respectively. These values are normalized by the absolute values of the x component of the swimming velocity vector |U∞| and the x component of the angular velocity vector |Ω∞| of the model swimming in free space. Note that, in the coordinate system defined in Fig. 8, the forward swimming direction is in the negative x direction.

Figure 9(a) shows that the swimming speed is unaffected if the cell swims at a distance of around ten times the diameter of the cell body (see d/2b=12.5)

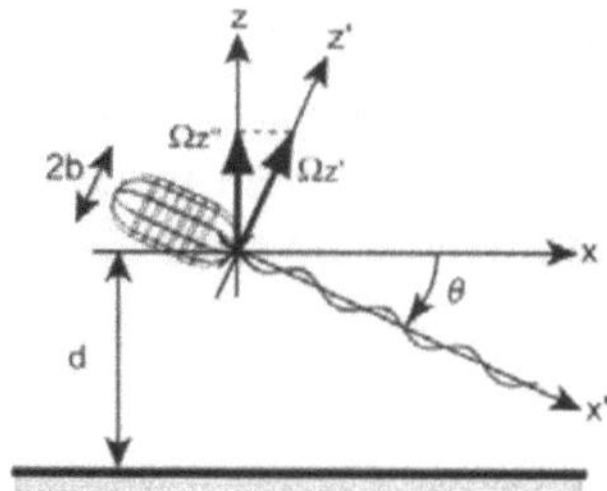

Fig. 8. Boundary element model applied to a bacterial cell with a body width of *2b* swimming at a distance *d* from a wall. The pitch angle θ is shown. From Goto et al. (2005) with permission.

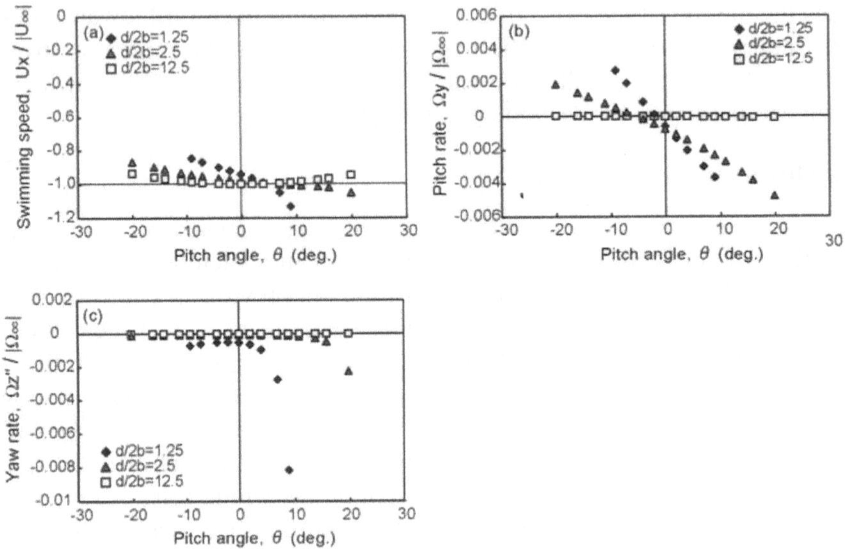

Fig. 9. Swimming speed and rotational rates in forward motion as functions of pitch angle. Parameter d/2b is a measure of the distance from the wall (see Fig. 8). These data should be inverted with respect to the abscissa for the case of backward motion. (a) Swimming speed normalized with respect to the swimming speed in free space, $|U\infty|$. (b) Pitch rate normalized using the angular velocity of the cell body in free space, $|\Omega\infty|$. (c) Yaw rate. $\Omega z''$ is the projection of $\Omega z'$ onto the z axis as defined in Fig. 8. From Goto et al. (2005) with permission.

from a surface. As the cell approaches a surface, the swimming speed varies with θ. When the cell swims parallel to the wall, namely θ=0°, the closer to the wall the cell is swimming, the slower it swims (the smaller the magnitude). When the cell swims in an orientation such that the flagellum is close to the wall and the cell body is away from the wall, namely θ>0°, the speed increases. Thus, the swimming speed increases either when the cell swims at a significant distance from the wall or when the flagellar filament interacts with the wall.

The pitch rate is shown in Fig. 9(b). If the cell swims at a significant distance from the surface, the pitch rate is independent of θ and is negligible. However, if the cell interacts with the wall, the pitch rate decreases almost proportionally with the pitch angle. Because the pitch rate is the time derivative of the pitch angle, the negative sign for the proportional constant indicates that the pitch motion associated with forward swimming has positive damping and is stable. For backward motion, all of the data will be inverted about the horizontal axis. Namely, their signs will invert from positive to negative and from negative to positive. Therefore, the pitch motion is unstable in backward motion.

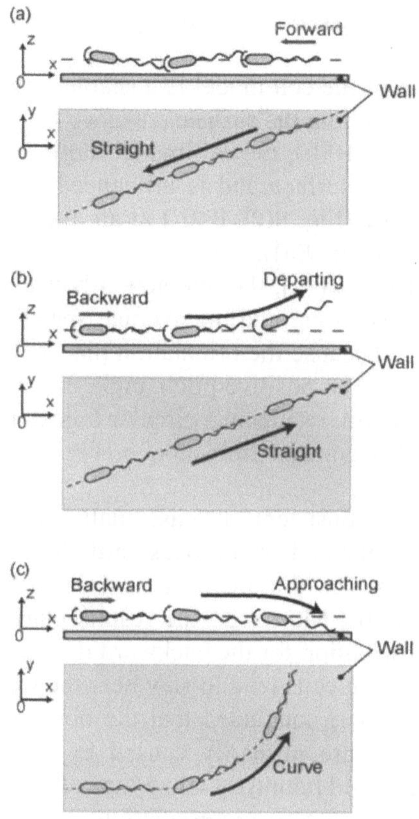

Fig. 10. Images of the motion of a bacterium close to a wall. Forward motion is stable: (a) the cell moves in a straight line maintaining a fixed distance from the wall. Backward motion is unstable: (b) the cell moves in a straight line when departing the wall, and (c) the cell trajectory is curved in a counterclockwise sense as viewed from above when approaching the wall. From Goto et al. (2005) with permission.

The relationship between the yaw rate and the pitch angle is shown in Fig. 9(c). The yaw rate is affected by the pitch angle only when the cell swims close to the wall and is oriented with the flagellum close to the wall. In such cases, for forward swimming motion, the yaw rate becomes negative and the resultant trajectory will curve in a clockwise direction when the motion is observed from above. For backward motion, similar to the pitch rate data, the signs of all data will be inverted with respect to the horizontal axis. Thus, the yaw rate becomes positive and the trajectory will curve in a counterclockwise direction if the motion is observed from above.

Using the numerical results, the image of the motion of the cell model swimming close to a rigid boundary shown in Fig 10 can be generated:

1. As drawn in Fig. 10(a), since the pitching motion is stable in forward motion (Fig. 9(b)), the cell model tends to swim parallel to the surface maintaining a certain distance. However, the swimming speed of the parallel motion is slower than that achieved in free space (Fig. 9(a)).

Significant circular motion is not observed since the yaw rate is negligible when $\theta \approx 0°$ (Fig. 9(c)).

2. In backward motion, the pitch motion of the cell model is unstable. The model is either departing from or approaching the surface.
3. In the departing motion depicted in Fig. 10(b), the cell model motion is steadily increasing its distance from the surface and is influenced very little. It swims almost in a straight line (Fig. 9(c), $\theta<0°$) at an average speed nearly equal to that in free space (Fig. 9(a)).
4. In the approaching motion shown in Fig. 10(c), the interaction between the flagellar filament and the surface is strong and the average speed is larger than that achieved in free space because the orientation places the flagellum close to the wall (Fig. 9(a)). The positive pitch angle induces rotation in the yaw direction which, in turn, results in a circular trajectory in the counterclockwise direction when viewed from above (Fig. 9(c), $\theta>0°$).

These diagrammatic representations are consistent, at least qualitatively, with the circular trajectories observed when the cells swim backward close to the surface. They are also consistent with the speed in the backward direction exceeding the speed in the forward direction. Moreover, the results clarify that the broadness in the residence time distribution for the backward direction is due to the unstable motion. The approaching cells tend to stay near the wall longer than the departing cells. Therefore, it is concluded that the observed asymmetrical characteristics of the motions are primarily caused by fluid-dynamic interactions between the cell and a rigid boundary when the cell has a non-parallel attitude relative to the boundary. These representations may be verified if the pitch angle is experimentally measured in conjunction with the swimming speed and trajectory, although the direct measurement of the pitch angle is very difficult.

5 Summary and remarks

This paper addresses the motion of a bacterium with a single polar flagellum swimming in a liquid medium close to a rigid surface. The unusual characteristics that appear only when a bacterial cell swims close to a surface were qualitatively explained as the result of the fluid-dynamic interaction between the cell and the surface.

These characteristics possibly explain the behavior of the bacterial taxis around surfaces, which may be animate or inanimate surfaces, or the formation of bio-films. A simulation suggests that the asymmetrical characteristics help the bacterial cell efficiently cover a searching area (Magariyama 2005). When a cell approaches a surface with its flagellum oriented toward the surface, the cell may have a higher tendency to attach to the surface.

In most observations of microorganisms, including bacteria, in a liquid medium, the medium is sealed between a glass slide and a cover slip. The studies presented here have revealed that the microorganism motion might be strongly influenced by the existence of such boundaries.

Microorganisms are considered examples of micro-machines. Their existence provides confirmation that such tiny micro-machines could be constructed without violating natural laws. From an engineering point of view, learning from microorganisms is essential to understanding the mechanism of swimming, propulsion, sensing, guidance, control, and structures suitable for the micro-scale realm. The mechanism of asymmetrical motion revealed here could be applied to build such tiny machines.

This work was partly supported by JSPS.

References

Berg HC, Turner L (1990) Chemotaxis of bacteria in glass capillary arrays., Biophys. J. 58, 919-930.

Blake JR (1971) A note on the image system for a stokeslet in a no-slip boundary., Proc. Camb. Phil. Soc. 70, 303-310.

DiLuzi WR, Turner L, Mayer M, Garstecki P, Weibel DB, Berg HC, Whitesides GM (2005) Right thinking bacteria., Nature 435, 1271-1274.

Frymier PD, Ford RM, Berg HC, Cummings PT (1995) Three-dimensional tracking of motile bacteria near a solid planar surface. Proc. Natl. Acad. Sci. USA 92, 6195-6199.

Goto T, Masuda S, Terada K, Takano Y (2001) Comparison between observation and boundary element analysis of bacterium swimming motion., JSME Int. J. Ser. C 44, 958-963.

Goto T, Nakata K, Baba K, Nishimura M, Magariyama Y (2005) A fluid-dynamic interpretation of the asymmetric motion of singly flagellated bacteria swimming close to a boundary., Biophys. J. 89, 3771-3779.

Kudo S, Imai N, Nishitoba M, Sugiyama S, Magariyama Y (2005) Asymmetric swimming pattern of *Vibrio alginolyticus* cells with single polar flagella., Microbiol. Lett. 242, 221-225.

Magariyama Y, Masuda S, Takano Y, Ohtani T, Kudo S (2001) Difference between forward and backward swimming speeds of the single polar-flagellated bacterium, *Vibrio alginolyticus*., FEMS Microbiol. Lett. 205, 343-347.

Magariyama Y, Ichiba M, Nakata K, Baba K, Ohtani T, Kudo S, Goto T (2005) Difference in bacterial motion between forward and backward swimming caused by the wall effect., Biophys. J. 88, 3648-3658.

Ramia MD, Tullok L, Phan-Thien N (1993) The role of hydrodynamic interaction in the locomotion of microorganisms., Biophys. J. 65, 755-778.

[illegible]

[illegible] and flagellar [illegible]

[illegible]

Chapter 2

Properties of a Semi-dilute Suspension of Swimming Micro-organisms

Takuji Ishikawa[1], Takami Yamaguchi[1], and T. J. Pedley[2]

[1] Department of Bioengineering and Robotics, Tohoku University, 6-6-01, Aoba, Aramaki, Aoba-ku, Sendai 980-8579, Japan
[2] Department of Applied Mathematics and Theoretical Physics, University of Cambridge, CMS, Wilberforce Road, Cambridge CB3 0WA, UK

Summary. In this paper, we review our recent researches on a suspension of swimming micro-organisms. Firstly, interactions between two swimming micro-organisms are discussed. We introduce our experimental works using *Paramecium caudatum*, and numerical works using a squirmer model. The results show that changes in direction between two swimming cells are induced mainly by hydrodynamic forces and that biological interaction is a minor factor. Secondly, the rheological and diffusive properties of a semi-dilute suspension of squirmers are discussed. We introduce our numerical works and show that a suspension of bottom-heavy squirmers has strong non-Newtonian properties. We also show that the spreading of squirmers is correctly described as a diffusive process after a sufficiently long time, even though all the movements of the squirmers are deterministically calculated.

Key words. Micro-organism, Suspension, Hydrodynamic Interaction, Stokesian dynamics, Viscosity, Diffusivity

1 Introduction

A suspension of micro-organisms appears, for instance, in a massive plankton bloom in the ocean, which is an integral part of the oceanic ecosystem. The phytoplankton absorb CO_2 from the water, and thus they play an important role in the global carbon cycle and hence in phenomena such as global warming. Blooms sometimes cause harmful red tides in coastal regions, which lead to economic damage to fish farms. In the engineering field, micro-organism suspensions are used for bioreactors to make chemicals and biomass. Therefore, the study of micro-organism behaviour is a proper subject for scientific investigation.

The size of individual micro-organisms is often much smaller than that of the flow field of interest, so the suspension is modelled as a continuum in which the variables are volume-averaged quantities. Various continuum models for suspensions of swimming micro-organisms have also been proposed for the analysis of phenomena such as bioconvection, Pedley & Kessler (1990) for instance. However, the continuum models proposed so far are restricted to dilute suspensions, in which cell-cell interactions are negligible. If one wishes to consider larger cell concentrations, it will be necessary to consider the interactions between micro-organisms. Then, the translational-rotational velocity of the micro-organisms, the particle stress tensor and the diffusion tensor in the continuum model will need to be refined.

In this background, we have investigated the properties of a suspension of micro-organisms both experimentally and numerically. In this review paper, we introduce our recent researches on this topic. Two-cell interactions are the basis of many cell interactions. Thus, in chapter 2, interactions between two swimming micro-organisms are discussed. We introduce our experimental works using *Paramecium* and numerical work using a squirmer model. In chapter 3, the continuum quantities, such as the rheological and diffusive properties, of a semi-dilute suspension of squirmers are discussed. We introduce our numerical works using a Stokesian-dynamics method.

2 Interaction between two micro-organisms

2.1 Experiments using *Paramecium*

Biological reactions of a solitary *Paramecium* cell to mechanical stimulations were investigated by Naitoh and Sugino (1984). Avoiding reactions occur when a cell bumps against a solid object with its anterior end. The cell swims backward first, then gyrates about its posterior end, and finally resumes normal forward locomotion. Escape reactions occur when the cell's posterior end is mechanically agitated. The cell increases its forward swimming velocity for a moment, then resumes normal forward locomotion. The change in the swimming motion is regulated by membrane potential charges, because *Paramecium* cells, as well as other monads, have no nerves for transmitting stimulative information nor synapses to determine transmission direction. Machemer (1974) clarified the directional responses of cilia to membrane potential changes in *Paramecium* cells. There are many Ca^{2+} channels in the anterior end, whereas there are many

K^+ channels in the posterior end. The locality of ion channels is the essential mechanism of controlling *Paramecium*'s biological reaction.

Interactions between two *Paramecium* were investigated by Ishikawa and Hota (2006). In this section, we briefly introduce the study using *P. caudatum*. Microscopic observation showed that the body length of an individual cell is in the range of approximately 200-250μm and the width is in the range of approximately 40-50μm. The swimming speed of an individual cell was approximately 1mm/s. The swimming motion of *P. caudatum* in a free space was not straight, but formed a left spiral. The pitch of the spiral was approximately 2mm and the width was 0.4mm.

The experimental setup was designed to measure the displacements of cells in a still fluid between flat plates as shown in figure 1. The experimental setup consisted of a digital video (DV) camera with a ×24 macrolens, light sources, and inner and outer dishes. The test fluid was placed between the top of the outer dish and the bottom of the inner dish. The gap between the two dishes was about 70μm, so that cells could not overlap three-dimensionally. The test fluid was same as the culture fluid, but the volume fraction of cells was adjusted to be in the range of about 0.5-1% so that three-cell interactions rarely occurred.

A typical hydrodynamic interaction, observed in the study is shown in figure 2. When one cell collides with the anterior end of the other cell, the two cells tend to swim side by side at first, then move away from each other with an acute angle, as shown in figure 2. When two cells are initially facing, they come close to each other at first, then they change their

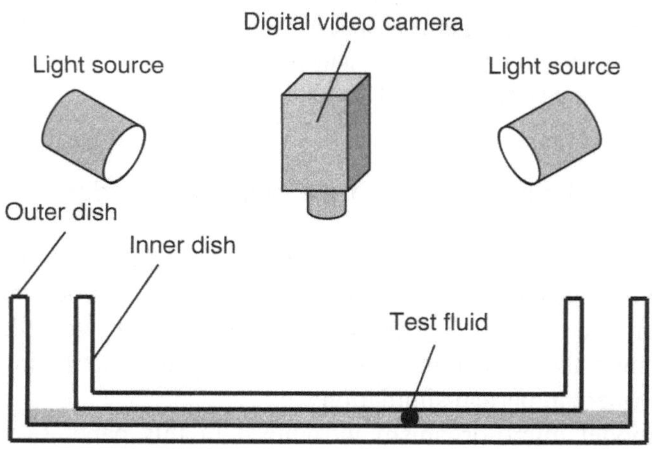

Fig.1. Schematics of the experimental apparatus. Test fluid was placed between the bottom of the inner dish and the top of the outer dish. (Ishikawa and Hota, 2006)

directions slightly in the near field, and finally move away from each other.

All the cells interactions were classified as hydrodynamic interactions, avoiding reactions or escape reactions. The total number of interacting cells recorded in the study was 602 (the total number of cases was 301). It was found that 84.7% of cells interact hydrodynamically. The ratio of escape reaction was slightly higher than that of avoiding reaction. We could conclude that the cell-cell interaction was mainly hydrodynamic and that the biological reaction accounted for a minority of incidents.

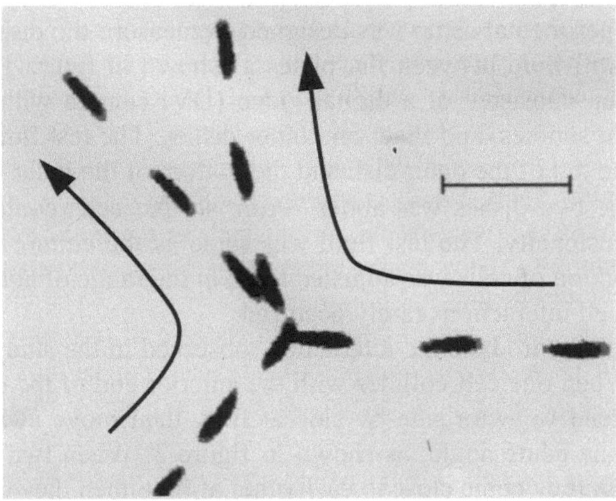

Fig.2. Sample sequences showing the hydrodynamical interactions when two swimming *P. caudatum* experience a near-contact. The time interval between each sequence is 1/3 sec. Long arrows are added to schematically show cell motion. Scale bars, 500 μm (Ishikawa and Hota, 2006)

2.2 Numerical simulation using a squirmer model

Although we assumed that the experimental data not satisfying the definition of avoiding reaction nor escape reaction were hydrodynamic interactions, as shown in figure 2, there was no evidence that two cells did not actively change their swimming motions during the interaction. Thus, we also performed numerical simulations.

The model micro-organism introduced in this paper (a squirmer) is used by Ishikawa *etal.* (2006). A squirmer is assumed to be neutrally buoyant, because the sedimentation velocity for typical aquatic micro-organisms is much less than the swimming speed. The centre of buoyancy of the micro-organism may not coincide with its geometric centre (*bottom-heaviness*). The model micro-organism is, therefore, force-free but may not be torque-free. Swimming speeds of micro-organisms such as *P. caudatum* range up to several hundred μm/s. However, the Reynolds number based on the swimming speed and the radius of individuals is usually less than 10^{-2}, so the flow field around the cells can be assumed to be Stokes flow with negligible inertia compared to viscous effects. Brownian motion is usually not taken into account, because typical locomotive cells are too large for Brownian effects to be important.

The model micro-organism was assumed to propel itself by generating tangential velocities on its surface. In fact, it is a reasonable model to describe the locomotion of ciliates, which propel themselves by beating arrays of short hairs (cilia) on their surface in a synchronised way. In particular, the so-called symplectic metachronal wave, in which the cilia tips remain close together at all times, employed by *Opalina*, for instance, can be modelled simply as the stretching and displacement of the surface formed by the envelope of these tips.

When there are N cells in an infinite fluid, the Stokes flow field external to the cells can be given in integral form as:

$$u_i(\mathbf{x}) - \langle u_i(\mathbf{x}) \rangle = -\frac{1}{8\pi\mu} \sum_{\alpha=1}^{N} \int_{A_\alpha} J_{ij}(\mathbf{x}-\mathbf{y}) q_j(\mathbf{y}) dA_y \tag{1}$$

where **u** is the velocity, **J** is the Green function, **q** is the single-layer potential, and A is the surface of a particle. The parentheses <> indicate the suspension average. The boundary condition is given by

$$\mathbf{u}(\mathbf{x}) = \mathbf{U}_m + \mathbf{\Omega}_m \times (\mathbf{x} - \mathbf{x}_m) + \mathbf{u}_{s,m} \text{ , } \quad \mathbf{x} \in A \tag{2}$$

where $\mathbf{U}_m$ and $\mathbf{\Omega}_m$ are the translational and rotational velocities of squirmer m, respectively. $\mathbf{x}_m$ is the centre of squirmer m, and $\mathbf{u}_{s,m}$ is the squirming velocity of squirmer m. In order to obtain $\mathbf{u}_s$, the velocity field around a swimming *P. caudatum* cell was measured by a PIV technique. Velocity vector was calculated for a time series of images, and approximately 80,000 velocity vectors around a swimming cell were obtained. These vectors were averaged by assuming that the velocity field is axisymmetric and time-independent. Then the surface velocity, $\mathbf{u}_s$, is interpolated from the surrounding flow field.

The boundary element method was employed to discretize equation (1). For the computational mesh, a maximum of 590 triangle elements per particle were generated, and the mesh was finer in the near-contact region. Time-marching was performed by 4th-order Runge-Kutta schemes. In simulating hydrodynamic interactions between two squirmers, we assume that the surface velocity is independent of the distance between the cells.

Thus, no biological reaction is modelled, and the interaction is purely hydrodynamic.

Figure 3 shows the interactions between two squirmers, in which one squirmer collides with the anterior end of the other. In this case two cells tend to swim side by side at first, then move away from each other with an acute angle. This tendency is the same as the experimental results shown in figure 2. We also performed simulations with various initial conditions, and found that the hydrodynamic interaction data in the experiments agreed well with the numerical results and that the interaction was purely hydrodynamic.

In most of the previous analytical studies on cell-cell interactions (Guell *etal*, 1988; Ramia *etal*, 1993; Nasseri and Phan-Thien, 1997; Jiang *etal*, 2002), two cells in close contact were not discussed. (Lega and Passot, 2003 included an *ad hoc* interactive force acting between cells, that in

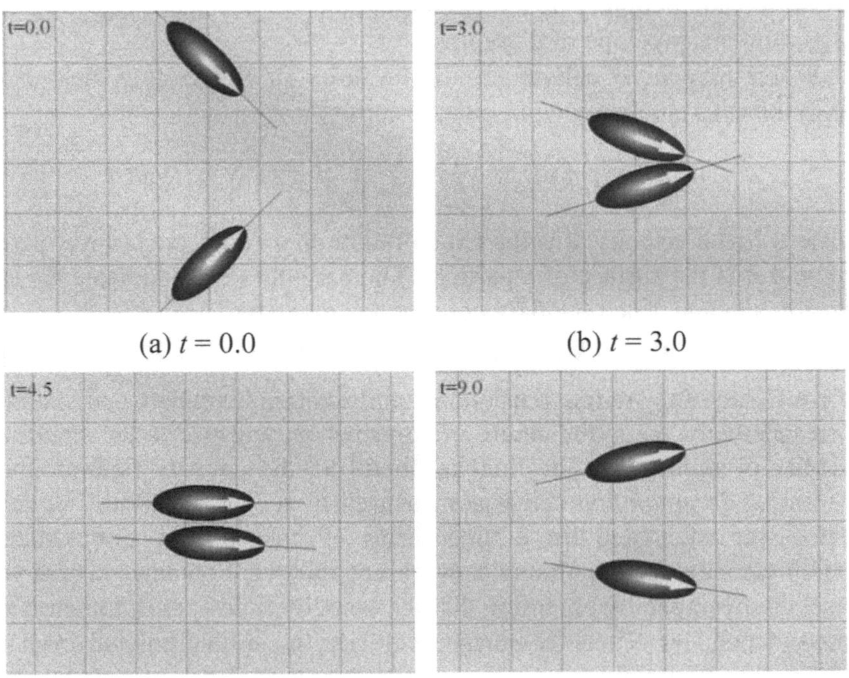

(a) $t = 0.0$ (b) $t = 3.0$ (c) $t = 4.5$ (d) $t = 9.0$

Fig.3. Sequences (a) to (d) showing the hydrodynamic interactions between two squirmers, where t is the dimensionless time. The orientation vectors of the squirmers are shown as large arrows on the ellipsoids, and a thin solid line is added so that one can easily compare the angle between the two squirmers. (Ishikawa and Hota, 2006) practice is unlikely to exist.) The present results show, however, that the near-field interaction dramatically changes the orientation of cells. Since orientation change affects the macroscopic properties of the suspension, such as the diffusivity, the near-field interaction has to be solved accurately.

3 A semi-dilute suspension of squirmers

3.1 Methods

In the following sections, a squirmer is assumed to have a spherical shape for simplicity. The surface velocities of a squirmer model was analyzed by Blake (1971), and we use a simplified version of his equation given by;

$$\mathbf{u}_s = \sum_{n=1}^{2} \frac{2}{n(n+1)} B_n \left(\frac{\mathbf{e.r}}{r} \frac{\mathbf{r}}{r} - \mathbf{e} \right) P'_n, \tag{3}$$

where $\mathbf{e}$ is the orientation vector, $\mathbf{r}$ is the position vector and $r = |\mathbf{r}|$. P_n is the nth Legendre polynomial, and B_n is the nth mode of the surface squirming velocity. (see details in Ishikawa *etal*, 2006)

A sample squirmer, which will be used in the following sections, generates velocity vectors, relative to a frame of reference in which the squirmer's centre is at rest, as shown in figure 4. The orientation vector of the squirmer is also shown, and the uniform flow of its swimming speed comes from the far right.

The numerical method for a semi-dilute suspension of squirmers can be found in Ishikawa and Pedley (2007a), based on the Stokesian-dynamics method (cf. Brady and Bossis, 1988). In the absence of Brownian motion and at negligible particle Reynolds number, the equation of motion for N identical squirmers suspended in a Newtonian solvent fluid otherwise at rest can be written as;

$$-\mathbf{R.U} + \mathbf{F}_{sq} + \mathbf{F}_{tor} + \mathbf{F}_{rep} = 0. \tag{3}$$

Here $\mathbf{U}$ consists of the translational-rotational velocities of the N particles evaluated at the squirmer centre. $\mathbf{R}$ is the grand resistance matrix, which is constructed by a pairwise superposition of exact two inert sphere results. The pairwise additivity is an approximation, but it is expected to be justified if the particle volume fraction is not too large (defining *semi-dilute*). $\mathbf{F}_{sq}$ is the force-torque due to the squirming motion, which is calculated from superposition of pairwise interaction between squirmers. $\mathbf{F}_{tor}$ is the external torques due to the bottom-heaviness. $\mathbf{F}_{rep}$ is the non-hydrodynamic interparticle forces introduced in order to avoid the prohibitively small time step needed to overcome the problem of overlapping particles. To model a suspension of infinite extent, periodic boundary conditions were employed.

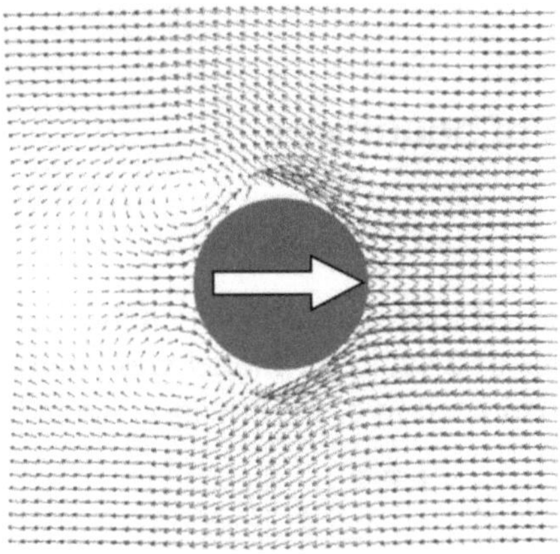

Fig.4. Velocity vectors relative to the translational velocity vector of a squirmer, in which $B_2/B_1 = 5$. The orientation vector is shown as a big arrow.

3.2 Rheological property

The rheological properties of a cell suspension play an important role in the flow field generated by populations of swimming micro-organisms (e.g. in bioconvection), which were discussed in Ishikawa and Pedley, (2005, 2007a). In these studies, the three-dimensional movement of 64 identical squirmers in a simple shear flow field, contained in a cube with periodic boundary conditions, was computed, for random initial positions and orientations, by the Stokesian-dynamics method. The instantaneous positions of the squirmers and their trajectories during one time interval are shown in figure 5, as an example.

The results for non-bottom-heavy squirmers showed that the squirming did not have a direct influence on the apparent viscosity. However, it did change the probability density in configuration space, and thereby caused a slight decrease in the apparent viscosity at order of c^2, where c is the volume fraction of squirmers.

In the case of bottom-heavy squirmers, on the other hand, the stresslet generated by the squirming motion directly contributed to the bulk stress at order of c, and the suspension showed strong non-Newtonian properties. When the background simple shear flow was directed vertically, the apparent viscosity of the semi-dilute suspension of bottom-heavy squirmers became smaller than that of inert spheres drawn by a dashed line (see figure 6). When the shear flow was horizontal and varied with the vertical coordinate, on the other hand, the apparent viscosity became larger than that of inert spheres. In addition, significant normal stress differences appeared for all relative orientations of gravity and the shear flow, in the case of bottom-heavy squirmers.

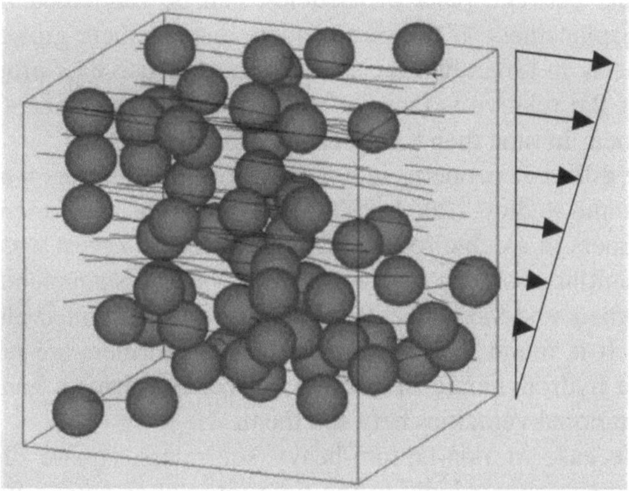

Fig.5. Instantaneous position of 64 identical squirmers. Solid lines are trajectories of the squirmers during one time interval. The background simple shear flow is exerted as shown by the arrows.

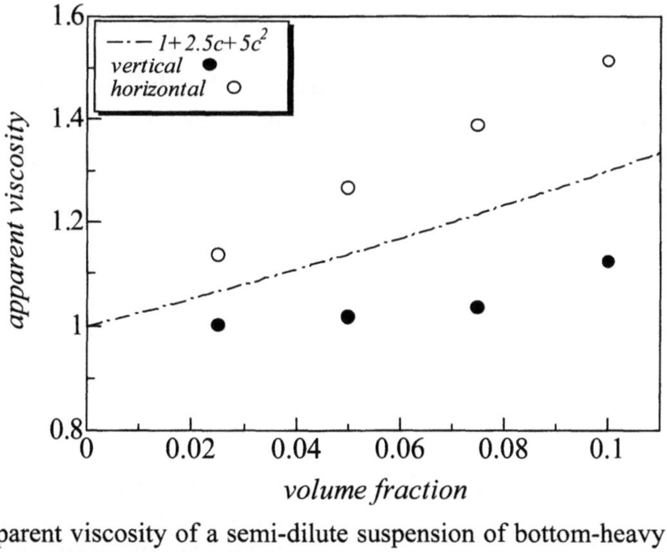

Fig.6. Apparent viscosity of a semi-dilute suspension of bottom-heavy squirmers. Where, *vertical* indicates that the background simple shear flow is directed vertically, whereas *horizontal* indicates directed horizontally

3.3 Diffusive property

Understanding the diffusive behaviour of swimming micro-organisms is important in order to obtain a better continuum model for a cell suspension. The translational diffusivity, which is a measure of the increasing displacements between pairs of particles, can be calculated from the mean square displacement. If the mean square displacement grows more rapidly than linearly in time, then the spread is not diffusive (if proportional to t^2, it is as if the relative velocity of two squirmers is constant), but if it becomes linear in time then the spread is diffusive.

The diffusive property in a suspension of squirmers was discussed in Ishikawa and Pedley, (2004, 2007b), in which the movement of 27 identical squirmers in a cubic region of fluid otherwise at rest was computed for random initial positions and orientations. The instantaneous positions of the squirmers and their trajectories during five time intervals are shown in figure 7. It is found that the trajectories of squirmers are not straight, because the hydrodynamic interaction between squirmers generates translational-rotational velocities between them.

In the case of non-bottom-heavy squirmers, square displacement of cells becomes proportional to the time interval as shown in figure 8. The results indicate that the spreading of squirmers was correctly described as a diffusive process over a sufficiently long time scale, even though all the movements of the squirmers were deterministically calculated. Scaling of the results on the assumption that the squirmer trajectories were unbiassed random walks was shown to capture some but not all of the main features of the results.

In the case of bottom-heavy squirmers, the diffusive behaviour in squirmers' orientations could be described by a biased random walk model, but only when the effect of hydrodynamic interaction dominated that of the

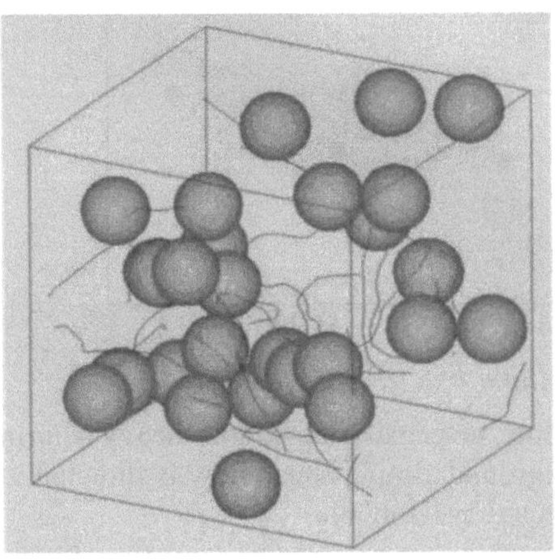

Fig.7. Instantaneous position of 27 identical squirmers. Solid lines are trajectories of the squirmers during five time interval. No background flow field is imposed.

bottom-heaviness. The spreading of bottom-heavy squirmers in the horizontal directions showed diffusive behaviour, and that in the vertical direction also did provided that the average upward velocity was subtracted. The rotational diffusivity in this case, at a volume fraction $c = 0.1$, was shown to be at least as large as that previously measured in very dilute populations of randomly swimming algal cells (*Chlamydomonas nivalis*).

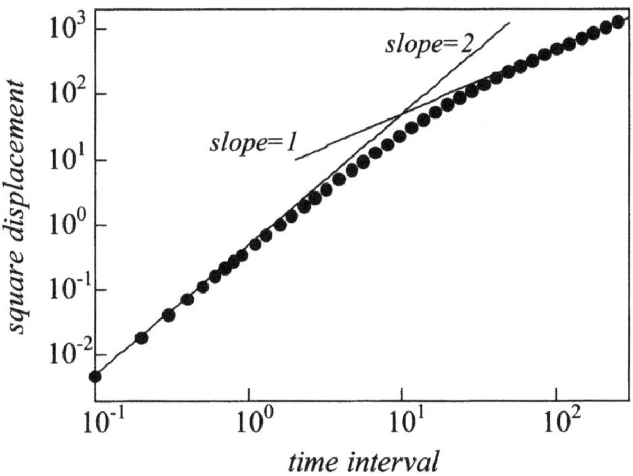

Fig.8. Translational diffusivity of non-bottom-heavy squirmers in a semi-dilute suspension with c =0.1. Where, lines with slope 1 and 2 are drawn for comparison.

4 Conclusion

In this paper, we have reviewed our fundamental studies on properties of a suspension of micro-organisms. Some of the continuum properties, such as the viscosity and diffusivity, were clarified in a semi-dilute regime. We expect to expand the present analysis to a denser suspension, which is the subject of current.

Taking into account biological characteristics in the squirmer model will be another important challenge. Though the biological reactions were a minor factor in the particular cell-cell interaction, discussed in section 2.1, they are likely to be important for many types of cell. Also most micro-organisms show some kinds of taxis, such as geotaxis, phototaxis, chemotaxis, rheotaxis or magnetotaxis, which dominate the preferred direction of cell's orientation. In order to develop a more realistic model of a suspension of micro-organisms, one may need to model both hydrodynamical and biological phenomena.

There are still plenty of unsolved phenomena in a suspension of micro-organisms. We expect that many researchers will be interested in this research field and will clarify the natural beauty exhibited by micro-organisms in suspension in the near future.

This work was partly supported by Grant-in-Aid for Scientific Research (S) 19100008 and Grant-in-Aid for Young Scientists (A) 19686016 by .

References

Blake, J. R. (1971), A spherical envelope approach to ciliary propulsion, J. Fluid Mech. 46: 199-208

Brady, J. F. and Bossis, G. (1988), Stokesian dynamics, Annu. Rev. Fluid Mech. 20: 111-157

Guell, D. C., Brenner, H., Frankel, R. B., Hartman, H. (1988), Hydrodynamic forces and band formation in swimming magnetotactic bacteria, J. theor. Biol. 135: 525-542

Ishikawa T, Pedley T. J. (2005), The rheology of a semi-dilute suspension of swimming cells, Bull. Amer. Phys. Soc. DFD 50: 124

Ishikawa T., Hota M. (2006), Interaction of two swimming *Paramecia*, J. Exp. Biol. 209: 4452-4463

Ishikawa T., Simmonds M. P., Pedley T. J. (2006), Hydrodynamic interaction of two swimming model micro-organisms, J. Fluid Mech. 568: 119-160

Ishikawa, T. and Pedley, T. J. (2007a), The rheology of a semi-dilute suspension of swimming model micro-organisms, J. Fluid Mech., in press

Ishikawa, T., Pedley, T. J. (2007b), Diffusion of swimming model micro-organisms in a semi-dilute suspension, J. Fluid Mech., in press

Jiang, H., Osborn, T. R., Meneveau, C. (2002), Hydrodynamic interaction between two copepods: a numerical study, J. Plank. Res. 24: 235-253

Lega, J., Passot, T. (2003), Hydrodynamics of bacterial colonies, Phys. Rev. 67: 1906

Machemer, H (1974), Frequency and directional responses of cilia to membrane potential changes in *Paramecium*, J. Comp. Physiol. 92: 293-316

Naitoh, Y., Sugino, K. (1984), Ciliary movement and its control in *Paramecium*, J. Protozool. 31: 31-40

Nasseri, S., Phan-Thien, N. (1997), Hydrodynamic interaction between two nearby swimming micromachines, Computational Mechanics 20: 551-559

Ramia, M., Tullock, D. L., Phan-Thien, N. (1993), The role of hydrodynamic interaction in the locomotion of microorganisms, Biophysical J. 65: 755-778

Pedley, T. J., Kessler, J. O. (1990), A new continuum model for suspensions of gyrotactic micro-organisms, J. Fluid Mech. 212: 155-182

Pedley, T. J., Ishikawa, T. (2004), Diffusion of swimming model micro-organisms in a semi-dilute suspension, Bull. Amer. Phys. Soc. DFD 49: 142

Chapter 3

Dynamics Modeling and Real-time Observation of Galvanotaxis in *Paramecium caudatum*

Naoko Ogawa[1,2], Hiromasa Oku[1], Koichi Hashimoto[3], and Masatoshi Ishikawa[1]

[1] Graduate School of Information Science and Technology, University of Tokyo, 7–3–1 Hongo, Bunkyo-ku, Tokyo 113–8656, Japan
[2] Japan Society for the Promotion of Science, 8 Ichibancho, Chiyoda-ku, Tokyo 102–8472, Japan
[3] Graduate School of Information Sciences, Tohoku University, 6–6–01 Aza Aoba, Aramaki, Aoba-ku, Sendai-shi 980–8579, Japan

Summary. Paramecium caudatum, a kind of ciliates, exhibits very strong galvanotaxis; when a DC electric field is applied, cells are made to swim toward the cathode. In this article, we propose a novel physical scheme for Paramecium galvanotaxis to provide a quantitative explanation, using a bottom-up approach based on systems theory. We analytically derived the torque produced with respect to the cell angle and constructed equations of translational and rotational motion. Using the proposed model, we performed numerical simulations. We also performed a preliminary evaluation of the proposed model by using real data. Experimental data were obtained by using a real-time galvanotaxis observation system developed in our laboratory. We found that the simulated data was approximately in agreement with the experimental results. [Arranged from (Ogawa et al. 2005) and (Ogawa et al. 2006)]

Key Words. Paramecium, galvanotaxis, dynamics model, tracking, microorganism

1 Introduction

Paramecium caudatum, a kind of ciliates, exhibits very strong galvanotaxis, a behavioral response to an electrical stimulus; when a DC electric field is applied, cells can be made to swim toward the cathode. Pioneers in this field of microbiology found that it is caused by a change in direction of the ciliary beating (Ludloff 1895), which is called the Ludloff phenomenon. However, there has been almost no quantitative discussion of the physical relationship between the microscopic ciliary beating pattern and the macroscopic behavior of a cell. Although several properties of Paramecium cells have been modeled (Jahn 1961; Cooper and Schliwa 1985), conventional models have mainly been physiological and biochemical ones, ignoring the physical properties. Moreover, the few physical models that have been presented have tended to disregard galvanotaxis. One rare physical model of galvanotaxis is that constructed by Roberts (Roberts 1970); however, its validity is uncertain because his assumptions were rough, and the accuracy of his model was not fully verified by comparing it with experimental data. Here in this article, we present a first attempt to construct a physical model of Paramecium galvanotaxis based on mechanics using a bottom-up approach, accompanied by experimental validation.

2 Dynamics Model of Paramecium Galvanotaxis

2.1 Paramecium and Its Galvanotaxis

In this article, we consider *Paramecium caudatum* because it has been extensively studied and its behavior is well known. *P. caudatum* is a unicellular protozoan with an ellipsoidal shape. It swims by waving cilia on its body; thousands of cilia beat the water backward to yield a forward reaction force (Naitoh and Sugino 1984). When an external electrical stimulus is applied, it modifies the membrane potential and alters the ciliary movements, thus affecting the cell motion. Viewed macroscopically, the cell is made to swim toward the cathode. This phenomenon is called negative galvanotaxis.

A Paramecium cell in an electric field shows a characteristic ciliary movement pattern. Assume an imaginary plane perpendicular to the electric field and located near the center of the cell, slightly closer to the cathodal end, dividing the cell into two parts, as illustrated in Fig. 1. The electric field causes cilia on the anodal end to beat more frequently (ciliary augmentation) (Kamada 1929), and the cilia on the cathodal end also beat more frequently but in the opposite direction (ciliary reversal) (Ludloff 1895). This is called the Ludloff phenomenon, and it provides a simple and qualitative explanation for galvanotaxis: the asymmetry in direction of the ciliary beatings generates a rotational force and orients the cell toward the cathode. Several previous works have reported existence of substantial boundary (Ludloff 1895; Statkewitsch 1903; Jennings 1923; Kamada 1931), and we also confirmed it in our wet experiments. The deviation of the boundary to the cathode is considered to be due to interaction between the medium and the negative resting potential of the cell (Görtz 1988).

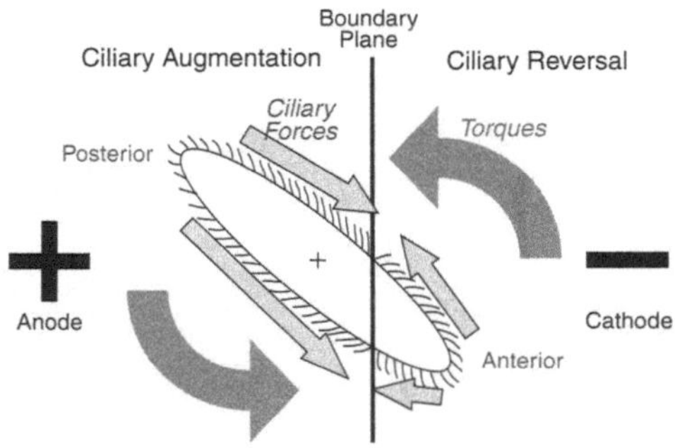

Fig. 1. Qualitative explanation for galvanotaxis. Reproduced from Ogawa et al. 2006 with permission.

Based on electrochemical and electrophysiological knowledge (Jahn 1961; Cooper and Schliwa 1985), the mechanism of the Ludloff phenomenon can be understood to be a combination of electrochemical, physiological, and physical factors. This article concentrates on physical factors, while regarding electrochemical and physiological factors as "black boxes".

2.2 Assumptions

By making several assumptions focusing only on those properties that are essential and dominant in galvanotaxis (Ogawa et al. 2006), we can describe the cell motion in a two-dimensional plane including the cell axis and the electric field vector. Hereafter, we consider cell motion only in this plane.

We define a global coordinate system (X,Y) and a local coordinate system (x,y) on the plane, as shown in Fig. 2. The global coordinate system is fixed with respect to the external world, with the X-axis parallel to the electric field $\boldsymbol{E}$. Let ϕ be the angle of the cell axis in the global coordinate system ($\phi < 0$ in Fig. 2(a), for the sake of convenience in deriving the model). The local coordinate system is fixed with respect to the cell.

Let the cell shape be an ellipsoid $\mathscr{E}$ with a major axis of length $2L$ and a minor axis of length $2R$. We assume that cilia are distributed uniformly around the edge of the ellipsoid with linear density n. In the presence of an electric field, imagine a plane perpendicular to the field (hereinafter referred to as "a boundary plane"). The shortest distance between the plane and the center of the cell is defined as l. The propulsion force yielded by one cilium is assumed to be f_0 in the absence of an electric field. In the presence of an electric field E, the force increases to $f = (1+\beta E)f_0$, where β is a positive parameter.

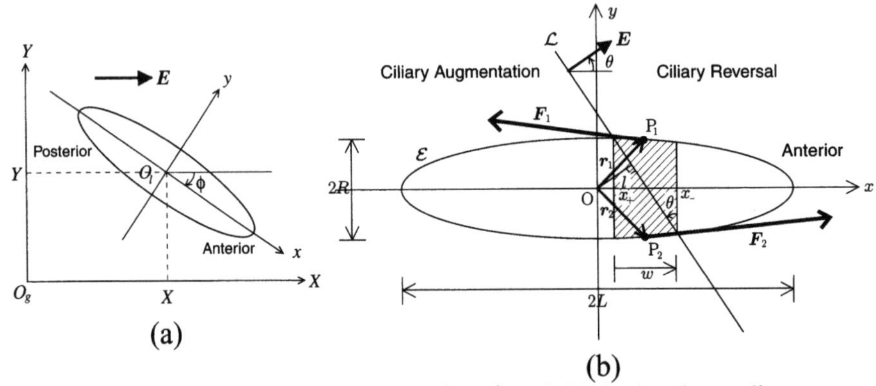

Fig. 2. (a) The global coordinate system (X,Y) and (b) the local coordinate system (x,y). Reproduced from Ogawa et al. 2006 with permission.

2.3 Model of the Torque

The phenomenon whereby a Paramecium cell swims toward the cathode is due to a torque caused by asymmetry of ciliary motion. In this section, we estimate this torque. Note that the following model is defined only if $\boldsymbol{E} \neq \boldsymbol{0}$.

First, consider an ellipsoid $\mathscr{E}$ in the local coordinate system (x,y), as illustrated in Fig. 2(b). In the substantially trapezoidal region formed by the intersection of the boundary plane and the ellipsoid shown as the hatched region in Fig. 2(b), beating direction of cilia on both sides differ, thus causing asymmetry in forces. The forces generated by the cilia on the cell other than the region do not contribute to the torque generation. Thus, we have only to consider the forces generated at this hatched region.

For convenience, let us introduce $\theta = -\phi$ as the angle of the electric field $\boldsymbol{E}$ in the local coordinate system.

Because it would be too complicated to consider the individual minute forces generated by each cilium, here we focus on the resultant forces for simplicity. We set sites of action, $\mathrm{P}_1(x_\mathrm{a}, y_\mathrm{a})$ and $\mathrm{P}_2(x_\mathrm{a}, -y_\mathrm{a})$, at the mid-points of the sides of the trapezoid and assume the directions of the forces to be tangential to the ellipsoid. We then define position vectors, $\boldsymbol{r_1} = \overrightarrow{\mathrm{OP}_1}$ and $\boldsymbol{r_2} = \overrightarrow{\mathrm{OP}_2}$. Next, let us suppose that the magnitude of the resultant force is proportional to the "height" of the trapezoid w, which is a signed value of the same sign as θ. Then, the propelling forces $\boldsymbol{F}_1$ and $\boldsymbol{F}_2$ at the points P_1 and P_2, respectively, are represented by $\boldsymbol{F}_{\{1,2\}} = \mp f w n \boldsymbol{m}_{\{1,2\}}$, where $\boldsymbol{m}_1$ and $\boldsymbol{m}_2$ are the unit tangent vectors at P_1 and P_2, and the $\mp$ symbol indicates the two directions of ciliary beating. Thus, we find the torques at the points $\mathrm{P}_{\{1,2\}}$, namely, $\tau_{\{1,2\}} = \boldsymbol{r}_{\{1,2\}} \times \boldsymbol{F}_{\{1,2\}}$. It should be noted that these vectors are treated as three-dimensional in calculating cross products. The total torque rotating the cell body is given by $\tau = \tau_1 + \tau_2$. Since its x and y components are obviously zero, hereafter we call its z component, τ_z, the "torque".

Finally, by substituting $\phi = -\theta$, the torque is described in the global coordinate system as:

$$\tau_z(\phi) = -\frac{4LR^2 fns\sqrt{L^2c^2 + R^2s^2 - l^2}}{\sqrt{L^4c^4 + 2L^2R^2c^2s^2 + R^4s^4 - L^2l^2c^2 + R^2l^2c^2}}, \tag{1}$$

where $s = \sin\phi$ and $c = \cos\phi$. This equation describes the torque generated in a cell oriented at angle ϕ.

2.4 Forward Propulsion Force

As mentioned above, only the cilia on the trapezoid contribute to the rotation. Other cilia are responsible for translational propulsion. Considering that the reversed effective strokes on the cathodal side produce a backward force, we can derive the magnitude of the force F as:

$$F = \begin{cases} 2fn|x_\mathrm{a}| & (|L\cos\theta| > l), \\ fn(2L - |w|) & (|L\cos\theta| < l). \end{cases} \tag{2}$$

2.5 Equations of Motion of Paramecium Cell

Using the torque estimated in section 2.3, we now discuss the equations of motion of the Paramecium cell.

In the micrometer-scale world that paramecia inhabit, the inertial resistance of the fluid is small enough to be negligible, and the viscous resistance becomes dominant. Hence, we can apply Stokes' law, derived from the Navier-Stokes equation by ignoring inertial force. Since a rigorous evaluation of the viscous resistance around an ellipsoid is quite complicated, here we approximate the viscous force by applying the formula for a sphere as a substitute. According to Stokes' law, the force exerted on a sphere with radius a, moving with velocity v in a viscous fluid is given by $F_\mathrm{s} = 6\pi\mu a v$, where μ is the viscosity of the fluid. From this equation, the viscous force around the ellipsoidal cell can be obtained by replacing the radius a by the cell radius R. Thus, the equation of motion for the translational motion of the cell can be approximated by $M\ddot{\boldsymbol{X}} + D\dot{\boldsymbol{X}} = \boldsymbol{F}$, where $\boldsymbol{X} = (X, Y)^\mathrm{T}$ is the cell position (the superscript T means the transposition), $\boldsymbol{F} = F\boldsymbol{e}$ is a forward propulsive force, $\boldsymbol{e} = (\cos\phi, \sin\phi)^{\mathrm{T}}$ is a unit vector along the body axis, $D = F_{\mathrm{s}}/|\dot{\boldsymbol{X}}| = 6\pi\mu R$ is the viscous friction coefficient, $M = \rho V$ is the cell mass, ρ is the cell density, and $V = 4\pi LR^2/3$ is the cell volume.

In addition, the equation of motion for the rotation is given by $I\ddot{\phi} + D'\dot{\phi} = \tau_z(\phi)$, where $I = \pi M(R^2 + L^2)/5$ is the moment of inertia for an ellipsoid, $D' = \delta\pi\mu L^3$ is the viscous friction coefficient, and δ is a coefficient to compensate for errors in the model.

Finally, integration of the equations of motion for the translational motion and the rotational motion leads to the following equations with a notation common in systems theory:

$$\dot{\boldsymbol{q}} = A\boldsymbol{q} + \boldsymbol{B}(\boldsymbol{q}), \tag{3}$$

$$A = \begin{pmatrix} 0 & 0 & 1 & 0 & 0 & 0 \\ 0 & 0 & 0 & 1 & 0 & 0 \\ 0 & 0 & -D/M & 0 & 0 & 0 \\ 0 & 0 & 0 & -D/M & 0 & 0 \\ 0 & 0 & 0 & 0 & 0 & 1 \\ 0 & 0 & 0 & 0 & 0 & -D'/I \end{pmatrix}, \quad \boldsymbol{B}(\boldsymbol{q}) = \begin{pmatrix} 0 \\ 0 \\ (P\cos\phi)/M \\ (P\sin\phi)/M \\ 0 \\ \tau_z(\phi)/I \end{pmatrix}$$

where $\boldsymbol{q} = (X, Y, \dot{X}, \dot{Y}, \phi, \dot{\phi})^{\mathrm{T}}$, and $P = 2fn|x_{\mathrm{a}}|$.

Practically, acceleration terms can be ignored because of the smallness of the cellsOgawa et al. (2007). In this paper, we included acceleration terms for accurate description.

3 Numerical Experiments

We performed some numerical experiments to verify the equations of motion using numerical analysis software (MATLAB, MathWorks Inc.).

Table 1 shows several physical parameters used in the numerical experiments. We obtained the cell size by observing cells incubated in our laboratory. The value of β, the increase in beating frequency with electric field, was estimated from the fact that the frequency increased to around 50 Hz under a stimulation of a few volts per centimeter from the frequency in the regular state of around 15-20 Hz (Naitoh and Sugino 1984).

The force yielded by cilia per unit length, fn, is still an unknown parameter. However, our model itself is based on approximation, and a strict evaluation of fn is not so critical. Hence, we estimated the order of fn by

Table 1. Parameters of the proposed model. Reproduced from Ogawa et al. 2006 with permission.

Parameters	Values	Comments
Major cell axis $2L$	$100\mu m$	our strain
Minor cell axis $2R$	$25\mu m$	our strain
Boundary plane offset l	$10\mu m$	(Jennings 1923), (Kamada 1931)
Viscosity of water μ	1.00×10^{-3} kg/(ms)	at 20 ℃
Cell density ρ	$1,000$ kg/m^3	same as water
Increase in beating freq. β	2.00×10^{-3} V^{-1}	

using the swimming velocity measured in past experiments. The terminal velocity of a cell was obtained by substituting $\ddot{\boldsymbol{X}} = \boldsymbol{0}$ into the equation of motion under the conditions $\phi = 0$: $\dot{\boldsymbol{X}} = \boldsymbol{F}/D = \boldsymbol{e}fn|x_a|_{\phi=0}/3\pi\mu R$. Since x_a equals l at $\phi = 0$, we can estimate fn from: $fn = 3\pi\mu R|\dot{\boldsymbol{X}}|/l$. Measurement of the cell velocity by using a high-speed tracking system, described later in section 4.3, gave a velocity of around 400μm/s. Using this value, we estimated fn to be 4.71×10^{-6} N/m. In addition, we adjusted the parameter δ to be 7.5 based on experimental data of the cell trajectories obtained by our system (Ogawa et al. 2005).

Next, we tested whether our proposed model could exhibit this phenomenon like real cells.

When an electric field is applied in the direction opposite to the swimming direction of a cell, the cell makes a U-turn motion. We chose this behavior for the test. Swimming trajectories for cells with eleven different initial orientations were calculated. Figure 3 shows all trajectories simultaneously. All cells were configured to have the same initial position, namely, at the origin $(0,0)$, but their initial angles ϕ differed by intervals of 30°($-150°$, $-120°$, ..., $150°$). An electric field of 5.0 V/cm was applied along the X-axis. The trajectory of each cell was calculated by solving ordinary differential equations. As shown in Fig. 3, all cells starting from the origin turned toward the cathode, like real cells. It is interesting that realistic macroscopic behavior emerged from a microscopic description of the ciliary motion.

4 Real-Time Observation System and Wet Experiments

In this section, we describe a novel system for real-time continuous observation of galvanotaxis using a visual tracking method. With this system,

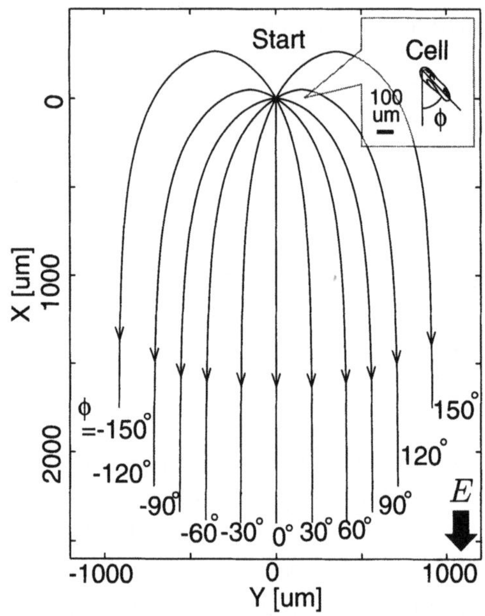

Fig. 3. Simulation of U-turn motions of cells. Reproduced from Ogawa et al. 2006 with permission.

we verified the validity of the model using these data.

4.1 Why Tracking?

To realize precise observation of freely swimming cells, it is necessary to perform continuous observation of rapidly swimming, non-immobilized cells, to provide a sufficiently large working area, and to perform detailed observation of a specific cell with high magnification to precisely control its actuation. We found, however, that most conventional microscope systems could not satisfy these demands.

To overcome these obstacles, we adopted a specially designed tracking method. In this article, we mainly use the term "tracking" in the sense that the camera pursues a target so as to keep it always in the center of the visual field (sometimes we also call it "lock-on tracking"). As shown in Fig. 4, a lock-on mechanism can be realized by moving the position of the specimen on a stage so that the camera always keeps the target at the center.

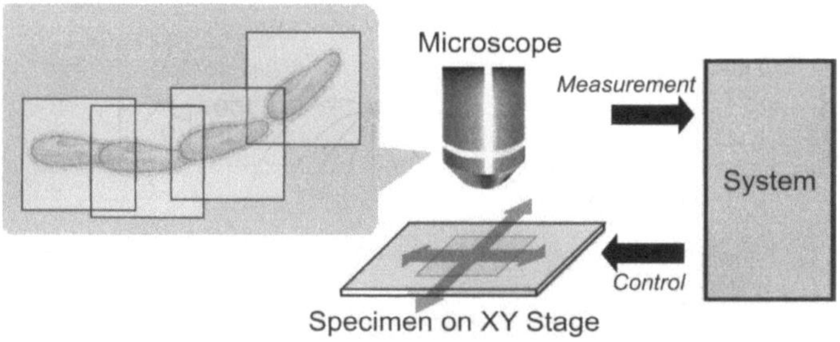

Fig. 4. Lock-on tracking scheme. The camera pursues a target so as to keep it always in the center of the visual field. Reproduced from Ogawa et al. 2005 with permission. ©2005 IEEE

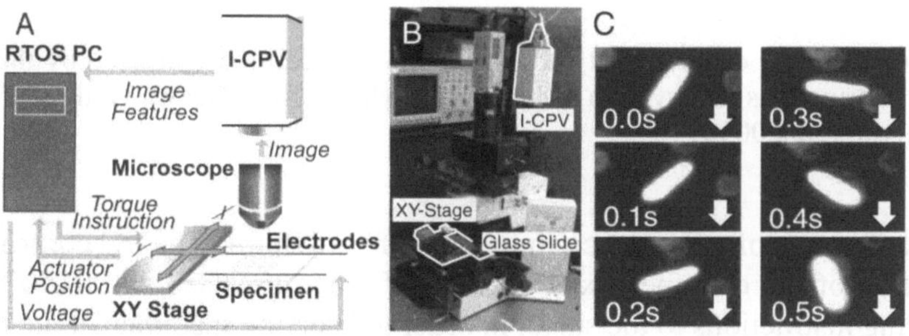

Fig. 5. Experimental set-up of our tracking system. A: Schematic overview of the system. B: Photograph. C: An example of U-turn motion monitored by our system. Reproduced from Ogawa et al. 2005 with permission. ©2005 IEEE

4.2 System Configuration

Our system measures the cell position and angle continuously at a 1-kHz frame rate, using a high-speed lock-on tracking method. The configuration of the overall system and its block diagram are illustrated in Fig. 5. An electrical stimulus is applied to cells swimming in a chamber placed on an electrical stimulus input device mounted on an XY stage. The stage is controlled by a high-speed vision system so as to keep a cell in the center of the field of view. The position and orientation of the cell are calculated from image features. More details of this system can be found in our previous work (Ogawa et al. 2005).

The I-CPV vision system (Toyoda et al. 2001) is mounted on an upright optical microscope (Olympus, BX50WI) and captures dark-field images at 1-kHz frame rate. From the captured images, the I-CPV calculates image moments and sends them to the PC. They are used for calculating the attitude and the position of the target. Figure 5 C shows an example of the

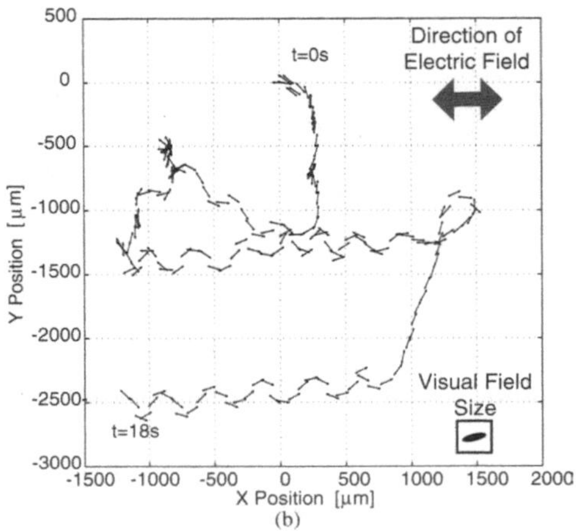

Fig. 6. Example of trajectory and orientation of a cell reconstructed by the real-time galvanotaxis observation system. Reproduced from Ogawa et al. 2005 with permission. ©2005 IEEE

tracked image of a Paramecium cell making a U-turn. The PC controls the position of the chamber fixed on the XY stage (SMC, LAL00-X070) by sending instructions to the stage, in order to keep the specimen at the center of the visual field.

In the electrical stimulus input device, two carbon electrodes of 0.5-mm diameter are placed 22-mm apart in parallel on a glass slide to allow control of the electrical stimulus in one direction. The specimen chamber, which is 0.17-mm deep, is placed between the electrodes. The PC provides a voltage in the range ± 10 V to the electrodes via a D/A converter board (Interface, PCI-3310).

The whole system is controlled at a frequency of 1 kHz by the PC running a real-time OS (800 MHz, ART-Linux). Figure 6 shows an example of the trajectory and orientation of a swimming Paramecium cell reconstructed by the real-time galvanotaxis observation system. In this experiment, the electric field was applied leftward during 0-6 s, then rightward during 6-12 s, and leftward again during 12-18s.

4.3 Experiment

Experimental data were obtained by high-speed measurement of the responses of a single cell to an electric field, using the real-time galvanotaxis observation system. Wild-type *P. caudatum* cells were cultured at 20-25°C in a soy flour solution. Cells grown to the logarithmic or stationary phase

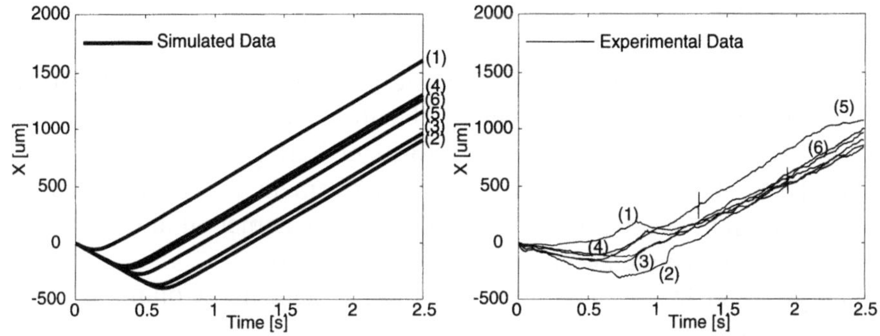

Fig. 7. Comparison between simulated data (left) and experimental data (right) in the U-turn motion. Only the X-coordinates (positions along the electric field) were extracted. Experimental and artificial data for three seconds from the application of a stimulus in six trials are overlaid. Each number indicates a couple of data sharing the same initial angle. The figure indicates that the simulated data was approximately in agreement with the experimental results.

(4-10 days after incubation) were collected together with the solution, filtered through a nylon mesh to remove debris, and infused into a chamber. A DC electric field with a step-like temporal profile rising to 4.1 V/cm was applied to the cells.

We compared simulated and experimental positions in the U-turn motion. We extracted positions along the electric field (X direction), because X-disposition is almost independent of fluctuations caused by spiral motions, which we disregarded.

Figure 7 shows experimental data (right) for three seconds from application of a stimulus (reversal of the electric field) in six trials. In the simulation (left), the initial angle in each trial was set to the same value in the measured data. The field strength was set to 4.1 V/cm, the same as in the wet experiment. The figure indicates that the simulated data was approximately in agreement with the experimental results. A more rigorous comparison will require three-dimensional tracking (Oku et al. 2006), and 3-D experiments are currently underway in our laboratory.

5 Conclusion

In this article, we have proposed a physical model of Paramecium galvanotaxis using a bottom-up approach to link the microscopic ciliary motion and the macroscopic behavior of a cell. We investigated the validity of the model by numerical experiments using a novel observation system.

References

Cooper MS, Schliwa M (1985) Electric and ionic controls of tissue cell locomotion in DC electric fields. J Neurosci Res 13:223–244

Görtz HD (Ed) (1988) Paramecium: Springer-Verlag

Jahn TL (1961) The mechanism of ciliary movement. I. Ciliary reversal and activation by electric current; the Ludloff phenomenon in terms of core and volume conductors. J Protozool 8:369–380

Jennings HS (1923) Behavior of the Lower Organisms: Columbia University Press

Kamada T (1929) Control of galvanotropism in Paramecium. J Fac Sci Imp Univ Tokyo, Sect. IV, Zoology 2:123–139

——— (1931) Polar effect of electric current on the ciliary movements of Paramecium. J Fac Sci Imp Univ Tokyo, Sect. IV, Zoology 2:285–298

Ludloff K (1895) Untersuchungen über den Galvanotropismus. Arch Ges Physiol 59:525–554

Naitoh Y, Sugino K (1984) Ciliary movement and its control in *Paramecium*. J Protozool. 31:31–40

Ogawa N, Oku H, Hashimoto K, Ishikawa M (2005) Microrobotic visual control of motile cells using high-speed tracking system. IEEE Trans Robot 21:704–712

——— (2006) A physical model for galvanotaxis of paramecium cell. J Theor Biol 242:314–328

——— (2007) Trajectory planning of motile cell for microrobotic applications. J Robot Mech 19:190–197

Oku H, Theodorus, Ishikawa M, Hashimoto K (2006) High-speed autofocusing of a cell using diffraction pattern. Opt Express 14:3952–3960

Roberts AM (1970) Motion of *Paramecium* in static electric and magnetic fields. J Theor Biol 27:97–106

Statkewitsch P (1903) Ueber die Wirkung der Inductionsschläge auf einige Ciliata. Le Physiol Russe 3:1–55

Toyoda H, Mukohzaka N, Nakamura K, Takumi M, Mizuno S, Ishikawa M (2001) 1ms column-parallel vision system coupled with an image intensifier; I-CPV. In: Proc Symp High Speed Photography and Photonics 2001 5-1:, pp 89–92, in Japanese

Chapter 4

Object Manipulation by a Formation-controlled *Euglena* Group

Akitoshi Itoh[1] and Wataru Tamura[2]

[1] Department of Mechanical Engineering, School of Engineering, Tokyo Denki University, 2-2 Kanda Nishiki-cho, Chiyoda-ku, Tokyo 101-8457, Japan
[2] Graduate School Student, Tokyo Denki University, 2-2 Kanda Nishiki-cho, Chiyoda-ku, Tokyo 101-8457, Japan

Summary. This study investigates how to use protist group as huge group of living micromachines. Motion control of protist is made by using the orientation phototaxis of Euglena. Blue laser scanning system is made to form the group of Euglena. Euglena gather around the blue laser irradiated area. The construction of the experimental system is that blue laser passed through two galvano-scanners to make the two dimensional positioning of laser possible, concentrated by a convex lens and then irradiated into the experimental pool. After that, blue light is attenuated by high cut filter, and recorded by CCD camera with macro lens. Since there is little response of phototaxis of Euglena, red LED light is used for background illumination. Experimental results show that this system can make any shape of Euglena group by gathering along the scanned laser beam by their positive orientation phototaxis. If the moving speed of the laser beam is under 4μm/s, Euglena group can follow to the laser. The formed Euglena group can transport objects by moving its group. The maximum transportation weight force transported by the Euglena group was about 70nN.

Key words. Micromachine, Micro Cyborg, Micro Manipulation System, Laser Scanning System and Automatic Motion Control.

1 Introduction

Recently, many studies have been done to investigate the moving mechanisms of microorganisms to make use of making future micromachines. Typical examples are studies of bacterial flagellar motor. When we see the recent great results of biotechnology, however, we can find that almost all of these results are applied the real biological systems of living things, such as gene manipulation, tissue engineering, etc. Therefore, it is very important to investigate the applicability of bio-applied mechanical system or mechano-bio fusion system in the micro-mechanical research field.

As far as the author have known, the idea of using micro organisms as living micromachines are firstly seen in the paper of R. Fearing et al. (Fearing 1991). They controlled Paramecium's motion along the square shaped guide route.

The author had the same idea independently with Fearing et al., and made automatic motion control system using negative galvano-taxis of Paramecium. They succeeded to control Paramecium automatically along the star shaped guide route. They could also rotate ϕ0.5mm micro impeller by using motion controlled Paramecium. This result showed that we can use micro organisms as micromachines (Itoh 2000).
The author et al. also developed the positioning control system of the downward flow of bioconvection using Tetrahymena. They can move 20mm span seesaw reciprocatedly by the position controlled downward flow of bioconvection. This is an example of the shapeless actuation system using huge groups of microorganism (Itoh et al. 2001).
Negative galvano-taxis was applied in all of these experiments. The taxis of microorganism, however there are a lot of variety of taxis like phototaxis, magnetotaxis, thermotaxis, gravitaxis, etc. If we can apply the other taxis to control micro organisms, we can choose so many varieties of micro organisms to use. From this point of view, the author et al. also developed the motion control system for mastigophoran using their phototaxis (Itoh 2004). The control method is that a shadow is made in the strong intensity bright field and a Euglena can be captured into the shadow by its negative photophobic response. And the position of the Euglena can be controlled by moving the shadow. The experimental results showed that the position of the Euglena can be controlled by this system. The flexibility of this system, however, is not so high. Therefore, it is necessary to modify the control system.
Based on these details, a phototaxis based motion control system for Euglena was constructed by using a blue laser and galvano scanners. Motion of a huge group of Euglena was controlled and objects are manipulated by using this Euglena group. Main results are reported as follows.

2 Experimental System and Specimen

The main features of laser beam are, high density distribution of energy, high directivity, mono wave length light and coherency. High density energy distribution and high directivity have advantages for trying various control methods. Laser light is mono wave length, therefore, if we choose the frequency which is very effective to Euglena, we can get high response rate of Euglena. Basic biological study showed that the wave length range between 460 and 500 nm is the most effective for Euglena (Egans et al. 1975). It was also confirmed by the basic reaction experiment using LED mono wave length light.
Based on these facts, the blue laser of which wave length λ=473nm (5mW output, CrystaLaser Inc. BIOL-005) was used to generate phototaxis. Red LED light (l=644nm) was used for background illumination since the light over 600nm is not affected to Euglena.
Basic construction of the laser motion control system is depicted in Fig.1. Laser beam is refracted twice by the mirrors which were positioned by galvano scanners.

They make laser beam positioned in any place in the x-y plane of the experimental pool. Then, beam was concentrated by the convex lens and irradiated into the experimental pool. After that, the beam was attenuated by a polarizing filter and a high cut filter. After attenuation, we can see the experiment by normal CCD camera with x200 macro lens. We cannot see the laser light during experiments. Therefore, an automatic calibration program was made to make the open-loop positioning of laser beam possible.
The experimental protists are typical photo-responsive mastigophora, "Euglena gracilis". The body length of euglena is about 40-50μm. A long flagellum is grown from the front side for a motor organ. The optical sensor of euglena is considered as an eyespot and a subflagellum body, which are both located at the base of the flagellum.

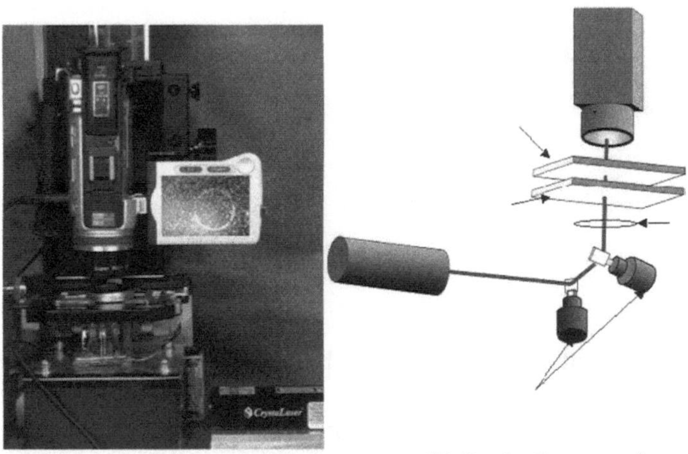

(a) External Appearance (b) Basic Construction

Fig. 1. Experimental System for Motion Control Experiments

3 Basic Reaction Properties to The Blue Laser

3.1 Reaction Behaviors to The Fixed Laser beam

First, the reaction properties of Euglena to the fixed blue laser spot were examined. Experimental pool is a shallow pool (0.13mm depth). The top of this pool is sealed by a cover glass.
The experimental results showed that every Euglena reacted to the fixed laser beam. There were two types of reaction. (1) Avoiding reaction: When the laser beam was irradiated to the Euglena, the Euglena turned its swimming direction.

This is the typical Euglena's negative photophobic response. Fig.2 (a) depicts this motion. (2)Rotating reaction: When the laser beam was irradiated to the Euglena, the Euglena started rotating and kept rotating. Fig.2 (b) depicts this motion. This mechanism is unknown; however, the possibility of continuous avoiding reaction is very high. The positive orientation phototaxis has not seen when these responses were observed. This fact seems that the observation magnification is too high and the energy density distribution is too keen to observe positive reaction.
The results of avoiding reaction and rotating reaction have counter side effects. Avoiding reaction means "escape". However, rotating reaction means "capture". Therefore, it is very important to predict which reaction will be appeared. It is also important to predict the turning angle when Euglena shows avoiding reaction. Therefore, the observation results are summarized and the effect of reacting distance, reacting direction and laser power were investigated.

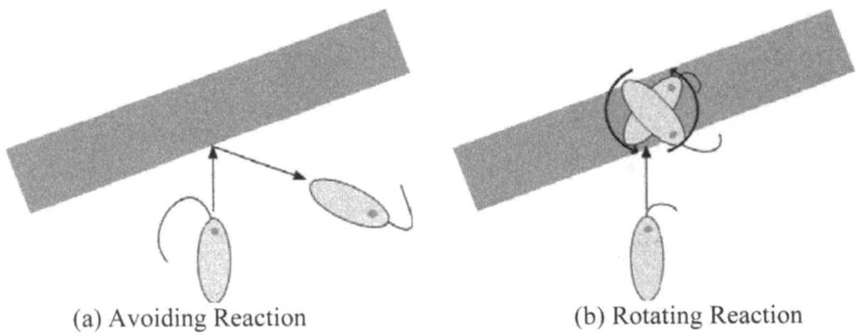

Fig. 2. Response Examples of Euglena to the laser beam

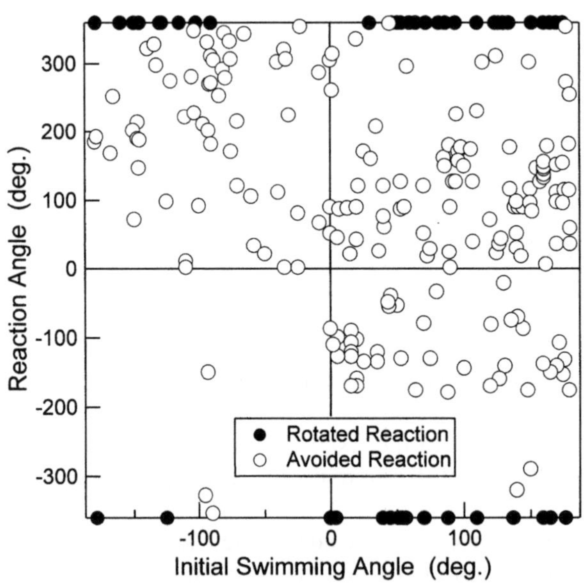

Fig. 3. Relationship between initial swimming angle and reacted turning angle

The results showed that the turning angles are basically at random, and there is no correlation among these conditions and turning angles. Fig.3 shows the relationship between initial swimming angle and reacted turning angle. Rotating reaction was indicated by plotted on 360 deg area in Fig.3. Therefore, we have to forgive to predict which reaction may occur.

3.2 Reaction Behaviors to The Scanning Laser beam

Next, the reaction properties of Euglena to the scanning laser beam were examined. The scanning frequency was varied during 1 - 20 Hz. Basic reactions on these experiments were, above mentioned avoiding reaction, rotating reaction, no reaction and swimming along the laser line with spiral rotation. The last reaction was considered that the rotating reaction was observed from the different (upward in Fig.2 (b)) angle.
The investigation results showed no correlation. Therefore, it is very difficult to predict which reaction will be appeared or to predict the turning angle.
The amount of not reacted examples has a correlation to the scanning frequency and the length of the scanning line. This is corresponded to the probability ratio that the laser beam is not irradiated to the sensory spot of the Euglena when the Euglena swims across the scanning laser line.
Above observation experiments showed that the response of Euglena is basically random, and it is very difficult to control the swimming direction to our aiming direction by irradiating the laser beam. The only possibility is to change the swimming direction from the present swimming direction. Therefore, the control efficiency is worse than the control method using galvano-taxis when we control an individual cell.

4 Mechanism of The Gathering and The Group Formation

The reaction of Positive phototaxis has not seen in the individual cell level observations. However, when a laser was irradiated into the high population density culture media, a gathering and group formation of Euglena was seen in the media. Fig.4 shows an example. In this experiment, fixed spot laser beam was irradiated to the high density (2600 cell/mm^3) culture medium. This photograph was taken after 10 minutes irradiation. Therefore, it is necessary for the motion control of Euglena group to investigate how the Euglena gather and form the group.
Therefore, the influence of cell population density of culture media and the power of the laser beam on the gathering and group formation process was measured when a laser spot light is irradiated. Fig.5 summarizes the changes of the density distribution around the laser spot light through the time passage. First, a ring-shaped group was formed around the spot laser beam, and then, the high density area enlarged and beam area was finally covered with Euglena.

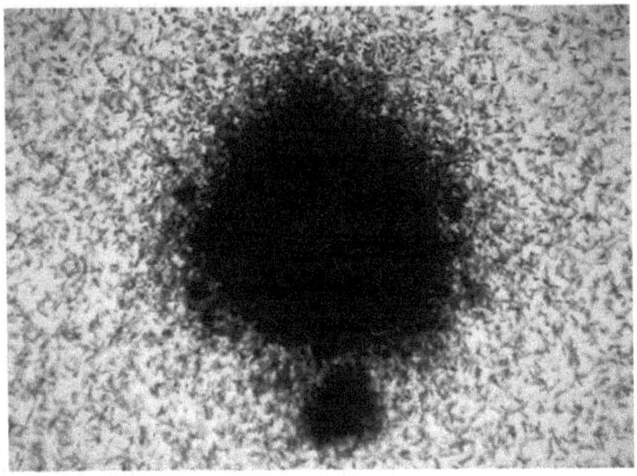

Fig. 4. An example of the Euglena group gathered to the fixed laser beam (Cell density of the culture medium:2600 cells/mm^3, After 10 minutes laser irradiation)

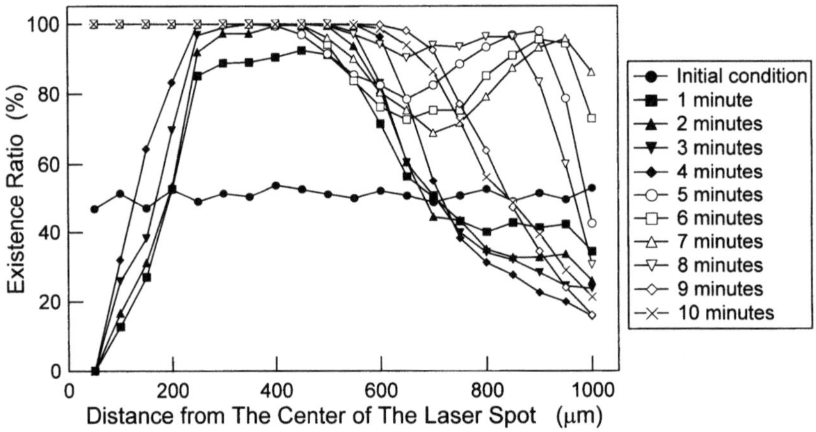

Fig. 5. Changes of the density distribution around the laser spot light through the time passage at the population density of 2600 cells/mm^3

These group formations are basically by the positive orientation phototaxis of Euglena. However, the reason why we cannot see positive orientation phototaxis at the individual cell is not clear. Therefore, low density culture media were used for the next experiments. Fig.6 is an example photograph that was taken after 90 minutes irradiation of spot laser beam to the 190 cell/mm^3 culture medium. The changes of the density distribution around the laser spot light were also summarized in Fig.7. The results showed that the Euglena group was formed in the final stage in the case of the low density media, it needs longer time. There is no group after 10 minutes irradiation. Group formation starts after 30 minutes irradiation.

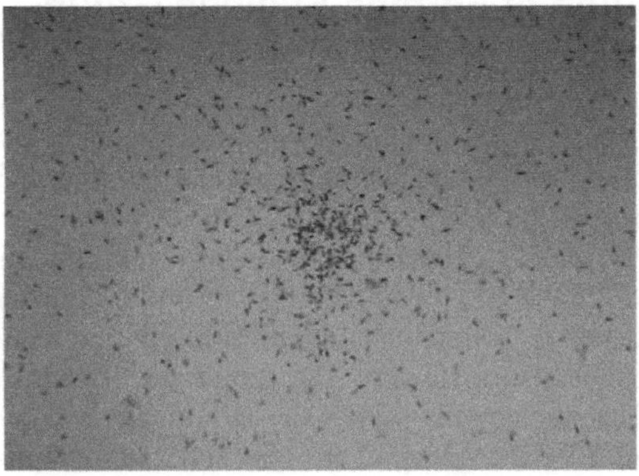

Fig. 6. An example of the gathering of Euglena to the laser beam at the population density of culture medium: 190 cells/mm^3, taken after 90 minutes laser irradiation

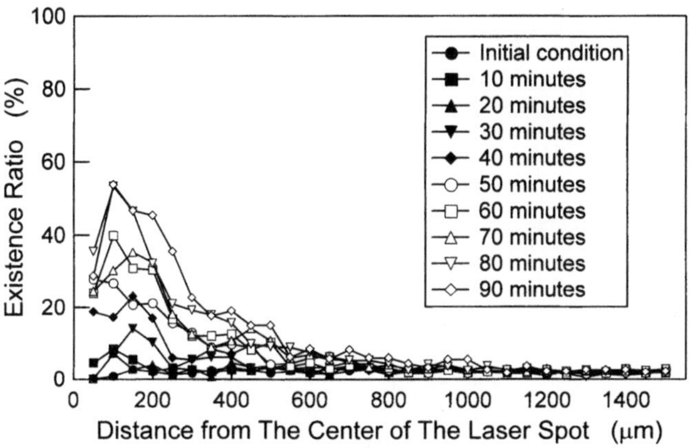

Fig. 7. Changes of the density distribution around the laser spot light through the time passage at the population density of 190 cells/mm^3

That is to say, the reasons why we have not observed positive orientation phototaxis are these two. (1) Positive orientation phototaxis is a statistic phenomenon based on the photophobic response, and it needs longer time to go to the comfortable illumination intensity area. Therefore, it has not appeared in one reaction on the above-mentioned observation. (2) The magnification for individual cell reaction is too large and we only observed too near area around the laser. As a result, the Euglena only showed the negative reaction.

The reason why group formation was accelerated by the population density is considered as follows. High density media have more probability ratio of collision between the Euglena that are swimming to the laser by positive phototaxis and that are swimming against the laser by negative phototaxis. This effect restricted the swimming course; hence, the group formation was accelerated.

5 Deformation and Transition of Formed Euglena Group

To apply the formed Euglena group, we have to move and deform the group. It is done by moving the scanning laser beam. If the moving speed of laser beam is too fast, the Euglena group can not catch up with laser. Therefore, we have to limit the moving speed of laser irradiation area.

The shape and the condition of the Euglena group to the moving laser spot irradiation were investigated to clarify the limit of the moving speed. The culture media that the population density is about 2000 cells/mm3 were used for this experiment. The depth of this experimental pool is also 0.13 mm. The results are shown in Fig.8. The left side photographs show the condition after the 10 minutes irradiation of spot laser beam to form the Euglena group. The right side photographs show the results after moving the laser beam to the right down area by the setting speed. The results show that if the moving speed is under 4mm/s, the Euglena group can catch up with the transition of laser beam.

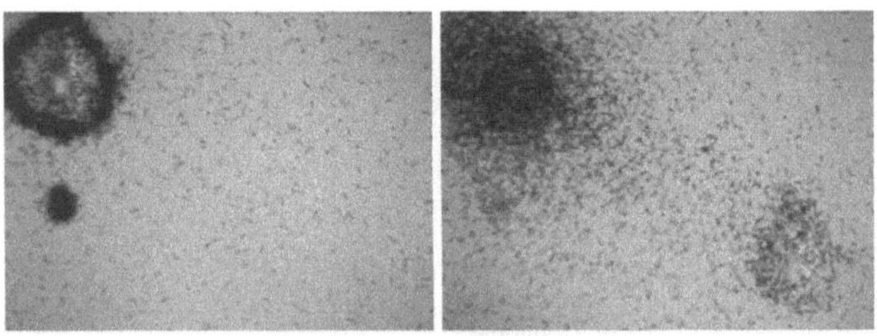

(a) Transition speed: 12.3μm/s

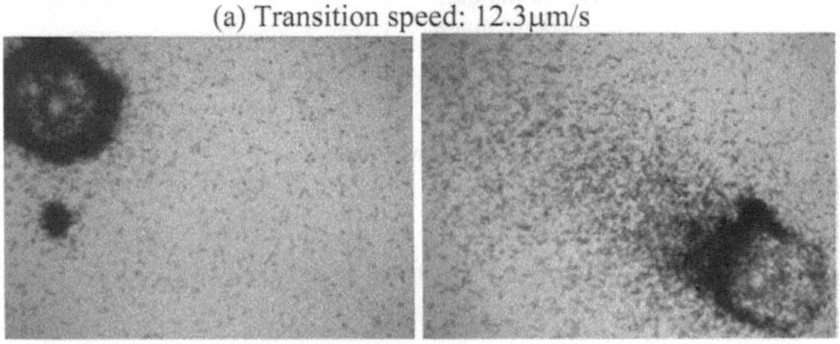

(b) Transition speed: 6.49μm/s

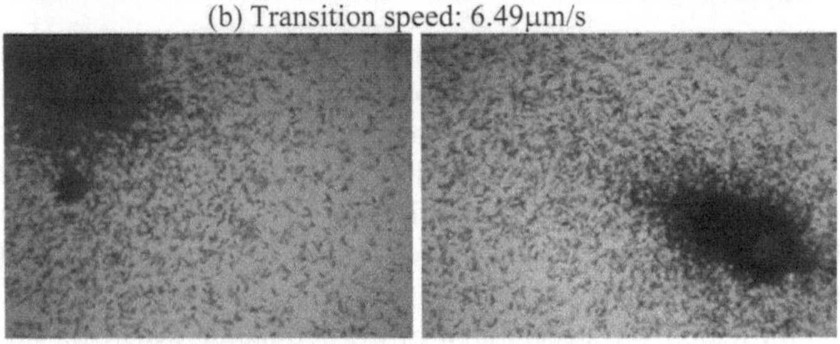

(c) Transition speed: 3.74μm/s

Fig. 8. Transition of Euglena group by moving the laser spot beam (Pool depth: 0.13mm, Population density: 2000 cells/mm^3)

Therefore, the deformation of the Euglena group was examined in the next experiments. The method is, first, a spot laser was irradiated in the pool. And then, the spot beam was deformed to the circle shape by controlling the galvano-scanner. Then the scanning circle line was gradually deformed to the line. Finally, the scanning line was deformed again to the circle shape. The transition speed is 3 mm/s. Fig.9 shows the continuous photographs from the experimental video image. This result shows that we can generate any shapes of the Euglena group if the transition speed of laser is under 4 mm/s.
However, to generate the shape of the Euglena group faithfully to the scanning laser, we have to prepare shallow pool (< 0.2mm) and high density culture media (> 1000 cells/mm3). If the experimental pool is too thick, the ground which is not irradiated by laser was covered by the Euglena and it is difficult to distinguish the shape of the Euglena group.

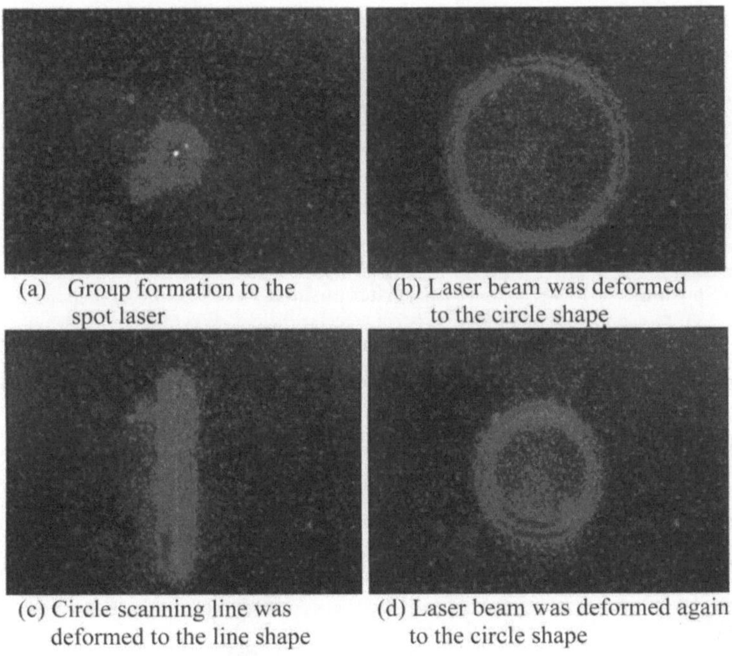

(a) Group formation to the spot laser

(b) Laser beam was deformed to the circle shape

(c) Circle scanning line was deformed to the line shape

(d) Laser beam was deformed again to the circle shape

Fig. 9. Deformation of the Euglena group by changing the scanning line of laser beam (Pool depth: 0.13mm, Population density: 2000 cells/mm^3)

6 Object Manipulation by Motion Controlled Euglena Group

Now we can make any shapes of Euglena groups and move them and deform them at our will by above mentioned technology. Therefore, the possibility whether

these controlled Euglena group can move and manipulate the object or not was investigated.
If the Euglena group was hit on the object, the unbalanced force from the collision of so many Euglena to the object might move the object. Therefore, object transporting experiments were done by using spot laser beam and high density culture media (Population density: 2000 cells/mm^3). Fig.10 shows an example. In this experiment, spot laser was moved to the upward direction. If the condition is within the range, we can manipulate the object by the motion controlled Euglena group.
It is very easy for this experimental system to generate the shape of the Euglena group that is fit to the target object. Therefore, a line shaped Euglena group was generated for long rectangle target object. Fig.11 is an example. Object transportation can be done more stably by fitting the shape of the Euglena group to the target object.

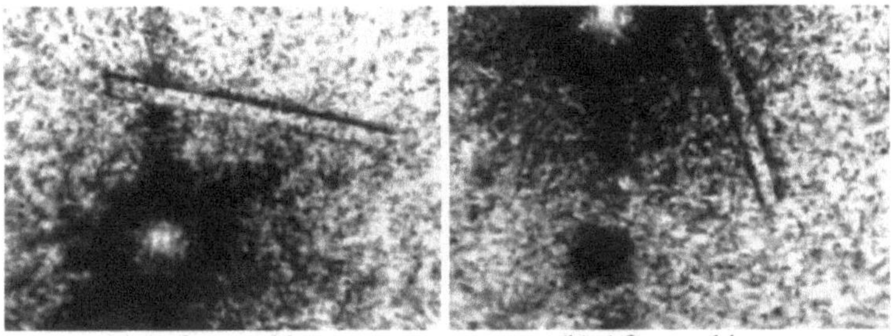

(a) Before pushing (b) After pushing

Fig. 10. Object manipulation experiment by position controlled Euglena group generated by spot laser beam (Object size: 1000x50x18μm, Fluorocarbon resin)

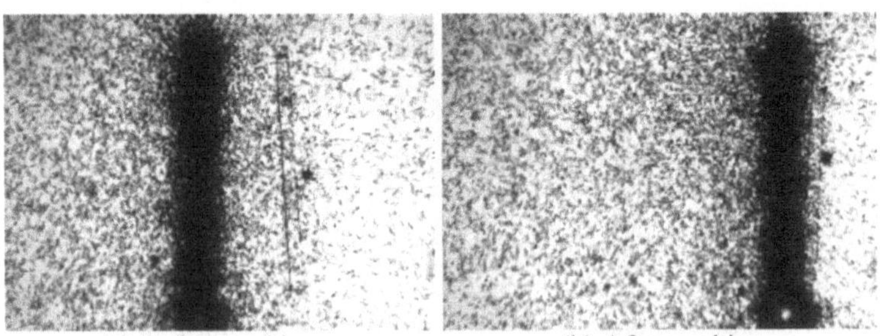

(a) Before pushing (b) After pushing

Fig. 11. Object manipulation experiment by position controlled Euglena group generated by 1mm line scanned laser beam (Object size: 1000x50x18μm, Fluorocarbon resin)

There is a limitation for the target object that can be moved by this method. Weight (force to the ground surface of the pool) and shape of the object dominate it. The densities of metals, such as aluminum or steel, are too large to push. Therefore, metal cannot be moved by the Euglena group. If the shape is the same to the

object of Fig.10, many of plastic films can be moved by this method, such as fluorocarbon resin, polyester, polyimide, polypropylene and polyethylene terephthalate.
The effect of the friction force and the adsorption force from the bottom surface is large. Therefore, it is very important to maintain the contact condition. The effect of the surface tension force is also large. Therefore, if the object is lighter than the water, sealed type experimental pool is better than the open air type pool.
Basically, the transportation of the object is getting easier as the contacting area to the Euglena is getting larger, and as the force to the bottom surface of the pool is getting smaller. In the case of thickness t=18μm fluorocarbon resin, this system can move within 500x500 μm rectangular size object. If the size of the object is near the limit, the transportation speed decreases and constant transportation becomes difficult.
Fig.12 summarizes the results of the object transportation experiments by Euglena group. In these experiments, the rectangle shaped objects made of fluorocarbon resin were prepared. The widths of the objects were 1mm constant and the lengths were varied between 50 and 500 μm to change the weights. The population density of the Euglena solution was 3210 cell/mm^3. The Euglena group was formed by the 500μm straight line scanning laser beam and the beam was moved between 0.68 and 6.5 μm/s. The results were estimated whether the objects were transported to the setting distance or not.
It was very difficult to maintain the friction (= shapes of the edge of the transported object), therefore, the results were varied widely. However, the maximum transported weight in these experiments was 70nN and the Euglena group can carry heavier object if the moving speed is slower.
It is necessary to improve the experimental conditions especially the shapes of the edges of the objects to maintain the friction force to measure the exact force generated by Euglena group.

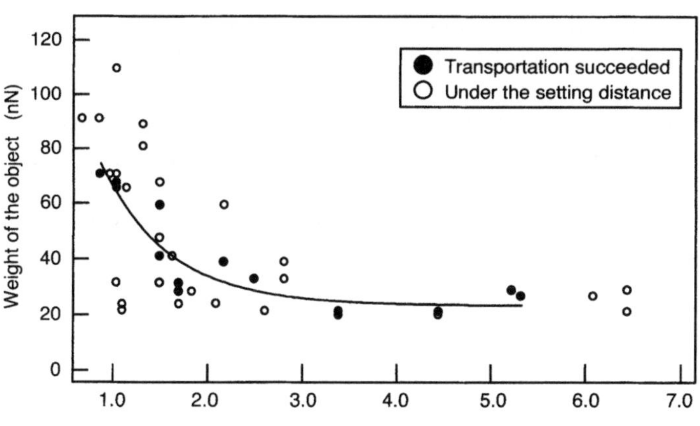

Fig. 12. Relationship between the moving speed of the scanning laser beam and the weight of the transported object moved by the line shaped Euglena group

7 Conclusion

Above-mentioned experimental results showed the possibility to use the protozoa for the living micromanipulators. One meaning, this system is an example of a shapeless mechanical system. If we need special shape, we can make any desired shape by gathering Euglena. This is a common dream of micromachine if we consider Euglena as living micromachines.

To the future, three dimensional motion control methods shall be developed. The application to the industrial field or the medical science has to be investigated.

Acknowledgements

The authors would like to their deep gratitude to Prof. Yuji Tsukii of Hosei University for giving them the seed of Euglena. Their deep gratitude is also to those graduating students, Yoji Toyoda, Takayuki Kawaguchi, Yasuo Kawai, Tetsuro Mishima, Wataru Sasaki, Hitoshi Yagi and Yoshitaka Matsumoto who joined them in the experiment and helped the authors.

This work was partially supported by the Research Institute of Technology of Tokyo Denki University under Grant Q03S-07.

References

Fearing R (1991) Control of a micro-organism as a prototype micro-robot. 2nd Int. Symp. on Micromachines and Human Sciences. pp 1-15

Itoh A (2000) Motion control of protozoa for bio-mems. IEEE/ASME Trans. on Mechatronics, vol.5, no.2. pp 181-188

Itoh A, Toida H (2001) Control of bioconvection and its mechanical application. IEEE/ASME Int. Conf. on Advanced Intelligent Mechatronics AIM'01. pp 1220-1225

Itoh A (2004) Euglena Motion Control by Local Illumination. Kato N, Ayers J, Morikawa H (Eds.) Bio-mechanism of Swimming and Flying. Chapt.2, Springer-Verlag Tokyo. pp13-26

Egans J. Jr., Dorsky D, Schiff JA (1975) Plant Physiology. vol.56, pp 318-323

Chapter 5

Passive Mechanisms Controlling Posture and Trajectory in Swimming Fishes

Malcolm S. Gordon, Dean V. Lauritzen, and Alexis M. Wiktorowicz

Department of Ecology and Evolutionary Biology, University of California, Los Angeles, CA 90095-1606, USA

Summary. Control of both posture and trajectory are important for swimming animals. The better controlled their movements are the greater their abilities to move as necessary.

Mechanisms contributing to posture and trajectory control are either passive or powered. Passive mechanisms operate automatically and continuously, requiring neither attention nor work. Powered mechanisms derive mainly from body and fin (or limb) movements. They require active coordination and control.

This paper focuses on passive mechanisms of posture and trajectory control in swimming fishes. It summarizes important results of comparative studies of fish swimming biomechanics, emphasizing our recent observations of four families of tetraodontiform bony fishes. All species studied are rigid-bodied multipropulsor swimmers. They are unusually dynamically well controlled while swimming.

We also discuss possible new insights concerning passive mechanisms of posture and trajectory control that derive from our work with tetraodontiform fishes and may also apply in other groups of fishes. Specifically, we consider possible roles of lateral keels and scutes near or on the caudal peduncles of a diverse variety of both elasmobranch (sharks) and bony fishes. Similar considerations may also apply to some groups of fossil fishes and to all living species of the mammalian cetaceans (porpoises, dolphins, whales).

Key words. fish swimming, posture control, trajectory control, passive mechanisms, bony fishes, sharks.

1 Introduction

Robust mechanisms for both posture and trajectory control are essential for actively swimming fishes moving and maneuvering at varying speeds through variably turbulent waters. The better these fishes can control their body postures, trajectories, stability and maneuverability the more easily they can move as they need to.

Mechanisms contributing to posture and trajectory control are either passive or powered. All fishes use varying combinations of both types of mechanisms while swimming, their relative importance changing continuously as types and levels of activity vary in response to changing motivational and environmental conditions. Passive mechanisms operate automatically and continuously, requiring neither work nor attention by the fish. Powered mechanisms require coordination and control. The relative quantitative importance of these two categories of control mechanisms is unknown for all fishes under all conditions.

The diversity of ways available for fishes to achieve high levels of posture and trajectory control while swimming is so great that analytically manageable levels of understanding of the patterns and processes used by fishes moving about in their own environments often elude us. Contributing to this diversity is the fact that most, if not all, of the morphological and integumentary features participating in these control processes are actually multifunctional. Many of these features also play roles in the generation of other flow-related phenomena, including directing flows and producing lift, drag, and thrust (Fish 1998).

Considering separately the passive and powered mechanisms of these controls can reduce this complexity. Fish locomotion generally is an active area of research; the relevant literature is increasing rapidly. This discussion is not a comprehensive survey of the broader field. Recent useful review articles include Weihs (1993, 2002), Webb (1997, 2002, 2006), Lauder and Drucker (2002, 2004), Drucker and Lauder (2005), Lauder (2006), and Fish and Lauder (2006).

This paper focuses on passive mechanisms for posture and trajectory control. Passive mechanisms are either internal and static or external and dynamic. A widespread internal/static mechanism is the metacentric height difference between the positions of the centers of mass and centers of buoyancy within fish bodies (Weihs 1993, 2002; Webb 1997, 2002, 2006). Many, if not most, fishes are hydrostatically unstable because their centers of buoyancy are below their centers of mass. This may be the case for both neutrally buoyant and denser than water species. Reasons for this situation are not well understood, but may relate to maneuverability properties.

External/dynamic mechanisms are based upon uses of hydrodynamic flows and forces. Hydrodynamically based mechanisms derive mainly from morphological features, including body shapes and sizes and fin placements, shapes, and sizes (Wilga and Lauder 1999, 2000; Fish and Shannahan 2000; Gordon et al. 2000; Hove et al. 2001; Bartol et al. 2003, 2005). Integumentary armoring, ornamentation, and skin compliance properties are also significant (Brainerd 1994; Gordon et al. 1996; Arreola and Westneat 1996; Plaut and Chen 2003).

Active control mechanisms derive mainly from patterns of body and fin movements (Gordon et al. 2000; Walker and Westneat 2002; Walker 2004;

Drucker and Lauder 2005; Lauder 2006; Drucker et al. 2006; Lauder and Tytell 2006). Fishes swimming in body and caudal fin (BCF) modes use different combinations of active control mechanisms as compared with fishes using median and paired-fin (MPF) modes. As speed through the water varies, and as different maneuvers are performed, substantial changes occur in the frequencies, amplitudes, coordination, and relative phasing of fin movements. Integrated sets of these variations, occurring consistently within particular ranges of speed, are recognized as changes in gaits (Alexander 1989; Drucker 1996; Gordon et al. 2000; Hove et al. 2001; Drucker et al. 2006; Lauder and Tytell 2006). Recent work with trout indicates that at least some subcarangiform swimmers may actually be using multipropulsor swimming modes that are best-termed 'M-BCF' gaits (Drucker and Lauder 2005).

This paper emphasizes work done in our laboratory in recent years focused on the varied passive mechanisms used for posture and trajectory control by selected tetraodontiform bony fishes. All species studied are rigid-bodied multipropulsor swimmers that show unusually high levels of control while swimming.

The paper concludes with a discussion of what we consider a plausible extrapolation of some of the principal findings of our work with tetraodontiform fishes to a wider diversity of other fishes, both elasmobranchs (sharks) and bony fishes, as well as to the mammalian cetaceans. The widespread, phylogenetically sporadic, occurrence of either lateral keels or rows of sharp bony scutes on and near the caudal peduncles of many groups of fishes probably indicates hitherto unrecognized hydrodynamically significant roles for these structures. We note that all living cetaceans also have well developed peduncular keels.

2 Materials and Methods

2.1 Fishes and their Morphologies

Tetraodontiform fishes are among the most highly evolved and specialized of the teleosts (Nelson 2006). We have studied eight representative species belonging to four families (Ostraciidae, Aracanidae, Tetraodontidae, Diodontidae). The former two families are comprised of armored fishes having varied body shapes and an assortment of keels and other ornaments on their bony carapaces (Fig. 1 A, B). The keels are major sources of passive control forces, at least in the ostraciids (Bartol et al., 2003, 2005). In the ostraciids the carapaces are closed posteriorly to the dorsal and anal fins; they are open posteriorly in the aracanids. This small difference results in ostraciids being MPF swimmers, aracanids BCF swimmers (D.V. Lauritzen, A.M. Wiktorowicz, and M.S. Gordon unpublished).

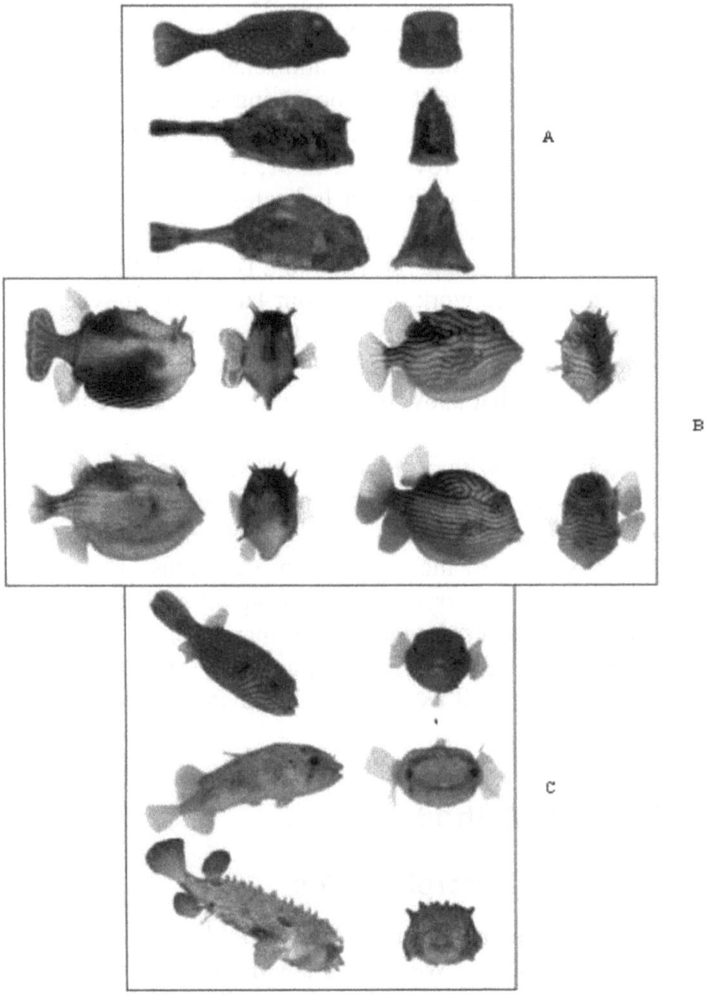

Fig. 1. Lateral and frontal photographs of eight species of tetraodontiform fishes we have studied. (A) Three species of ostraciid fishes [Bartol et al. (2005)]. Top: spotted boxfish (Ostracion meleagris); middle: scrawled cowfish (Acanthostracion quadricornis); bottom: buffalo trunkfish (Lactophrys trigonus). (B) Dimorphic males (left) and females (right) of two species of aracanid fishes [Lauritzen et al. (in preparation)]. Upper pair: ornate cowfish (Aracana ornata); lower pair: Shaw's cowfish (Aracana au-

rita). (C) One species of tetraodontid fish and two species of diodontid fishes (Wiktorowicz et al. in preparation). Top: reticulated puffer (Arothron reticularis); middle: porcupinefish (Diodon holocanthus); bottom: striped burrfish (Chilomycterus schoepfi).

Another difference between these two families relates to sexual dimorphism. In the ostraciids the sexes are very similar in terms of body shapes, though they may differ markedly in coloration. In the aracanids body shapes are strongly different between males and females (Fig. 1 B).

The tetraodonts (smooth puffers) and diodonts (spiny puffers) are not armored. Their body shapes also vary, but within narrower limits than occur in the armored forms (Fig. 1 C). Their bodies do not have keels; sources of their passive control forces remain to be determined (Wiktorowicz AM Lauritzen DV and Gordon MS unpublished).

Due to space limitations the rest of this paper discusses only our work with ostraciids and aracanids and extrapolations from that work.

2.2 Study Methods

Data on the biomechanics and kinematics of swimming in these fishes derive from analyses of high speed digital video recordings of animals swimming straight ahead at various speeds in low turbulence water tunnels (Gordon et al. 2000; Hove et al. 2001).

Data on passive mechanisms of posture and trajectory control in ostraciids were obtained using dimensionally exact rapid prototyped models of these fishes with their fins removed. These models were studied in various water and wind tunnels using 2-dimensional digital particle imaging velocimetry (DPIV) for flow visualization, force balances for determination of global forces on the models, and surface pressure measurements to supplement and verify the DPIV results (Bartol et al. 2003, 2005).

3 Results and Discussion

3.1 Swimming modes and passive controls – ostraciid and aracanid fishes

Gordon et al. (2000), Hove et al. (2001), and Bartol et al. (2003, 2005) give detailed descriptions of ostraciid swimming modes and passive controls on swimming postures and trajectories. Living ostraciids are rigid-bodied multi-propulsor MPF swimmers. They are impressively dynamically stable over wide ranges of speed (they glide smoothly through the water), and they are both fast and maneuverable. Their upper critical swimming speeds are high compared with most other bony fishes.

As shown in Fig. 1A, the bony carapaces of different ostraciid species are structurally diverse. Shared features within that diversity contribute significantly to passive control; these include:

a) Carapaces cover about 65% of body length;
b) Carapaces have bony closures behind the insertions of the dorsal and anal fins;
c) Carapace surfaces are smoothly curved in ways that control water flow patterns so they contribute to trim control;
d) Carapace corners are shaped into keels that are powerful generators of continuous leading edge vortices that are major contributors to trim control;
e) Various ornaments on carapaces slow separation of vortices from surfaces, enhancing their contributions to trim control.

The aspect we emphasize here is the role of the keels and other ornaments as leading edge vorticity generators. The sizes and intensities of the continuous vortices produced increase with increased speed through the water. The positions of the vortices vary with changing angles of attack in ways that are consistent with their generation of forces on the carapaces that assist significantly with automatic passive postural and trajectory control by the fishes.

Bartol et al. (2003, 2005) give detailed quantitative descriptions of the results of these studies. Important points relevant here are: i) The three independent experimental approaches used all gave results consistent with this description; ii) Although three very different carapace morphologies were studied, the patterns and mechanisms of flow direction were closely similar in all cases; iii) The patterns of change of lift coefficients with changing angles of attack, the quantitative values of those coefficients, and the absence of stall at high pitch angles (measurements based on force balance studies), were consistent with those characteristic of delta wings. A broad characterization of the hydrodynamic properties of the carapaces of ostraciid fishes is that they resemble carefully sculpted boxes mounted on delta wings.

The relative importance of these automatic passive forces as compared with the powered forces (lift, drag, and thrust forces, also turning moments) generated by the complex movements of the fins remains to be determined. Both flow visualization measurements on living boxfishes and computational fluid dynamic modeling will be needed to produce a complete picture and analysis.

Gordon et al. (2000) published a qualitative physical model that takes into account many of the issues involved. That model postulates as a working hypothesis that high precise levels of active posture and trajectory control result from continuously varying, near exact balances of turning moments generated by all five fins. Specific patterns of balancing vary as swimming speeds and gaits vary. The model was developed before the discovery of the vorticity generating roles of the carapace keels and ornaments. The model is due for refinement.

We have studied two species belonging to the evolutionary side branch of the boxfishes that includes the cool water cowfishes (family Aracanidae). This relatively unstudied family differs from all ostraciids in having the carapaces of all species (which also cover about 65% of body lengths) terminating just forward of the insertions of both the dorsal and anal fins, rather than enclosing the bases of those fins. They also have a different pattern of development of keels on their carapaces (their carapaces have only a single mid-ventral longitudinal keel, rather than paired ventrolateral keels and, often, smaller keels in other locations) (Fig. 1B).

We have studied swimming in two species of these fishes, the ornate cowfish (*Aracana ornata*) and Shaw's cowfish (*Aracana aurita*) (Lauritzen DV Wiktorowicz AM and Gordon MS unpublished). Their carapaces make their bodies almost completely rigid, but the unenclosed bases of their vertical fins provide sufficient additional flexibility at the bases of their tails that their swimming mode is a unique combination of both BCF and MPF modes. This mode is sufficiently unique that it should be designated as aracaniform swimming.

Measurements of upper critical swimming speeds indicate that the maximum performance levels of aracanids are significantly lower than those of the ostraciids [about 3 body-lengths/s (BL/s) as compared with 5-6 BL/s]. We have a substantial database on morphometrics, physical properties, and rectilinear swimming biomechanics and kinematics in both sexes of each species. The only significant recoil movements they make as they swim are yawing motions. In all other respects they are as dynamically stabilized as are the ostraciids. As we complete the analyses of these data we anticipate they will permit us to make fully informed inferences as to why this difference in maximum performance levels exists.

The resolution of this question will require additional flow visualization studies with living fishes and computational fluid dynamic calculations. The energy lost to yawing movements may be sufficient to slow them significantly. The shapes of their carapaces may also not be as hydrodynamically streamlined as are those of the ostraciids studied.

3.2 Some Implications

All members of the four families of tetraodontiform fishes we have studied are rigid-bodied multipropulsor swimmers that move through the water in ways that are unusually well controlled with respect to posture and trajectory. The morphological, biomechanical, kinematic, and hydrodynamic bases for these abilities are presently partially understood. It is clear, however, that diverse mixtures of passive and powered mechanisms of control contribute.

The research literature is silent with respect to what we believe may be one of the most important probable sets of broader implications of our work. The issues relate to the possible functions in passive control of the widely occurring, morphologically varied, lateral keels and rows of sharp-edged bony scutes that occur on and near the caudal peduncles of many groups of active swimmers. We believe that these structures, whatever other functions they also may serve, act as continuous generators of leading edge vortices. We postulate that those vortices play significant roles in generating forces important in passive posture and trajectory control in the animals having the structures.

Table 1. Phylogenetic distribution of well-developed lateral keels and rows of scutes on and near the caudal peduncles of unarmored, non-tetraodontiform living fishes

Class	Order	Family (Genera, common names)	Keels/scutes
Chondrichthyes			
	Orectolobiformes	Rhincodontidae (*Rhincodon,* whale shark)	3 keels
	Lamniformes	Cetorhinidae (*Cetorhinus,* basking shark)	1 keel
		Lamnidae (*Carcharodon*, great white shark; *Isurus*, mako shark; *Lamna*, salmon and porbeagle sharks)	1 keel
	Carchariniformes	Carcharhinidae (*Galeocerdo* only, tiger shark)	1 keel
	Squaliformes	Squalidae (*Squalus*, all dogfish sharks)	1 keel
Actinopterygii			
	Beloniformes	Belonidae (*Belone*, *Platybelone* only, needlefishes)	1 keel
	Perciformes	Carangidae (*Caranx* and 22 related genera in subfamily Caranginae, jacks)	1 row of scutes
		Bramidae (*Brama*, few species, pomfrets)	1 keel
		Acanthuridae (*Naso*, unicornfishes)	1-2 keels
		Luvaridae (*Luvarus*, louvar)	1 keel
		Gempylidae (*Lepidocybium* only, escolar)	1 keel
		Scombridae (*Scomberomorus,* Spanish mackerels; *Sarda,* bonitos; *Thunnus* and four related genera in Tribe Thunnini, tunas)	1 keel
		Xiphiidae (*Xiphias*, swordfish)	1 keel
		Istiophoridae (*Istiophorus*, sailfishes; *Tetrapterus*, spearfishes; *Makaira*, marlins)	2 keels

Numbers of keels/rows of scutes are for one side of the fishes; the structures are bilaterally symmetrical. See text for details.

Phylogenetic sequence, taxonomic terminology, and common names follow Nelson (2006).

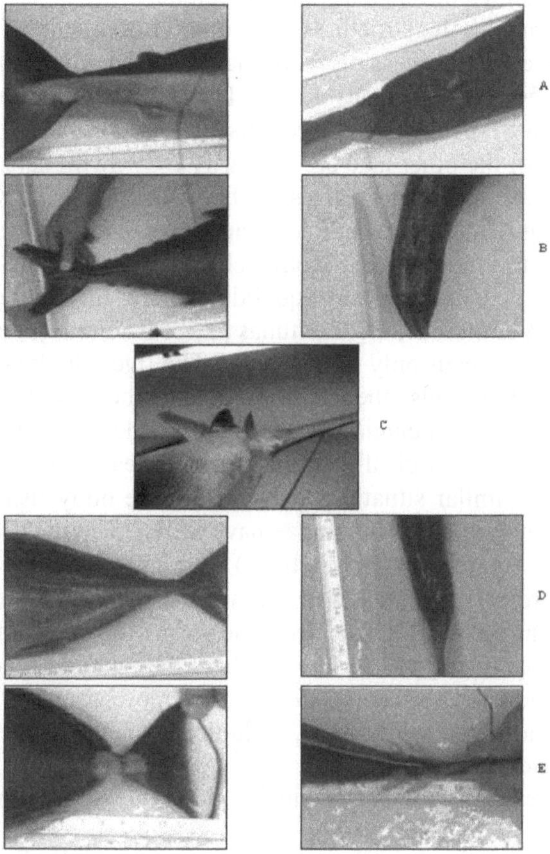

Fig. 2. Lateral and dorsal photographs of caudal peduncles and parts of the tails of preserved museum specimens [Los Angeles County Museum of Natural History (LACM)] of five selected species of marine fishes symmetrically having either one or two well developed lateral keels or a row of well developed lateral scutes. (A) salmon shark (*Lamna ditropis*), LACM 47306-1, 93 cm TL (total length), one keel. (B) escolar (*Lepidocybium flavobrunneum*), LACM 36230-1, 74 cm SL (standard length), one keel. (C) black marlin (*Makaira nigricans*), LACM uncataloged, 3.5 m TL, two keels (oblique lateral view only). (D) horse jack (*Caranx caballus*), LACM 9946-1, 33 cm SL, row of scutes. (E) lined unicorn fish (*Naso lineatus*), LACM 54199-1, 21 cm SL, two keels. See text for additional information.

Table 1 summarizes the phylogenetic distribution of lateral keels and rows of scutes among the sharks and bony fishes. There are at least several hundred species represented. They are found in at least five families of sharks, representing four Orders. They occur in at least nine families of bony fishes, representing two Orders. Figure 2 shows five selected examples of these structures.

Interesting features of this distribution include:

Most fishes having the structures have one on each side of their peduncles. Two of the three taxa possessing more than one (the whale shark and the billfishes) reach very large sizes as adults. The third taxon (species of unicornfishes, which are specialized surgeonfishes) are primarily labriform MPF swimmers. All of the other taxa are primarily BCF swimmers.

The morphologies and relative sizes of the keels vary greatly within and between taxa. The relative sizes of keels vary substantially allometrically within species. Keels are important taxonomic characters, occurring in all species, in multiple taxa (e.g. Lamnidae and Squalidae among the sharks; subfamily Caranginae in the Carangidae, the true tunas (Tribe Thunnini), and the billfishes). In other taxa keels occur only sporadically. The tiger shark is the only species in its group having keels; there are about 50 other species of carcharhinid sharks (including other pelagic species) that lack keels. The eight species of the closely related, ecologically similar hammerhead sharks (family Sphyrnidae) lack keels. Similar situations exist within the bony fishes. Only two of the ten genera in the family Belonidae have keels. The tuna-like escolar is the sole snake mackerel (family Gempylidae) having keels.

The single rows of scutes that characterize the carangin jacks are also greatly varied in size and extent. In some forms the scutes are restricted entirely to the sides of the peduncles, in others they extend forward almost half the length of the body. Still other forms have more than half of the length of the lateral line armored with scutes, including portions of the lateral line high up on the sides of the fishes.

These features raise a series of biologically and hydrodynamically interesting questions:

What are the implications for swimming speeds, hydrodynamic properties, and maneuverability of having or not having keels? The research literature is silent on this subject. Fish and Lauder (2006) and Lauder and Tytell (2006) discuss the roles of peduncular finlets in flow control and caudal fin function in scombrid fishes; the only experimental data now available are for fishes lacking lateral keels (Nauen and Lauder 2001, 2002).

Among the sharks, what are the implications in these subject areas relating to both the presence and absence of keels and to the interactions of keels with the varied morphologies of heterocercal tails present in different groups? Wilga and Lauder (2002, 2004) provide relevant experimental data on this subject, but only for sharks lacking keels. The multiple convergences between swimming mechanisms and properties in lamnid sharks and tunas that have been demonstrated by Donley et al. (2004) and Shadwick (2005) are relevant here.

How important, in both absolute and relative terms, are the putative vorticity generating functions of keels/scutes? How important are their possible contributions to posture and trajectory control with respect to the other functions that have been suggested for these structures? At least three other functions have been suggested: in swimming tunas peduncular keels far aft of the center of mass act as small low aspect ratio wings that generate lift that damps pitching movements (Magnuson 1978); they act as mechanical struts that strengthen and stiffen the bases of the tails of tunas (Hebrank 1982; Dewar and Graham 1994; Donley and Dickson 2000); they act to streamline the transverse structures of the caudal peduncles in both tunas and lamnid sharks, thus reducing turbulence and therefore drag generated by side to side tail movements (Aleev 1963; Collette 1978; Sepulveda and Dickson 2000; Shadwick 2005).

For non-biologists we point out that the structures we describe are almost all solid, rigid, and often reinforced internally by bone, cartilage, or very stiff ligaments and tendons. They are not soft and deformable, therefore unlikely to produce important hydrodynamic effects. While some are relatively small in size, their locations, far aft of the centers of mass and buoyancy of the animals, imply substantial leverage for the forces they generate (Weihs 1993). Their effects are most likely greatest on both pitching and yawing recoil movements.

References

Aleev YG (1969) Function and gross morphology in fish. Keter Press, Jerusalem.

Alexander RM (1989) Optimization and gaits in the locomotion of vertebrates. Physiol Rev 69: 1199-1227.

Bartol IK, Gharib M, Weihs D, Webb PW, Hove JR, Gordon MS (2003) Hydrodynamic stability of swimming in ostraciid fishes: role of the carapace in the smooth trunkfish *Lactophrys triqueter* (Teleostei: Ostraciidae). J Exp Biol 206: 725-744

Bartol IK, Gharib M, Webb PW, Weihs, D, Gordon MS (2005) Body-induced vortical flows: a common mechanism for self-corrective trimming control in boxfishes. J Exp Biol 208: 327-344

Breder CM (1926) The locomotion of fishes. Zoologica N.Y. 4: 159-256

Collette BB (1978) Adaptations and systematics of the mackerels and tunas. In: Sharp, GD, Dizon, AE (eds) The Physiological Ecology of Tunas. Acdemic Press, New York, pp. 7-39

Dewar H, Graham JB (1994) Studies of tropical tuna swimming performance in a large water tunnel. III. Kinematics. J Exp Biol 192: 45-59

Donley JM, Dickson KA (2000) Swimming kinematics of juvenile kawakawa tuna (*Euthynnus affinis*) and chub mackerel (*Scomber japonicus*). J Exp Biol 203: 3103-3116

Donley JM, Sepulveda CA, Konstantinidis P, Gemballa S, Shadwick RE (2004) Convergent evolution in mechanical design of lamnid sharks and tunas. Nature 429: 61-65

Drucker EG (1996) The use of gait transition speed in comparative studies of fish locomotion. Amer Zool 36: 555-566

Drucker EG, Lauder GV (2005) Locomotor function of the dorsal fin in rainbow trout: kinematic patterns and hydrodynamic forces. J Exp Biol 208: 4479-4494

Drucker EG, Walker JA, Westneat MW (2006) Mechanics of pectoral fin swimming in fishes. In: Shadwick RE, Lauder GV (eds) Fish Biomechanics. Elsevier and Academic Press, Amsterdam, pp. 369-423

Fish FE (1998) Imaginative solutions by marine organisms for drag reduction. Proc Intl Symp Seawater Drag Reduction, Newport, RI: 443-450.

Fish FE, Shannahan LD (2000) The role of the pectoral fins in body trim of sharks. J Fish Biol 56: 1062-1073

Fish FE, Lauder GV (2006) Passive and active flow control by swimming fishes nd mammals. Annu Rev Fluid Mech 38: 193-224.

Gordon MS, Hove JR, Webb PW, Weihs D. (2000) Boxfishes as unusually well-controlled autonomous underwater vehicles. Physiol. Biochem. Zool. 73: 663- 671

Hebrank MR (1982) Mechanical properties of fish backbones in lateral bending and in tension. J. Biomech. 15: 85-89

Hove JR, O'Bryan LM, Gordon MS, Webb PW, Weihs, D (2001) Boxfishes (Teleosteii: Ostraciidae) as a model system for fishes swimming with many fins: kinematics. J Exp Biol 204: 459-1471

Lauder GV (2006) Locomotion. In: Evans D, Claiborne J (eds) The Physiology of Fishes, 3rd edn. CRC Taylor and Francis, Boca Raton, pp. 3-46

Lauder GV, Drucker EG (2002) Forces, fishes, and fluids: hydrodynamic mechanisms of aquatic locomotion. News Physiol Sci 17: 235-240

Lauder GV, Drucker EG (2004) Morphology and experimental hydrodynamics of fish fin control surfaces. IEEE J Ocean Engr 29: 556-571

Lauder GV, Tytell ED (2006) Hydrodynamics of undulatory propulsion. In: Shadwick RE, Lauder GV (eds) Fish Biomechanics. Elsevier and Academic Press, Amsterdam, pp. 425-468

Magnuson JJ (1978) Locomotion by scombrid fishes: hydromechanics, morphology, and behavior. In Hoar WS, Randall DJ (eds) Fish Physiology, vol. VII, Locomotion. Academic Press, New York, pp. 239-313

Nauen JC, Lauder GV (2001) Locomotion in scombrid fishes: visualization of flow around the caudal peduncle and finlets of the chub mackerel *Scomber japonicus*. J Exp Biol 204: 2251-2263

Nauen JC, Lauder GV (2002) Hydrodynamics of caudal fin locomotion by chub mackerel, *Scomber japonicus* (Scombridae). J. Exp. Biol. 205: 1709-1724.

Nelson JS (2006) Fishes of the World, 4th edn. John Wiley and Sons, New York.

Sepulveda C, Dickson KA (2000) Maximum sustainable speeds and cost of swimming in juvenile kawakawa tuna (*Euthynnus affinis*) and chub mackerel (*Scomber japonicus*). J Exp Biol 203: 3089-3101

Shadwick RE (2005) How tunas and lamnid sharks swim: an evolutionary convergence. Amer. Scientist 93: 524-531

Walker JA (2004) Dynamics of pectoral fin rowing in a fish with an extreme rowing stroke: the threespine stickleback (*Gasterosteus aculeatus).* J Exp Biol 207: 1925-1939

Walker JA, Westneat MW (2002) Kinematics, dynamics, and energetics of rowing and flapping propulsion in fishes. Integr Comp Biol 42: 1032-1043

Webb PW (1997) Designs for stability and maneuverability in aquatic vertebrates: what can we learn? In: Tenth International Symposium on Unmanned Untethered Submersible Technology. Autonomous Undersea System Institute, Durham, NH, pp. 86-108

Webb PW (2002) Control of posture, depth, and swimming trajectories of fishes. Integr Comp Biol 42: 94-101

Webb PW (2006) Stability and maneuverability. In Shadwick RE, Lauder GV (eds) Fish Biomechanics. Elsevier and Academic Press, Amsterdam, pp. 281-332

Weihs D (1993) Stability of aquatic animal locomotion. Contemp Math 141: 443- 461

Weihs D (2002) Stability *versus* maneuverability in aquatic locomotion. Integr Comp Biol 42: 127-134

Wilga CD, Lauder GV (1999) Locomotion in sturgeon: function of the pectoral fins. J Exp Biol 202: 2413-2432

Wilga CD, Lauder GV (2000) Three-dimensional kinematics and wake structure of the pectoral fins during locomotion in leopard sharks *Triakis semifasciata*. J Exp Biol 203: 2261-2278

Wilga CD, Lauder GV (2002) Function of the heterocercal tail in sharks: quantitative wake dynamics during steady horizontal swimming and vertical maneuvering. J Exp Biol 205: 2365-2374

Wilga CD, Lauder GV (2004) Hydrodynamic function of the shark's tail. Nature 430: 850

Chapter 6

Mechanical Properties of the Caudal Fin Resulting from the Caudal Skeletal Structure of the Bluefin Tuna

Hirohisa Morikawa[1], Konami Yusa[1], and Shunichi Kobayashi[1]

[1] Department of Functional Machinery and Mechanics, Faculty of Textile Science and Technology, Shinshu University, 3-15-1 Tokida, Ueda-shi, Nagano 386-8567, Japan

Summary. The objective of our study was to investigate the effect of caudal fin behavior resulting from the caudal skeletal structure of tuna on the propulsive performance. In this paper, the caudal skeletal structure and mechanical properties on the caudal fin of bluefin tuna (Thunnus thynnus) were examined. The Young's modulus of fin rays composing the caudal fin was obtained. A two dimensional oscillating wing theory was applied to the caudal fin sinusoidally oscillating in a uniform flow. Then the pressure distribution over the fin was calculated and the fin deformation was obtained. It was found that a phase delay existed between the leading edge and the middle part of the fin.

Key words. Biomechanics, Mechanical properties, Tuna, Caudal fin, Skeletal structure

1 Introduction

It is important to understand the mechanisms that enable biological creatures to move effectively and efficiently through their environment, and to analyze the characteristics of their various forms of behavior from the engineering viewpoint. It is known that a fish such as tuna with a lunate caudal fin can swim at high speed and the caudal fin plays an important role in the efficient generation of propulsive force. Studies on oscillating fins of fish have been theoretically and experimentally carried out by a number of researchers including Lighthill (1975) and Chopra (1977). These studies have shown that the fins have the potential for high propulsion efficiency. Recently, a robo-tuna (Triantafyllou et al. 1995) and a dolphin robot (Nakashima 2003) were developed, and the experiments on their propulsive performance were carried out. However, their caudal fins were rigid or flexible wings though they were similar to real caudal fins in shape. The mechanical properties of the fins and the effect of the caudal fin deformation resulting from the caudal muscle-skeletal structure on the performance of propulsion were not taken into account. While the study on tail kinematics of scombrid fishes indicated that the movement of the caudal fin was complex and affected force generation (Gibb et al. 1999).

From a biomechanical point of view, we focused on the muscle-skeletal structure and the mechanical properties of a caudal fin of tuna related to the deformation of the fin for further improvement of the propulsive performance (Morikawa 2001). The objective of our study was to investigate the effect of the caudal fin behavior resulting from the caudal muscle-skeletal structure of tuna on the propulsive performance.

This article describes the caudal skeletal structure and the mechanical properties on the caudal fin of a bluefin tuna (Thunnus thynnus). The basic knowledge on the elastic modulus of fin rays composing the caudal fin was provided in order to analyze the relationship between the deformation of caudal fin and the fluid dynamic force generated by oscillating the fin.

2 Muscle-Skeletal Structure of Caudal Fin of Tuna

We examined the muscle and the skeletal structure of the caudal fin of a tuna. Figure 6.1 is a sketch which shows the caudal skeleton of a bluefin tuna (Thunnus thynnus) of body length 1.55 m. The caudal vertebral column has process-like plate bones of which both neural and haemal spines are deformed, and are connected to each other. The up and down motion of the caudal vertebral column is restricted because each column is inserted into the space between the process-like plate bones. Some caudal vertebrae have lateral flanges at both the sides. It is thought that the lateral flanges reinforce bending rigidity of the caudal peduncle. A caudal fin is composed of a number of fin rays and collagen. A fin ray is composed of spinous rays from the leading edge to the middle part of the fin and soft rays in the rear part of the fin. Urostyles or hypurals of the fin are located between pairs of fin rays. The sweepback angle of the caudal fin is larger than the angle of each spinous ray composing the fin shown in Fig.6.1.

Each muscle's tendon reaches the hypural and connects the ends of each fin ray (Fig.6.2). It is suggested that this tendon-fin ray structure enables the caudal fin to deform itself and generates the wave motion of the fin. As the caudal part and fin of tuna are flexed by the alternative contraction of muscles arranged on both sides of the vertebrae, a phase difference between caudal vertebrae transmits motion to the fin rays. It is suggested that the phase delay exists between the leading edge and the middle part of the fin, and therefore the caudal fin becomes concave in the direction of oscillation.

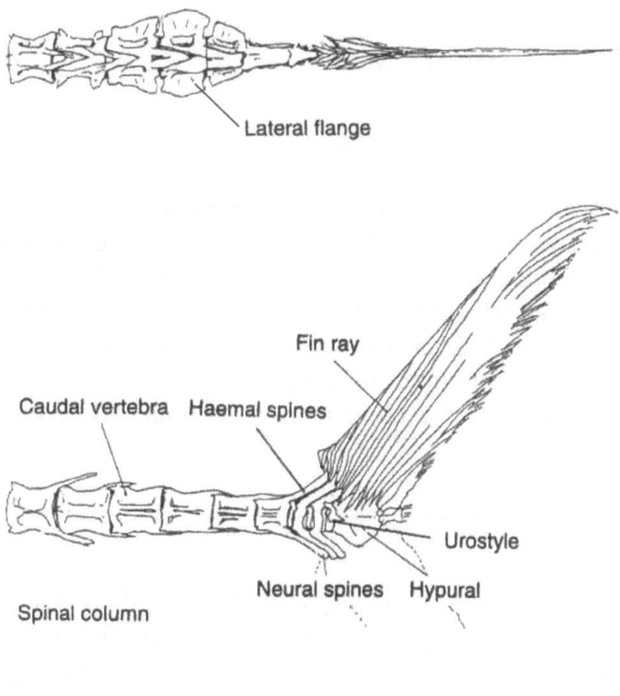

Fig.6.1. Caudal skeleton of bluefin tuna (Thunnus thynnus)

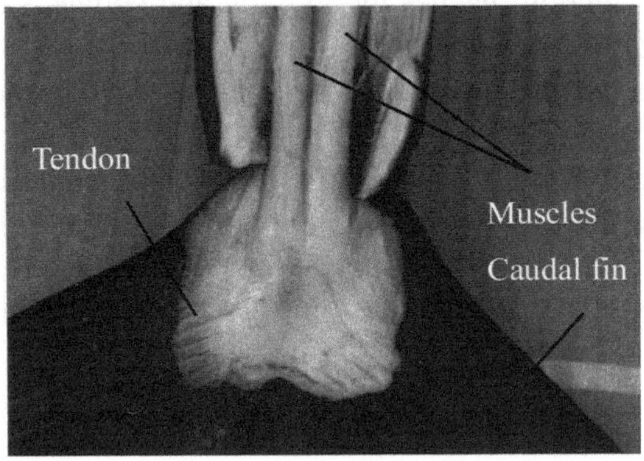

Fig.6.2. Caudal muscles and tendons of bluefin tuna (Thunnus thynnus)

3 Analysis Using FEM

We analyzed the deformation of the caudal fin using Finite Element Method (FEM) as an anisotropy plate of composite material according to Kirchhoff assumption. A model of the upper part of the caudal fin with mesh is shown in Fig.6.3. The number of triangle element is 284 and the number of nodal point is 170. The restriction condition of the model of the fin is fixed at the root of the upper part of fin, which is indicated by hatched lines in Fig.6.3. Constants, expressing the mechanical properties on the fin, are required to perform a structural analysis using FEM. Young's modulus for each element E_c is obtained by the following equation.

$$E_c = E_f V_f + E_m V_m \tag{6-1}$$

where E_f and E_m are the Young's modulus of the fin ray and the medium such as collagen, respectively, and V_f and V_m are the content volume rate of the fin ray and the medium composed the fin respectively. In this report, the Young's modulus on the medium of the caudal fin E_m was assumed to be equal to zero because the effect of the medium on the deformation of the caudal fin can be thought to be small. In order to investigate the mechanical properties of the caudal fin, we first examined the flexural rigidity of fin ray and obtained the Young's modulus of the fin ray through the following experiment. The content volume rate of fin ray was estimated by measuring a area of the fin ray from photographs of cross sections of the caudal fin cut parallel to the body axis. A sample of the photograph of cross section of the caudal fin of tuna is shown in Fig.6.4.

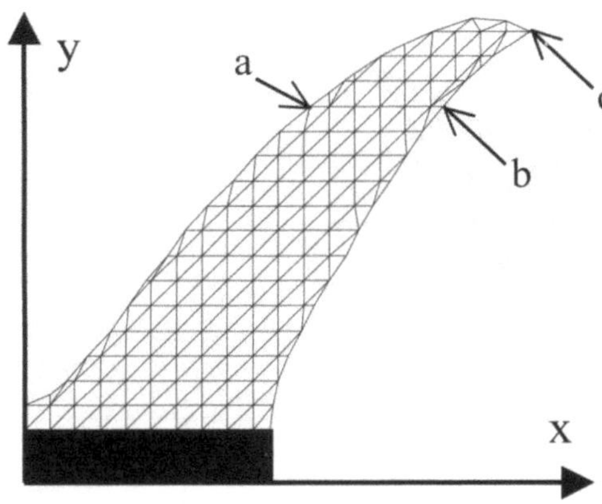

Fig.6.3. Model of the upper part of caudal fin with mesh

Fig.6.4. Cross section of the caudal fin of tuna

4 Experimental Method

Six fin rays were removed from the caudal fin of a bluefin tuna (Thunnus thynnus) of body length 1.45 m and weight of 52.4 kg. Each fin ray was numbered from 1 to 6 (Fig.6.5). Each fin ray was tested for its bending characteristic with bending test equipment (Shimazu Co.Ltd Autograph AG20KND). Figure 6.6 shows a schematic diagram of the device for measuring the bending deformation of a fin ray. The fin ray was marked at intervals of 20 mm. The bending experiment on fin rays was carried out at the room temperature 15 - 18 °C taking care to keep fin rays moist as follows: First the fin ray was fixed at the root. Weight was put on the first marked point by pushing a rod with a ϕ 6 ball. The deflection of the fin ray was measured while applying the load at the displacement rate of 2 mm/min. Next the fixed part was changed to the first marked point of the fin ray. The weight was placed on the second marked point of the fin ray and the deflection of the point was measured. This measurement procedure was repeated for the other marked points of the fin ray. That is why the fin ray has a variation in cross section (Morikawa et al. 2004).

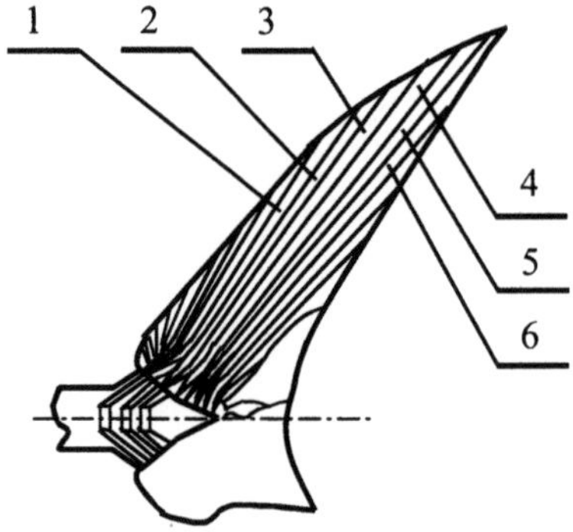

Fig.6.5. Fin rays removed from the caudal fin for bending test and numbered from 1 to 6

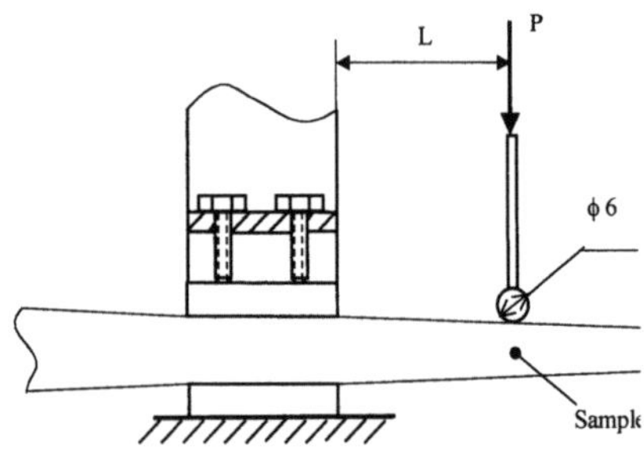

Fig.6.6. Schematic diagram of the setting device for measurement of bending deflection

Figure 6.7 shows an example of the relationship between the load P and the deflection δ measured at the first point of fin ray 2 shown in Fig.6.5. It was shown that they have a simple proportional relation. Other fin rays showed almost the same relationship. The flexual rigidity EI in a measured section was obtained by the following equation,

$$E I = \frac{P L^3}{3\delta} \tag{6-2}$$

where E is the Young's modulus, I is the second moment of area of the fin ray and L is the span of the measured part. The average second moment of area was obtained by assuming that the sectional shape neighboring the root of the fin ray was a ellipse and the shape neighboring the tip of the fin ray was a rectangle by observing photographs of the cross sections in the fin ray. Young's modulus was obtained by calculation.

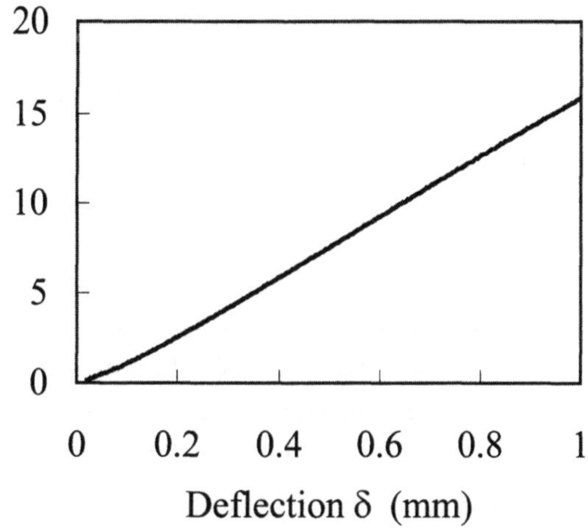

Fig.6.7. Relationship between load P and deflection δ of the fin ray 2

5 Results and Discussion

As an example of Young's modulus on each part of the fin ray obtained by use of Eq. (6-2), the distribution of Young's modulus along the fin ray 2 is shown in Fig.6.8. Here, the abscissa represents the longitudinal distance from the root of fin ray and the ordinate represents Young's modulus. Figure 6.9 shows the average Young's modulus of each fin ray indicated in Fig.6.5. The average Young's modulus of the fin ray composing the caudal fin was almost constant at 4.59 GPa (Deviation; 0.24 GPa).

Here, we estimated the mechanical properties on the caudal fin of tuna composed of fin rays and the medium. The Young's modulus of the longitudinal direction of the fin ray E_L was 8.72 GPa, perpendicular to the fin ray E_T was 0.65 MPa, and both the Poisson ratio and the shear modulus of the caudal fin were almost zero.

An example of the result calculating the deformation of the caudal fin by use of FEM when the caudal fin was uniformly loaded is shown in Fig.6.10. A model of the caudal fin used here was a flat plate with a shape similar to the caudal fin. The dimensions were 440 mm in span and 100 mm in chord length of the root in the upper part of the caudal fin. The Young's modulus obtained by the experiment was applied and the value of uniform load was 1 kPa in this case. The leading part of the fin was thick and hard to bent, while the trailing part and the tip of the fin was thin and flexible. The deflection of the tip of the fin was about 28 mm.

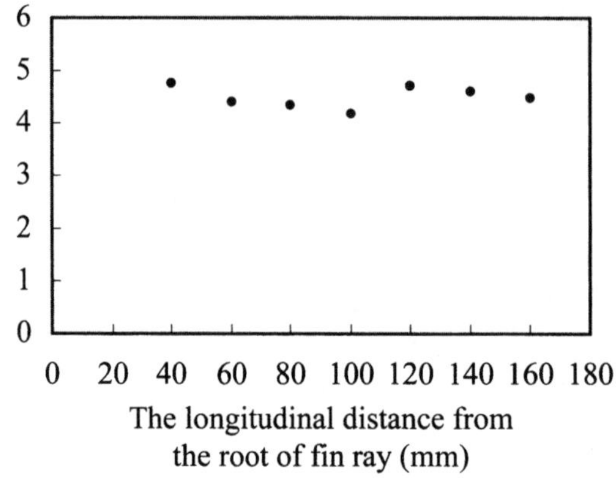

Fig.6.8. Distribution of Young's modulus along the fin ray 2

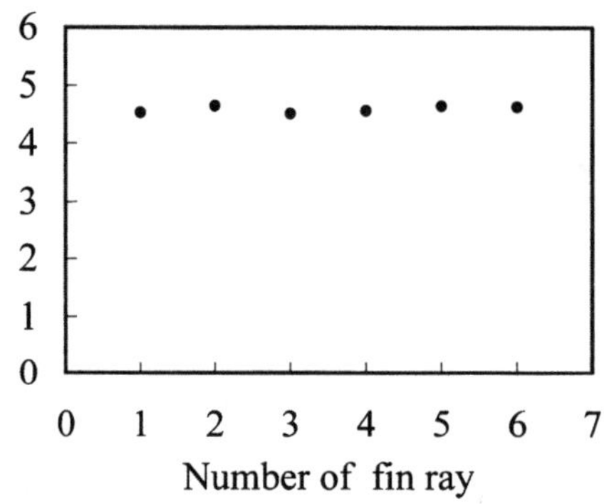

Fig.6.9. Average Young's modulus of fin ray

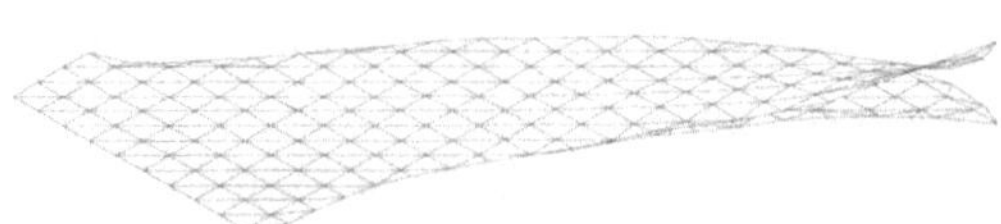

Fig.6.10. An example of FEM analysis results

Next, a caudal fin was divided into rectangular elements (Fig.6.11). Each element was assumed to be a two dimensional plate wing. The two dimensional oscillating wing theory (Bisplinghoff, Ashley and Holfman 1957) was applied to the caudal fin sinusoidally oscillating with the pitching and heaving motions in an uniform flow of which fluid is inviscid and imcompressible. The coordinate of the oscillating wing is shown in Fig.6.12. The assumption of Kutta's condition for the trailing edge of the wing is satisfied. A thin plate wing is replaced with a vortex sheet. The vorticity on the wing $\gamma(\xi)$ was calculated in consideration of vorticity in the wake γ_w. The pressure difference δp between the upper and the lower surface of the wing was obtained by the following equation:

$$\delta p(\xi,t) = \rho U \gamma(\xi,t) \tag{6-3}$$

where U is the velocity of the uniform flow and ρis the density of water. The pressure distribution over the fin was calculated and the deformation of the fin was obtained.

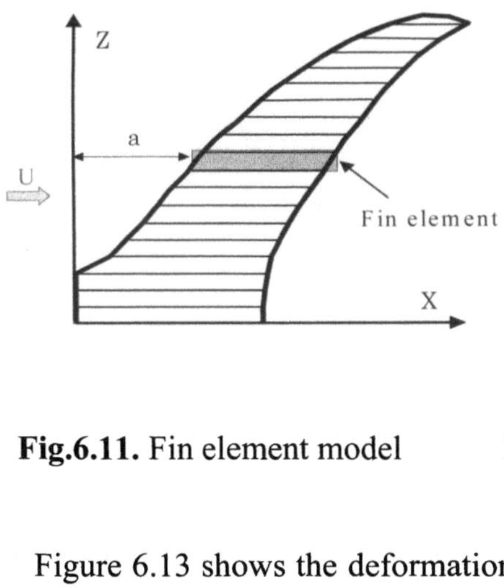

Fig.6.11. Fin element model

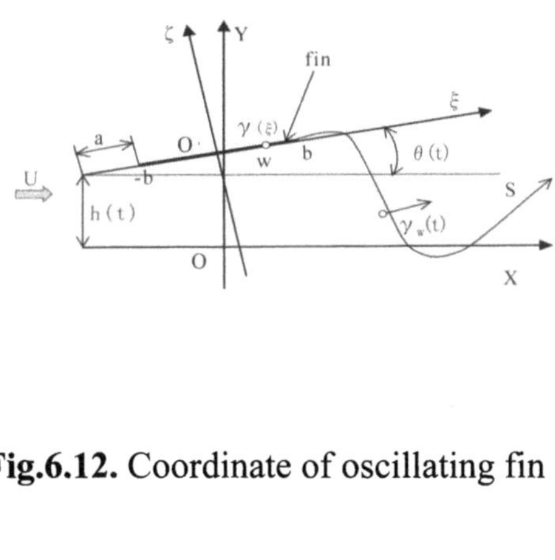

Fig.6.12. Coordinate of oscillating fin

Figure 6.13 shows the deformation of the oscillating fin over time. Calculation conditions were as follows: The maximum amplitude of the heaving motion was 30 mm, the maximum angle of the pitching motion was 40 degree, the phase delay between the heaving and the pitching motions was 90 degree, and the frequency of fin oscillation was 1 Hz. The velocity of uniform flow was assumed to be 25 m/s. The calculation was carried out in the quasi steady state of the fin behavior. Figure 6.14 shows the deflection of both the point a of the leading edge and the point b of the trailing edge on the model fin shown in Fig.6.3. The fin was found to have the difference and the phase delay of deflection between the leading edge and the trailing edge. Therefore the static mechanical properties on the caudal fin is thought to cause the concave deformation of the caudal fin to the direction of oscillation in consideration of the caudal skeletal structure.

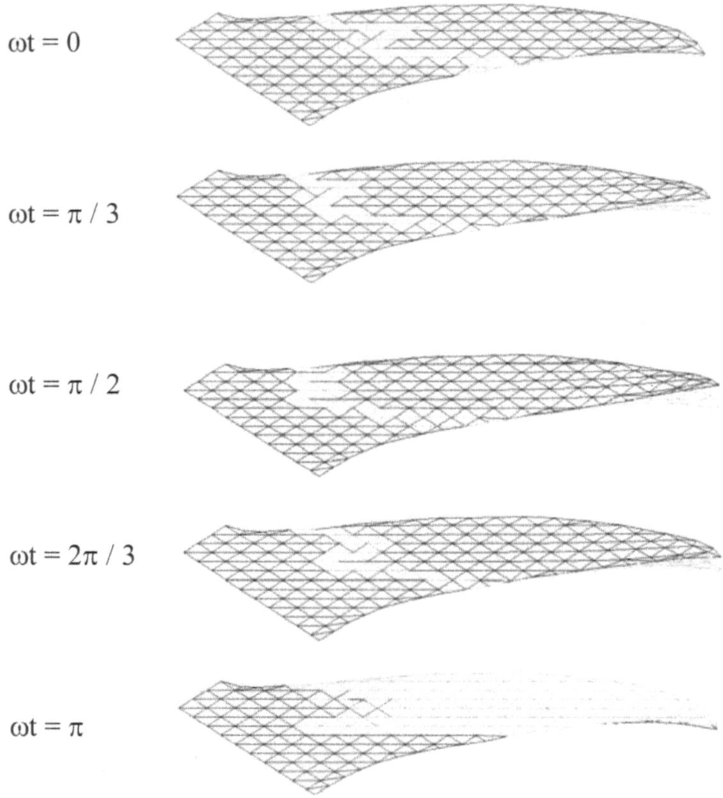

Fig.6.13. Deformation of the oscillating fin (ωt: phase of oscillation cycle)

6 Conclusions

The skeletal structure and mechanical properties on the caudal fin of bluefin tuna were examined. As a result, it was found that each fin ray of the caudal fin has almost the same value of Young's modulus, and a phase delay exists between the leading edge and the middle part of the fin. As the caudal fin achieves active movement during the tail beat, the function of the muscle-skeletal structure in producing the complex caudal fin movement and the effect of the deformation of the caudal fin on the propulsive performance would be investigated.

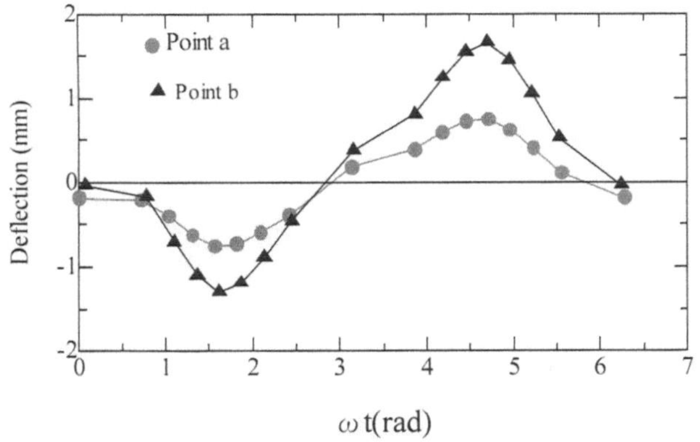

Fig.6.14. Deflection of the leading edge (a) and the trailing edge (b) of the caudal fin

References

Bisplinghoff R, Ashley H, Halfman R (1957) Aeroelasticity. Adison-Wesley, 251-281

Chopra M and Kambe T (1977) Hydrodynamics of lunate-tail propulsion. Part 2, J. Fluid Mech, Vol 79, 49-69

Gibb AC, Dickson KA, and Lauder GV (1999) Tail kinematics of the chub mackerel Scomber japonicus: testing the homocercal tail model of fish propulsion, J. Exp. Biol. 202, 2433-2447

Lighthill MJ (1975) Mathematical Biofluiddynamics, Soc.for Industrial and Applied Mathematics

Morikawa H, Nakao S, and Kobayashi S (2001) Experimental study on oscillating wing for propulsor with bending mechanism modeled on caudal muscle-skeletal structure of tuna, JSME Int J, Series C, Vol 44, No 4, 1117-1124

Morikawa H, Yusa K and Kobayashi S (2004) Mechanical characteristics of caudal fin resulting from caudal muscle-skeletal structure of bluefin tuna, Proc 14th Int Offshore and Polar Engineering Conference, Toulon, France, 280-283

Nakashima, M, Takahashi, Y, Ono, N, and Tubaki, T (2003) Three-dimensional maneuverability of a dolphin robot (Roll control and loop-the-loop motion), Proc 2nd Int Symp on Aqua Bio-Mechanisms, Honolulu

Triantafyllou, MS and Triantafyllou, GS (1995) An efficient swimming machine, Scientific American, Vol 272, No 3, 64-70

Chapter 7

Design of Artificial Tail Flukes for a Bottlenose Dolphin

Wataru Seki[1], Shingo Kato[2], Shinji Saito[3], and Takashi Yokoi[4]

[1] Diversified Products Research and Development Division, Bridgestone Corporation, 3-1-1 Ogawahigashi, Kodaira, Tokyo 187-8531, Japan
[2] Chemical and industrial Products Technology Division, Bridgestone Corporation, 1 Kashio, Totsuka, Yokohama, Kanagawa 244-8510, Japan
[3] Chemical Products Development Department, Bridgestone Corporation, 1 Kashio, Totsuka, Yokohama, Kanagawa 244-8510, Japan
[4] Research Center, Bridgestone Corporation, 3-1-1 Ogawahigashi, Kodaira, Tokyo 187-8531, Japan

Summary. A female bottlenose dolphin (*tursiops truncatus*), kept in an aquarium, Okinawa, Japan, lost most part of its tail flukes by diseases. Due to the loss of the flukes, the dolphin was not able to swim well. To regain the swimming ability, artificial tail flukes were designed. After several prototypes were tested, the latest prototype, made of silicon rubber and fiber reinforced plastics, achieved the goal. The flukes were attachable to the dolphin safely and easily. The field tests showed that the dolphin was able to swim as before with the help of the flukes.

Key words. dolphin, artificial flukes, biomechanics, fluid dynamics, composite materials

1 Introduction

1.1 About Fuji and Kick-Off of the Project

A female bottlenose dolphin (*tursiops truncatus*), named Fuji (Fig.1), has been under human care in Okinawa Churaumi aquarium since 1975. She suffered progressive necrosis in both distal parts of the tail flukes in 2002. To save Fuji, the necrotic parts, approximately 75% of the flukes, were excised, as reported by Ueda (2005). After the surgery, Fuji had recovered, but was not able to swim well due to the loss of the flukes. Then, Bridgestone Corporation, collaborating with the aquarium and the Japanese institute of cetacean research, started a joint project on designing prosthetic flukes for Fuji.

Fig.1. Fuji (Photo given by Churaumi Aquarium)

1.2 Key Factors of the Design

The key factors for designing the artificial flukes are as follows. Firstly, it should be safe. The flukes should not hurt Fuji or other dolphins in the pool. If the flukes break up into pieces in water, Fuji or other dolphins may possibly swallow the pieces. So, the flukes should have enough strength for hydrodynamic loading acting on them.

Secondly, it can be attached to and detached from Fuji easily. For the safety reasons, Fuji was expected to wear the flukes only daytime, so the flukes were attached to and detached from Fuji once a day.

Thirdly, the flukes should help Fuji regain swimming ability. For this point, hydrodynamic studies were needed. The flukes should generate more propulsive forces without producing large resistant drag forces in forward swimming. The flukes should be handled easily for several swimming patterns.

Fourthly, flexibility for possible design changes was desired. Since there had been no previous works to learn, it was expected that there should be several design changes as the project proceeded.

Training was also essential in the project. Usually dolphins do not like wearing foreign objects, so trainers of the aquarium started to train Fuji to get used to wearing towels on her tail while swimming. From the training, it was expected that Fuji would probably accept the artificial tail flukes.

2 First Prototype (*Boots* Type)

As shown in Fig.2(a), the remaining part of Fuji's flukes were almost symmetric. A replica, shown in Fig.2(b), was made for dimensional coordination. We adopted a pull-on type structure. Silicon rubber was used because it is easy for shaping, relatively harmless for animals' bodies, and tough against weathering. Foamed rubber sheets were pasted inside of the fin to prevent scratching the skin of Fuji.

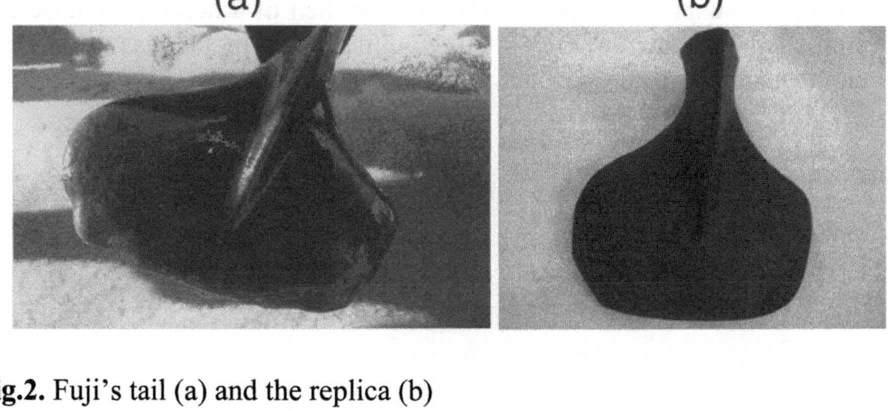

Fig.2. Fuji's tail (a) and the replica (b)

The first prototype, or so-called the *Boots* type is shown in Fig.3. The flukes were attached to the dolphin by a nylon belt.

Fig.3. First Prototype (*Boots Type*)

From the field tests, we found Fuji did not reject the flukes and Fuji's swimming ability was improved to a certain extent. However, we also found problems in the boots type flukes. Firstly the shape was somewhat dull, the flukes generated larger hydrodynamic drag forces comparing to a streamline shaped body. Secondly, it was too soft to generate enough propulsive forces.

3 Improved Prototypes

3.1 Revision of Shape, Materials and Structure

As for the shape, we copied from the specimen of a dead Bottlenose dolphin's tail, as shown in Fig.4. A three-dimensional model was made based on the geometry of the specimen.

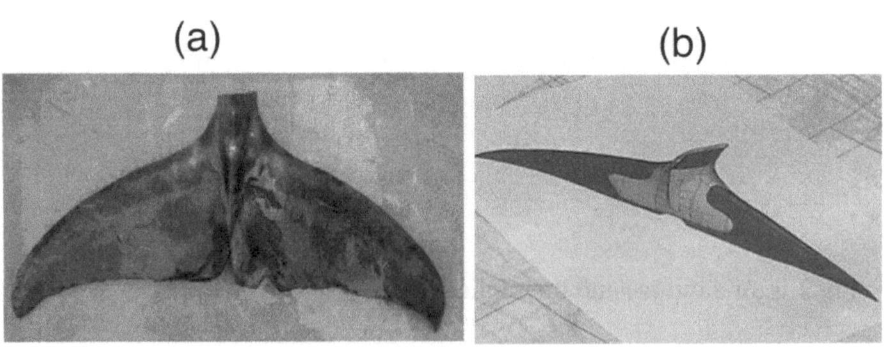

Fig.4. Specimen of the flukes (a) and the 3D model (b)

To adjust the stiffness of the flukes, we adopted a core plate, made of CFRP (carbon fiber reinforced plastics). By inserting it to the flukes, as shown in Fig.5, the tail flukes can be properly reinforced. The core plate should have enough robustness to support hydrodynamic loading. The choice of materials and structure of the core plate was quite important so that the flukes can bear hydrodynamic loads acting on them. The field tests showed the first prototype of the core plate, made of CFRP, did not have sufficient strength, especially when the flukes underwent impulsive loads. The materials and structure of the core plates, such as number of layers and fiber directions, were revised using finite element methods so that the maximum stresses in the core plates, caused by hydrodynamic loads, were minimized.

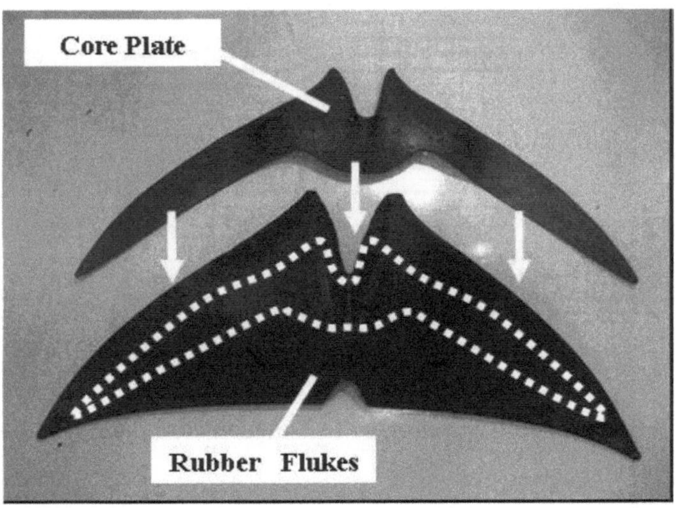

Fig.5. Core plate and the flukes

Fig.6 showed the schematic configuration of the revised core plate. The carbon fiber was used only for the layers near the neutral plane and glass fiber was used for layers near the surface. The configuration and thickness were also different to place to place as shown in Fig. 6. The plate was the thinnest at the edge (part A), and the thickest around the center (part E), in Fig.6. It was thicker in front and thinner at the other side. The setup was intended to simulate the real dolphin flukes so that the flukes had proper flexibility and robustness as well.

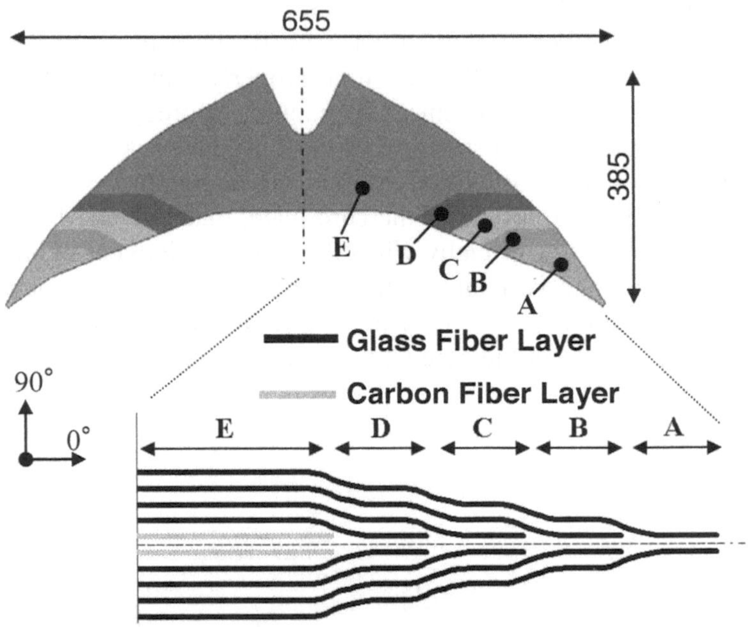

	A	B	C	D	E
glass layers	8	10	12	14	14
carbon layers	0	0	0	0	4

thichness of each layer is 0.2mm
direction of fibers are 0° and 90°、alternately

Fig.6. Schematic configuration of the revised core plate fiber layers

3.2 *Bandage* and *Cowling* Prototypes

As for the attachment methods, we tried the *bandage* type (Fig.7) and the *cowling* type (Fig.8). For the bandage type, the flukes were fastened by X-crossed nylon bands to the dolphin's tail. For the cowling type, the flukes were gripped by the cowling, a fastening plate made of CFRP. Two bolts were used, at the circles in Fig.8, to fasten the cowling to the flukes. In either way, we confirmed that the flukes were attached to the dolphin's tail firmly without any harm to the dolphin.

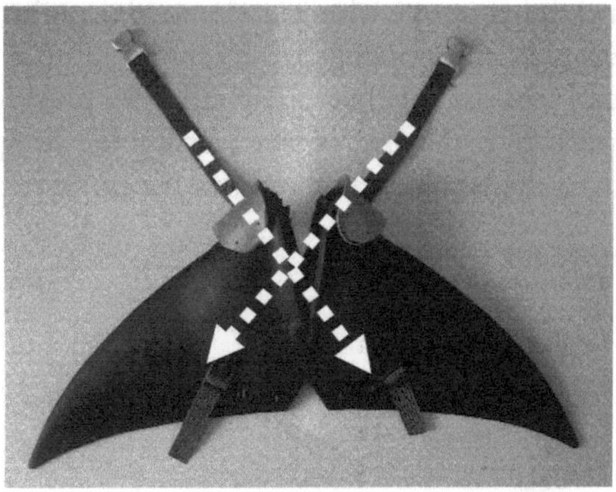

Fig.7. *Bandage* prototype

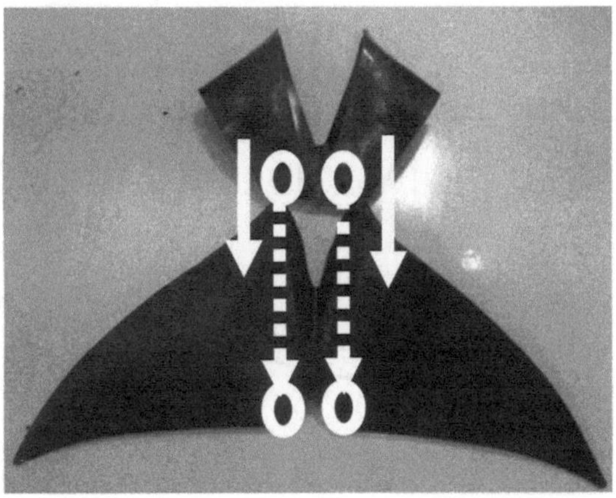

Fig.8. *Cowling* prototype

4 Water Tank Tests

4.1 Theoretical Backgrounds

A series of water tank tests were carried out to investigate the hydrodynamic performances of the several prototypes of the flukes. Nagai (2002) explained how dolphins generate swimming propulsion based on the oscillating wing theory. As shown in Fig.9, the tail flukes of a dolphin move along a sinusoidal path in forward swimming. Then two components of hydrodynamic forces – Lift and Drag Forces – act on the flukes. Usually the drag force is negligibly small comparing to the lift, so the horizontal component of the lift force works as the thrust, which makes the dolphin swim forward.

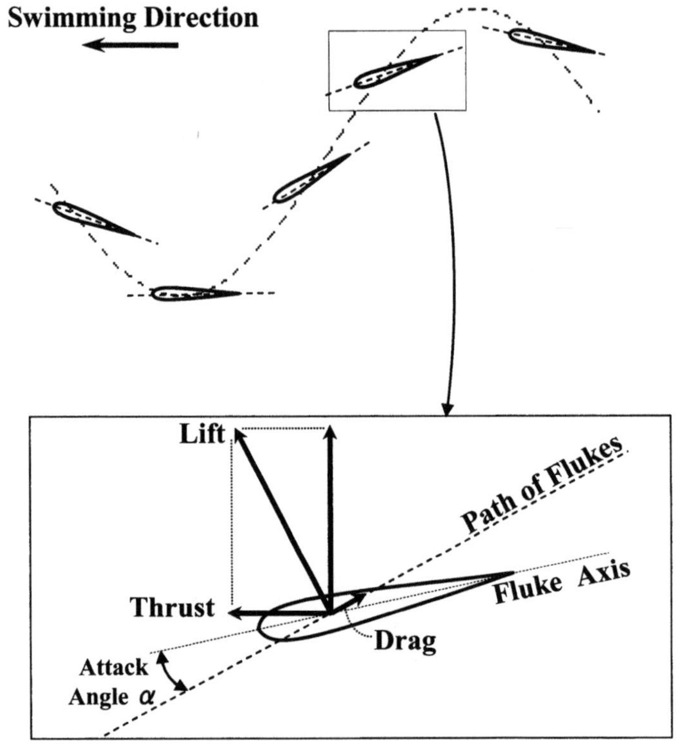

Fig.9. Oscillating wing theory

From the hydrodynamic point of view, it is desirable to design the tail flukes in such a way that the drag force is minimized and the lift force is maximized. To minimize the drag, the shape of flukes and the surface should be smooth as much as possible. As for the shape, we assumed that the current shape of dolphins has been naturally optimized through the process of evolution. From the field tests using the boots type, we found that the flukes were not able to generate enough propulsion if the flukes were too soft. To investigate the hydrodynamic property of the flukes, we conducted the following water tank tests.

4.2 Experiments

The tests were carried out at the water tank shown in Fig.10, at the Department of Environment and Ocean Engineering, the University of Tokyo. Figure 11 shows the setup of the experiment. The lift and drag forces were measured while the flukes were towed with the speed of 2.5m/s.

Fig.10. Water tank at the University of Tokyo

Fig.11. *Bandage* type flukes attached to the test equipment

4.3 Results and Discussion

The Fig.12(a) and (b) show the drag and lift coefficients, C_D and C_L, respectively, calculated from;

$$F_D = \frac{1}{2} C_D \rho V^2 S \tag{1}$$

$$F_L = \frac{1}{2} C_L \rho V^2 S \tag{2}$$

where, F_D; drag force, F_L; lift force, ρ; density of water, V; forward speed of the flukes, S; plane area of the flukes. From the experimental results, we found;

1) As for the attachment methods, the cowling type was superior to the bandage type since the former generates less drag forces than the latter.

2) The flukes reinforced by the core plates generate larger lift force than the one without core plates.

Therefore, we finally adopted the *Cowling* type flukes reinforced by a core plate, among the prototypes tested.

(a) (b)

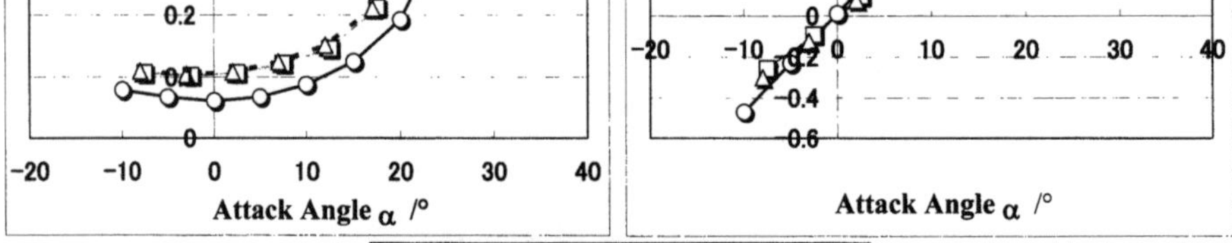

A	Cowling Type with Core Plate
B	Bandage Type with Core Plate
C	Bandage Type without Core Plate

Fig.12. Coefficient of Drag (a) and Lift (b)

5 Field Tests

The developed fluke prototypes were checked by field tests at Okinawa Churaumi Aquarium. We let Fuji try several swimming patterns with the flukes. The swimming patterns included ordinary ones with constant speed, turning round, backstroke, twisting, treading water, continuous jumping, and high jumping. In high jumping, Fuji was able to reach 1.5m above the water level, as shown in Fig.13. The field tests showed that the latest fluke prototype –Cowling type- had enough robustness. We confirmed that Fuji regained her swimming ability she had had before.

Nagai (2002) proposed the *Swimming Number (Sw)* defined by;

$$Sw = U/fl \qquad (3)$$

Where f is the tail beat frequency, U is the swimming velocity, and l is the body length. The swimming number means how far the animal can go forward per body length during one tail beat. It is a dimensionless number and can be interpreted as a measure of the animal's swimming ability.

Fig.13. Field tests – Fuji (the dolphin) jumped into the air with the artificial Flukes

Average ***Sw*** for bottlenose dolphins is about 0.8-0.9, according to Nagai (2002). Otani (2005) found that ***Sw*** of Fuji was 0.4-0.5 and 0.6-0.7 when she was not wearing and wearing the artificial flukes respectively. Although the ***Sw*** of Fuji did not match the values of non-impaired dolphins, her swimming ability was clearly improved by the artificial tail flukes.

6 Conclusions

Artificial tail flukes for a dolphin that lost its own tail by diseases were developed. The key points for the fluke design were to assure safety and robustness for swimming. Laboratory tests were conducted to evaluate and improve the basic hydrodynamic performances of the flukes. Based on finite element analyses, the flukes were fabricated. The latest fluke prototype showed to have enough performance and robustness for most of the dolphin's swimming patterns.

Acknowledgements

We would like to express our sincere thanks to Dr. Uchida, Mr. Miyahara, and Mr. Ueda (DVM) of Okinawa Churaumi Aquarium, Professor Miyata and Mr. Tsuchiya of the University of Tokyo, Dr. Kanbe, Professor Nagai of Ryukyu University, Dr. Otani of the Institute of Cetacean Research for their cooperation.

References

Nagai M (2002), Thinking Fluid Dynamics with dolphins, IOS Press Inc.: 7, 43-44

Ueda K, Miyahara H and Uchida S (2005), Artificial Tail Flukes 1 – Symptoms and Medical Treatments, Proc. of the 16th Biennial Conference on the Biology of Marine Mammals, San Diego, December.

Otani S, Ueda K, Miyahara H and Uchida S (2005), Artificial Tail Flukes 3 – Swimming Behavior, Proc. of the 16th Biennial Conference on the Biology of Marine Mammals, San Diego, December.

Chapter 8

Changes in Drag Acting on an Angled Wavy Silicon-rubber Plate as a Model of the Skin Folds of a Swimming Dolphin

Hui Zhang[1], Naoki Yoshitake[1], and Yoshimichi Hagiwara[1]

[1] Department of Mechanical and System Engineering, Graduate School of Science and Technology, Kyoto Institute of Technology, Matsugasaki, Sakyo-ku, Kyoto 606-8585, Japan

Summary. We have conducted experiments of turbulent water flow over an angled wavy plate on the bottom of an open channel. The plate consists of a silicon-rubber sheet, a thin adhesive film and a metal base. The wave amplitude and wavelength are 0.7mm and 20mm, respectively. We have measured velocity by applying a particle-tracking-velocimetry technique to the images of particles in flow. The local value of time-averaged wall-shear stress has been calculated from the local mean velocity. The total drag is estimated from the measured strains of thin metal strips, which support the plate. The experimental results show that the streamwise mean velocity decreases and the turbulence intensities increase in the near-plate area in and around the valley region. This is due to the flow separation. The time-averaged wall-shear stress is lower than that in the case of the flat wall in and around the valley region. The 15% reduction is obtained for the overall wall shear stress compared with that in the case of the flat wall. The increasing rate of the total drag is lower than that estimated for a non-angled wavy wall.

Key words. Dolphin, Skin folds, Angled wavy plate, Experiment, Drag

1 Introduction

The swimming of dolphins has been discussed widely in various research fields. Among many topics of dolphin swimming, the reduction of friction drag has been focused on recently not only in marine biology and bio-mimicry but also in fluids engineering and naval engineering. This is due to the necessity of reducing the fuel consumption of boats, ocean liners and submarines. Fish (2006) discussed the investigation of drag-reduction mechanisms in his review. He pointed out the following mechanisms; (1) viscous damping by compliant skin, (2) ridges at the skin surface, (3) skin folds, (4) induced turbulent boundary layer, (5) boundary layer acceleration, (6) secretions of polymers from eyes, and (7) boundary-layer heating. The effectiveness of the last two mechanisms is considered to be limited or insignificant. The fourth and fifth mechanisms are mainly associated with flow around the pectoral fins and caudal fins of dolphins, respectively. Thus, these two mechanisms are not related to the friction-drag reduction in a wide region of a dolphin's body. The review states that mobile skin folds (the third mechanism) were observed on accelerating dolphins by some researchers. In this case, the force acting on the skins of dolphins is stronger than that of dolphins swimming with the original speed. The folds are much bigger than the ridges. Actually, one of the present authors took photos of skin folds around the base of the pectoral fins and the side of the abdominal area of the bottlenose dolphins, which were swimming upward in a huge seawater tank. Figure 1 shows the typical photos. Their speed of swimming was at its fastest. Therefore, the skin folds are one of the most probable drag-reduction mechanisms in the body of fast-swimming dolphins.

It is considered that the deformation of the skin surface to generate folds is gradual and that the motion of the skin folds is slow because of the structure of the skin. Thus, the skin folds can be modeled with a wavy wall of a water channel as the first-order approximation. Our research group carried out experiments using several kinds of wavy walls: Nagamine et al. (2004) obtained an increase in the wall shear stress at the hilltop of an aluminum wavy wall. Yamahata et al. (2005) measured the flow over the deforming wall synchronized with the flow acceleration and obtained a decrease in the wall shear stress at the valley region. They made the wall by using silicon rubber because it bears the most similarity to dolphin skin. Note that Choi et al. (1997) obtained turbulent drag reduction of a long slender body by using the silicon-rubber surfaces.

The ridgelines were perpendicular to the main flow in our previous studies. However, the angles between the body axes of dolphins and the ridgelines in Fig. 1 are not 90 degrees. Therefore, how this angle affects the drags has not yet been clarified. Recently, we measured the velocity of turbulent water flow over an angled wavy wall covered with a silicon-rubber sheet in a preliminary experiment (Zhang, et al., 2006). Although we obtained a slight reduction in friction drag, we could not measure the total drag.

In the present study, we conduct an experiment for turbulent water flow over an angled wavy plate. We measure velocity profiles near the plate and calculate the wall shear stress from the profiles. Also, we estimate the total drag from the measured strains of thin metal strips, which support the angled wavy plates.

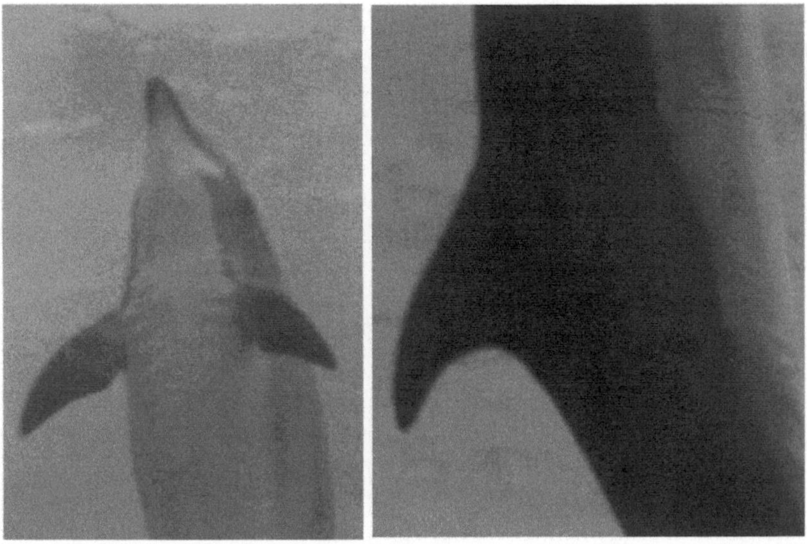

Fig. 1. Photos of swiftly swimming bottlenose dolphins

2 Experimental methods

2.1 Apparatus

Figure 2 shows the open channel. Water in the constant-head tank was introduced into the contraction chamber through a valve. The water in the chamber flowed into an open channel of 2000 mm in length and 270 mm in width. A tripping wire and an emery paper were attached to the bottom wall at the inlet of the channel in order to enhance the development of a turbulent boundary layer along the bottom wall. The perforated weir was located at the exit of the channel. The height of the weir was adjusted in order to control the water depth and velocity distribution. The x, y and z axes were positioned in the axial, upward and transverse directions, respectively.

The Reynolds number Re, based on the bulk-mean velocity and the water depth H (=35 mm), was 8.8×10^3. The Reynolds number Re_τ, based on the friction velocity of the flat plate $u_{\tau 0}$ and H, was 5.6×10^2. The Froude number was 0.41. We estimated the Kolmogorov length scale l_K and the Kolmogorov time scale t_K from DNS result of an open channel flow obtained by Yamamoto et al. (2001). These scales are shown in Table 1.

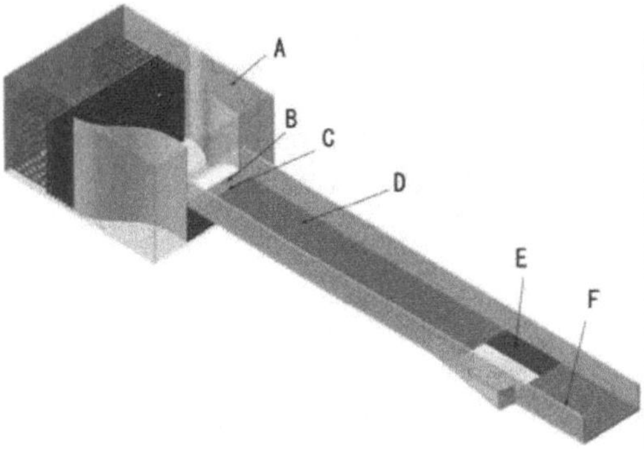

A: Contraction chamber
B: Emery paper, C: Tripping wire
D: Rubber sheet, E: Test section
F: Side wall

Fig. 2. Open channel

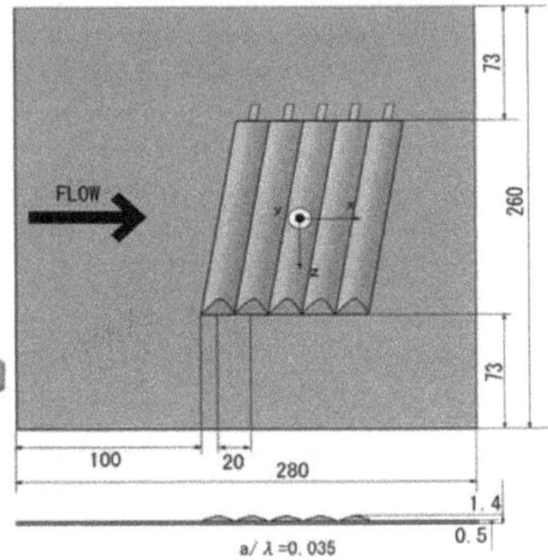

Fig. 3. Test plate

Table 1. Kolmogorov scales

l_K[mm]	l_K^+	t_K [s]	t_K^+	$1/t_K$[Hz]	$1/t_K^+$	f_r [Hz]	f_r^+
0.10	1.6	0.010	2.4	100	0.42	300	1.2

2.2 Test Plate

The test section was located in a developed region of the channel. Figure 3 shows the front and top views of the test plate. The plate consists of a stainless-steel flat plate (0.5 mm in thickness), a thin adhesive film and a thin silicon-rubber sheet. The angled wavy surface was made by inserting metal pipes into the slots of the film and sheet. The angle between the channel axis (flow direction) and the ridgeline of the wavy plate was set at 80 degrees. This angle was decided based on the skin folds seen in Fig. 1. The wave amplitude a, the wavelength λ, their non-dimensional values, the ratios of a/λ, a/H, a/l_K and λ/l_K are shown in Table 2. It is found from this table that the dimensions of the wavy plate are much larger than the smallest eddies. The area of the wavy surface was one-sixth of the total area of the test plate. This fraction will be used in the discussion of the total drag in Subsection 3.4. Note that the test plate without the insertion of pipes was used as a reference flat plate.

Table 2. Parameters of wavy plate

a [mm]	a^+	λ [mm]	λ^+	a/λ	a/H	a/l_K	λ/l_K
0.70	11	20	310	0.035	0.019	7.0	200

2.3 Flow Visualization

The flow was visualized with laser-induced fluorescent particles. We made the particles by adsorbing Rhodamine B to the particles (Mitsubishi Chemicals, SEPABEADS SP20SS: the specific weight is 1.02). The diameters of the particles were in the range of 0.063 - 0.075 mm. Thus, the particles are smaller than the Kolmogorov length scale. The upper limit of the frequency f_r, beyond which these particles cease to respond to fluid sinusoidal fluctuation completely, was estimated to be 300 Hz from the equations obtained by Hjelmfelt Jr. and Mockros (1966). The non-dimensional value of this frequency, $f_r^+ = f_r\nu/u_\tau^2$, is shown in Table 1. The reciprocal of the Kolmogorov time scale is the frequency corresponding to the smallest eddies. It is found from the table that f_r is much higher than the reciprocal. This shows that the particles can follow flow fluctuation caused by the smallest eddies. Therefore, the particles were appropriate.

2.4 Image Capturing

Figure 4 exhibits the optical system and image-capturing system. A green light of Nd: YVO_4 laser was used as the light source. The light was expanded with a plano-convex cylindrical lens and a plano-concave lens to obtain a light sheet. This light sheet passed through a slit of 5 mm in width above the free surface. The light sheet thus obtained illuminated the flow.

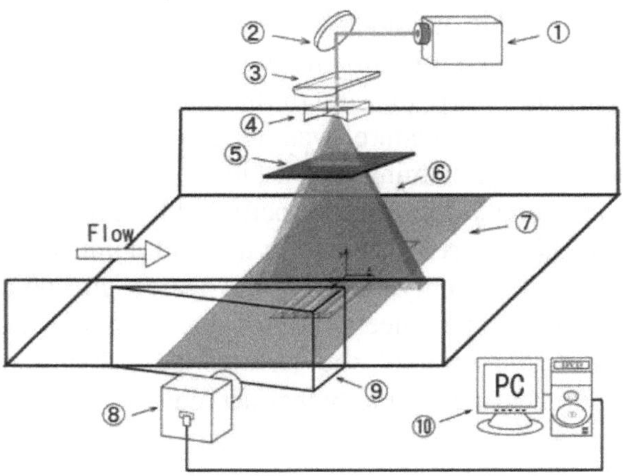

1 YVO_4 Laser
2 Mirror
3 Convex lens
4 Concave lens
5 Slit
6 Laser sheet
7 Test plate
8 CCD camera
9 Water prism
10 PC

Fig. 4. Velocity measurement system

Table 3. Image-capturing condition

Frame rate [frames/s]	300	Shutter speed [ms]	0.4
Pixel-covering area [mm^2]	0.035 × 0.035	Gray level [-]	256
Total area [mm^2]	22.4 × 11.2	Numbers of frames [-]	4000×3

Fluorescence was induced from the particles by the illumination of the laser light sheet. The fluorescence was captured by a CCD camera (IMPERX, Vga210-L) through an optical filter in front of the camera. The filter was used to reduce the reflected laser light at the sidewalls. The camera was located at a side of the channel as shown in Fig. 4. The optical axis of the camera was set parallel to the ridgelines of the wavy plate so that the valley of the plate was completely observed. In order to reduce refraction at the channel sidewall, a water prism was attached to the outer surface of the wall. Since the angle between the laser light sheet and the camera axis was 80 degrees, the captured images were distorted slightly. We captured a reference image in order to compensate for the distortion of images. A plate with grids was used for this reference image. This plate was placed parallel to the channel axis inside the laser light illumination area. The compensation of the captured images was carried out before the preprocessing of the images mentioned below. The image-capturing condition is summarized in Table 3. The captured images were directly recorded into the memory of a PC through a frame grabber board.

2.5 Image Processing

The particle-mask correlation method, developed by Etoh et al. (1999), was used to remove weak scattered light of particles from the images. The PTV technique based on the velocity gradient tensor method, proposed by Ishikawa et al. (2000), was applied to the preprocessed images for obtaining velocity vectors. In this method, the matrix including the velocity gradient tensor was calculated for pairs of neighboring particles in a specific region around a particle. Then, the sum of the square of errors in the matrix was evaluated. This procedure was repeated for all the candidate particles until the sum took its minimum value. This method has the advantage of reproducing strongly deformed velocity fields accurately.

The velocities of the particles were redistributed to the grid points of 14 × 50. The velocity of a particle was simply shifted to the nearest grid point in the redistribution procedure. The uncertainty of the velocity was estimated to be 5.6 mm/s in the case of the streamwise velocity of 250mm/s. This shows that the measured velocity was accurate.

2.6 Total Drag

Figure 5 indicates the measurement system for the streamwise component of total drag acting on the test plate. The test plate was steadied by phosphor bronze strips. The strain gages were adhered to the strips. The outputs of the gages were recorded through bridge circuits to a PC. Many particles of 0.3 mm diameter were inserted between the channel bottom and the lower surface of the test plate as bearings.

3 Results and discussion

Hereafter, we divide the whole area of each image into the following six regions as shown in Fig. 6 in order to clarify difference in results: valley, uphill1, uphill2, hilltop, downhill2 and downhill1.

3.1 Hilltop and Valley Regions

Figure 7 exhibits the profiles of mean velocities in the streamwise and vertical directions. It is found that the streamwise mean velocity in the area of $y < 0.5$ cm in the valley region shown with solid triangles is lower than that in the case of the flat wall. The velocity adjacent to the plate takes a negative value. Since the base point of the valley is the same as that of the flat wall, the velocity gradient in the valley region is lower than that of the flat wall.

On the other hand, the streamwise mean velocity in the hilltop region shown with solid circles increases sharply from zero at the base point (shown with a black dot on the horizontal axis). The velocity in the region 0.5 mm higher from the base point is slightly higher than the velocity at the same region in the case of the flat wall. This indicates that the velocity gradient is steeper in the hilltop region than that of the flat wall. The velocities in the vertical direction take positive values in the valley.

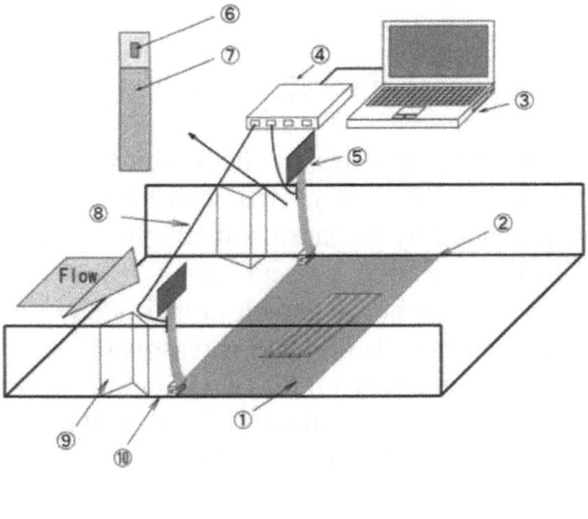

1 Angled wavy plate
2 Wire
3 PC
4 Bridge circuits
5 Clamped plate
6 Strain gage
7 Phosphor bronze strip
8 Lead wire
9 Fence
10 U-brass

Fig. 5. Drag measurement system

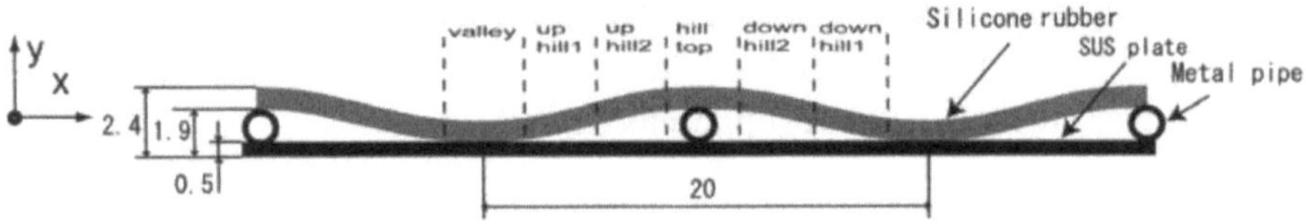

Fig. 6. Definitions of regions

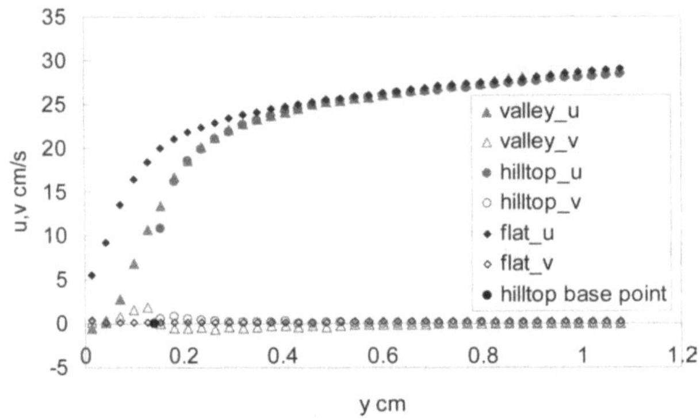

Fig. 7. Mean velocity profiles in the regions of valley and hilltop

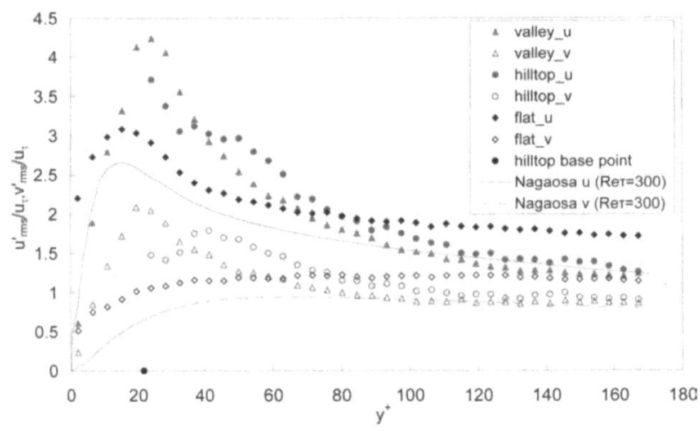

Fig. 8. Turbulence intensities in the regions of valley and hilltop

Figure 8 depicts turbulence intensity profiles in the streamwise and vertical directions. The intensities in the case of flat wall shown with lozenges are higher than those predicted by Nagaosa and Handler (2003). Re_τ in their computation is nearly half the present case. Since the intensities increase with the Reynolds number as shown in their study, the difference between their results and our results in Fig. 8 is reasonable.

In Fig. 8, the streamwise turbulence intensity in the valley region shown with solid triangles is higher than that in the case of the flat wall in the area of $y^+ < 70$. The vertical turbulence intensity in the valley region shown with open triangles is higher than that in the case of the flat wall in the area of $y^+ < 60$. Similarly, the streamwise and vertical turbulence intensities in the area of $25 < y^+ < 80$ in the hilltop region, which are shown with the solid circles and open circles respectively, are higher than those of the flat wall. All the turbulence modifications indicate that the flow structures near the valley and hilltop are different from that near the flat wall.

3.2 Uphill and Downhill Regions

Figure 9 shows the velocity profiles in the regions of downhill1 and uphill1. These regions are positioned symmetrically with respect to the valley. The velocities in the streamwise and vertical directions increase adjacent to the plate in the uphill1 region, while they do not increase adjacent to the plate in the downhill1 region.

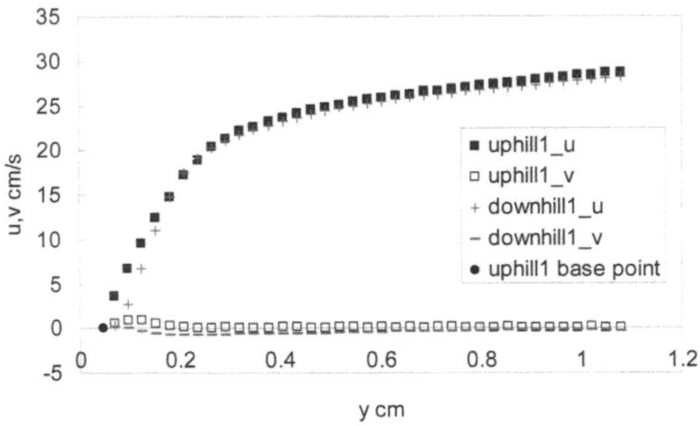

Fig. 9. Mean velocity profiles in the regions of uphill1 and downhill1

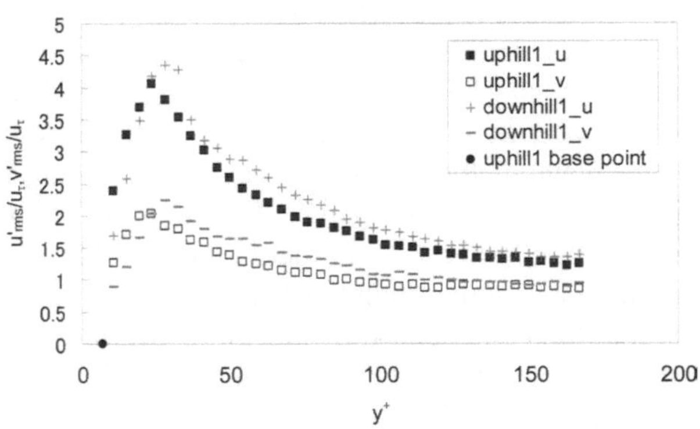

Fig. 10. Turbulence intensities in the regions of uphill1 and downhill1

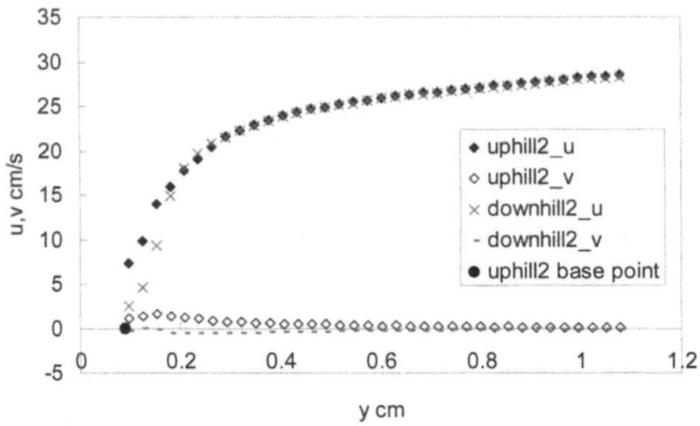

Fig. 11. Mean velocity profiles in the regions of uphill2 and downhill2

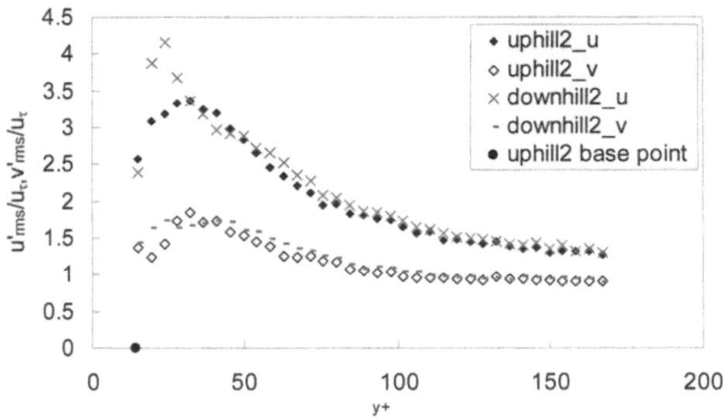

Fig. 12. Turbulence intensities in the regions of uphill2 and downhill2

The turbulence intensities in the two regions are indicated in Fig. 10. The intensities in the downhill1 region are slightly higher than those in the uphill1 region except for the near-plate region.

Figure 11 shows the profiles of mean velocities in the regions of uphill2 and downhill2. These regions are positioned symmetrically with respect to the hilltop. The velocities in the streamwise and vertical directions increase adjacent to the plate in the uphill2 region, while they do not increase adjacent to the plate in the downhill2 region. The turbulence intensities in these regions are shown in Fig. 12. The streamwise intensity in the downhill2 region is slightly higher than that in the uphill2 region.

The value of a/λ in Table 2 is higher than the critical value, over which a flow separation appears in the valley and downhill1 regions, in the case of flow over non-angled rigid wavy walls (Tuen et al., 2006; De Angeles et al., 1997). We observed the flow separation in our preliminary experiment. Thus, the asymmetry of the turbulence statistics with respect to the valley and hilltop mentioned above is due to the flow separation.

3.3 Wall Shear Stress

We calculated the wall shear stress in the streamwise direction from the gradient of u shown in Figs. 7, 9 and 11. The wall shear stress in each region is shown in Table 4. The wall shear stresses in the hilltop and downhill2 regions are higher than that in the case of the flat wall, while the shear stresses in the other regions are lower than that in the case of the flat wall. The mean value over the whole surface was calculated from the average of the values in Table 4. The mean value is 0.21 N/m^2 and thus, the friction drag was reduced by 15 percent by the angled wavy plate.

We measured the velocity in a horizontal plane just above the buffer region (Zhang et al, 2006). The ratio of velocity in the z direction to that in the x direction was in the range of 2.0 – 2.9%. Thus, flow along the ridgelines is weak. Note that the wall shear stress in the z direction is only 0.05 – 0.06% of the wall shear stress in the x direction.

Table 4. Wall shear stress in each region

	flat	valley	uphill1	uphill2	hilltop	downhill2	downhill1
τ_w [N/m^2]	0.25	-0.008	0.19	0.23	0.40	0.37	0.08

3.4 Total Drag

In the case of the flat wall, the total drag should be equal to the friction drag. The measured total drag was about 13 % higher than the estimated friction drag from the mean velocity gradient. The difference may be due to the friction of bearing particles or errors of the strain measurement.

In the case of the angled wavy plate, we considered that the transverse component of total drag is negligible. The streamwise component of total drag was 17 % higher than that of the flat wall. If the whole surface of the test plate is angled-wavy, the increase would be 17×6 = 102 %. This is approximately 20 % lower than that estimated from the result of the non-angled rigid wavy wall (Tuen et al. 2006). It is considered that the weak flow along the ridgelines causes instability of the flow separation, and that the unstable separation leads to the changes in the drag.

4 Conclusions

The experiment was conducted for water flow over an angled wavy plate covered with silicon rubber. The main conclusions obtained are as follows:

1. The streamwise mean velocity decreases and the turbulence intensities increase in the near-wall area in and around the valley regions. This is due to the flow separation.
2. The local wall shear stress is lower than that in the case of the flat wall except for the stress in the hilltop and downhill2 regions. The 15% reduction is obtained for the overall wall shear stress compared with that in the case of the flat wall.
3. The increasing rate of the total drag is reduced by the silicon-rubber angled wavy plate. This is probably due to the instability of flow separation caused by the weak secondary flow along the ridgelines.

The authors acknowledge Mr. T. Saito at Port of Nagoya Public Aquarium for his comments on dolphins.

References

De Angelis V, Lombardi P, Banerjee S (1997) Direct numerical simulation of turbulent flow over a wavy wall. Physics of Fluids. vol. 9, pp 2429–2442

Choi, KS et al. (1997) Turbulent drag reduction using compliant surfaces. Proc. Royal Soc. London A vol. 453, pp 2229-2240

Etoh T, Takehara K, Okamoto K (1999) Performance evaluation of the PMC and the KC methods for particle extraction and tracking through their application to standard particle images (in Japanese). Trans. JSME vol. 65B, pp 1688-1695

Fish FE (2006) Review of dolphin hydrodynamics and swimming performance. J. Biomimetics and Bioinspiration vol.1, pp. R21-R26

Hjelmfelt Jr AT, Mockros LF (1966) Motion of discrete particles in a turbulent fluid. Applied Scientific Research vol. 16, pp 149-161

Ishikawa M et al. (2000) A novel algorithm for particle tracking velocimetry using the velocity gradient tensor. Experiments in Fluids vol. 29, pp 519-531

Nagamine H et al. (2004) Turbulence modification by compliant skin and strata-corneas desquamation of a swimming dolphin. J. Turbulence vol. 5 no.018, 25p

Nagaosa R, Handler RA (2003) Statistical analysis of coherent vortices near a free surface in a fully developed turbulence. Physics of Fluids. vol. 15, pp 375–394

Tuan HA et al. (2006) Immersed boundary method for simulating turbulent flow over a wavy channel. Extended Abstracts of Whither Turbulence Prediction and Control. pp 116–117

Yamahata K et al. (2004) Turbulence modification by compliant wall as a model of dolphin skin. Proc 2nd Int. Symp. on Seawater Drag Reduction, pp 535-544

Yamamoto Y, Kunugi T, Serizawa A (2001) Turbulence statistics and scalar transport in an open-channel flow. J. Turbulence vol. 2, no. 010, 16p

Zhang H, Nakamura S, Hagiwara Y (2006) Reduction of friction drag by angled wavy wall simulating deformed skin of dolphins (in Japanese). Proc. 2006 Annual Meeting, Japan Soc. Fluid Mechanics (CD-ROM), no. AM06-12-003, 4p.

Chapter 9

Central nervous system underlying fish swimming [A review]

[1]Kazumasa Uematsu

[1] Laboratory of Fish Physiology, Graduate School of Biosphere Sciences, Hiroshima University, Kagamiyama 1-4-4, Higashi-Hiroshima, Hiroshima 739-8528, Japan

Summary. In the swimming of fish, thrust is generated by undulatory movements of the trunk muscles. The trunk muscles are composed of myomeres innervated by corresponding spinal segments, which are in turn composed of neural circuits which range longitudinally and interconnect. In each spinal segment there are various types of interneurons and motor neurons, which together form a central pattern generator (CPG). Although each CPG can potentially generate its own swimming rhythm, the activity of each CPG is regulated by excitatory commands from the brain and by mutual excitatory and inhibitory interactions between interneurons to produce controlled locomotion.

The nucleus of the medial longitudinal fasciculus (Nflm) has been identified as a locomotor region in the midbrain of the carp. Nflm neurons extend axons to the spinal cord and are connected to intraspinal motor neurons and interneurons. Physiological studies have demonstrated that there are neurons in Nflm which fire tonically only during swimming. These neurons probably drive spinal the CPGs to initiate and maintain swimming. Swimming of unrestrained goldfish can be induced by wired and wireless electric stimulation of Nflm, also supporting arguments that Nflm controls swimming in teleost fishes. The involvement of the fish cerebellum in swimming is also discussed.

Key words. central pattern generator, fish, locomotion, midbrain, spinal cord

1 Introduction

Animals move in their environment to search of food, to find prey, to avoid predators, and to locate partners for reproduction. This behavior is defined as locomotion, and includes various types of movement such as walking, running, flying, creeping, and swimming (Gray 1968, Day 1981, Biewener 2003). In this short review, I will describe previous and recent findings on the neuronal mechanisms underlying swimming by fish. Most fish swim using trunk muscles (axial muscles) and fins. The patterns of fish locomotion using both trunk muscles and fins have been classified into undulation and oscillation (Lindsey 1978). In this chapter, I will focus on the central mechanisms controlling undulatory swimming using trunk muscles.

Most fish swim by alternating contractions of the axial muscles traveling in a rostra-caudal direction (Lindsey 1978, William et al. 1989). The body musculature of fish is usually organized segmentally (Romer and Parsons 1977). Spinal motor neurons send their axons via ventral roots to innervate specific parts of each segmental myotome (Wallán et al. 1985). In vertebrates, this alternating rhythmic contraction of the axial muscles is produced by central pattern generators (CPGs) located in each spinal segment along the entire cord (Grillner et al. 1988, Roberts et al. 1986). For the production of swimming, excitatory drive, either from the brain or other CPGs, is required to activate the spinal CPG in the lamprey (Grillner et al. 1988). It has been demonstrated that spinal CPGs are driven by excitatory inputs from the brainstem in the carp *Cyprinus carpio* (Kashin et al. 1974). The nucleus of the medial longitudinal fasciculus (Nflm), located in the midbrain tegmentum, has been identified as a brain region initiating and maintaining swimming in the carp (Uematsu and Ikeda 1993, Uematsu and Todo 1997). Neurons in the Nflm have been labeled retrogradely from the spinal cord in the salmon *Oncorhynchus nerka* (Oka et al. 1986), the goldfish *Carassius auratus* (Prasada Rao et al. 1987), and the zebrafish *Danio rerio* (Kimmel et al. 1982, Becker et al. 1997), indicating that these neurons project to the spinal cord. However, the physiology and cytoarchitecture of neurons in the nucleus have yet to be determined in detail.

In this review, I describe the neuronal activities of Nflm neurons during swimming, the connections between Nflm neurons and neurons in spinal CPGs, and the cytoarchitecture of Nflm neurons, including their shapes, numbers, and distributions.

I will also refer to the efferent system of the cerebellum, an important structure in regulating movement and locomotion in tetrapods (Middleton and Strick 1998). Although the fish cerebellum is quite well developed, its functions are still poorly understood (Ikenaga et al. 2006). For example, cerebellar lesions did not significantly affect swimming by fish (Roberts et al. 1992). There have been many studies on the afferent connections of the cerebellum, and most sensory modalities are known to reach this brain region (Finger 1983). In contrast, our knowledge concerning the efferent connections of cerebellum is minimal. We have therefore attempted to identify the cerebellar efferent neurons and their transmitters and relationships with Purkinje cells using retrograde labeling and immunohistochemical techniques.

2 Fish Swimming

2.1 Fish Swim with use of the Trunk Muscles

Generally, fish can only swim forwards using the undulatory movement of the trunk muscles, though they sometimes swim backward by movements of the pectoral fins and not the trunk muscles themselves (Lindsey 1978). An exception to this is seen in a special fish group represented by the eels (*Anguilla* sps.), which can move backward using the trunk muscles (D'Août and Aerts 1999) and, interestingly, can even move their body after transection between the brain and spinal cord (spinal fish, Gray 1936). It is still unclear how they are able to move backward with the trunk muscles, though the answer to this question might be provided in the chapter by Dr. Ayers. In the future, we will also try to clarify the unique mechanisms of the eel CNS.

2.2 Muscular and Spinal Systems Involved in Undulatory Swimming by the Trunk Muscles

A morphological characteristic of the trunk muscles of principal importance in establishing undulatory movements like fish swimming is their segmental structure. The fish body is comprised of muscle segments, termed myomeres (myotomes), which function as power sources and are arranged longitudinally, and connective tissue myosepta, which act as flexible, articulated couplings between the myomeres (Figure 1). This configuration is essential for smooth and forceful undulatory swimming by the trunk muscles. For detailed descriptions of the structure of the fish locomotor muscles, see Nursall (1955) and Bone (1978).

During normal forward swimming, muscles in the left and right myomeres of the same segments contract alternately, with rostral myomeres contracting earlier. Exceptional usages of the trunk muscles are of great interest and include the startle reflex (C-start) mediated by the Mauthner neuron system and spawning of salmon (see below).

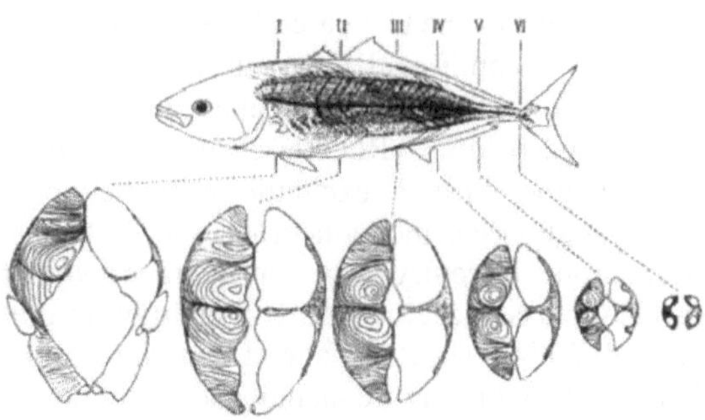

Fig. 1. A schematic drawing of the trunk muscle of the yellowtail Seriola quinqueradiata. Below, appearances of transverse sections of the body at the levels of indicated by I-VI in the top drawing. Shaded areas indicate the dark muscles. (Adapted from Tsukamoto (1984a) with permission)

The trunk muscles consist of dark muscle (red muscle) and ordinary muscle (white muscle) (Rayner and Keenan 1967, Tsukamoto 1984a). The former is located mostly superficially, and in the lateral portion of the trunk along the mid-horizontal axis around the lateral line (Figure 1). Migratory marine fish like the skipjack tuna *Katsuwonus pelamis* and the yellowtail *Seriola quinqueradiata* are unusual in having large volumes of deep dark muscle in addition to superficial dark muscle (Rayner and Keenan 1967, Tsukamoto 1984a). The dark muscles are adapted to aerobic metabolism and are used for sustainable swimming at relatively low speeds like that in cruising. The white muscle occupies large portions of the trunk and uses energy produced by anaerobic metabolic pathways (enzymes). Unlike the dark muscles it tires easily due to the accumulation of lactic acid and is usually used when fish needs to produce powerful thrust to escape predators or capture prey. It can thus be said that, muscularly speaking, most fish are equipped with a small but vital main-engine and a large and powerful sub-engine. For aerobic and anaerobic swimming performance by the yellowtail and the cod *Gadus morhua*, see Tsukamoto (1984b) and Reidy et al. (2000), respectively.

In the zebrafish, Westerfield et al. (1986) demonstrated that the white muscles are innervated by both a few larger primary and a larger number of smaller secondary motor neurons, while the dark muscles are innervated only by the secondary motor neurons. The two types of motor neuron have been identified in the larval zebrafish (Myers 1985). There are only three primary motor neurons in each spinal CPG of the adult zebrafish (Westerfield et al. 1986). The three uniquely innervate different dorsoventral portions of the myomeres of the same body segments, while the field of distribution of axons of the secondary motor neurons is not confined to within their own segment. Other fishes larger than the zebrafish most likely have many more motor neurons in each spinal segment (Fetcho and Faber 1988).

The basic, common neuronal circuit in the spinal CPG required for establishment of undulatory movements has been thoroughly characterized in the frog embryo (Roberts et al., 1986 and 1998) and the lampreys (Grillner et al., 1988 and 1998). The major components of the CPG are glutamatergic excitatory descending and glycinergic inhibitory commissural interneurons, and motor neurons. The connections and functions of these neurons are quite similar to those of the Mauthner system described below. Roberts et al. (1998) noted that the motor neurons in CPGs receive three different excitatory synaptic inputs: cholinergic and electrical inputs from neighboring motor neurons and glutamatergic inputs from the descending interneurons. Recentry, it was also suggested that some excitatory spinal interneurons also released acetylcholine in addition to glutamate as well as motor neurons (Li et al. 2004). They speculated that summation of all inputs between CPGs might control swimming frequency and direction. Although it has not been directly demonstrated that teleost fish are also equipped with the same central systems for generating swimming rhythm, I am convinced of the existence of circuits similar to those found in the tadpoles and lampreys, since Fetcho (1992) and Higashijima et al. (2004ab) have found the same types of neurons and circuits in the spinal cord of the goldfish and zebrafish, respectively.

Most fish can exhibit remarkable escape behavior when they perceive danger in the surrounding environment, usually by the auditory sense. The affected fish bends its body away from the sound source and moves off with a quick tail flip following the body bend within tens of milliseconds (Faber et al. 1989). This behavior is termed the C-start, based on the body shape produced when bending. This behavior is initiated by a pair of huge Mauthner cells (M-cell) in the tegmentum of the medulla oblongata below the cerebellum and just caudal to the Nflm (see below). One important characteristic of M-cells is that they each have one thick commissural axon projecting the entire length of the contralateral spinal cord, which directly or indirectly contacts all the interneurons and motor neurons on that side. M-cells also have two large dendrites which receive sensory inputs. Once stimuli are perceived by the fish, the M-cell closer to the sound source fires immediately, sending neural impulses to the spinal cord. In each spinal segment, there are excitatory descending interneurons, inhibitory commissural interneurons, and the two types of motor neurons noted above. The excitatory interneurons contact the two types of motor neurons in the spinal segment on the same side and excite them. The axons of the inhibitory interneurons cross the midline and contact all the neurons including interneurons and motor neurons in the contralateral spinal segment and completely inhibit their activity. A remarkable feature of this neuronal circuit is that the axon of an M-cell makes synapses to excite both the two types of interneuron and the primary motor neurons directly. As a result, every muscle in every myosegment on one side contracts simultaneously while those of the other side are not activated (Fetcho and Faber 1988). This is the neural basis of the C-start. Note, the spinal neuron types in the network for swimming and the C-start in fishes are very similar to those found in the spinal cord of the tadpole and the lamprey.

2.3 Spawning of Salmons, and a Hypothesis

In Pacific salmon *Oncorhynchus keta* it has been demonstrated by observations of electrical muscle activity that eggs are laid by tonic contraction of the trunk muscles, which are usually used for undulatory swimming (Uematsu et al. 1980, Uematsu and Yamamori 1982, Matsushima et al. 1986). However, the spawning center has yet to be identified in the salmon brain.

In the spinal cord of teleosts there are many inhibitory and excitatory interneurons along with the two types of motor neurons (Uematsu et al. 1993, Fetcho 1998, Yoshida et al. 1999, Higashijima et al. 2004ab). These neurons are components of the CPGs situated in each spinal segment, and are activated by commands from the brain or by sensory signals, as already described. I would like to suggest the hypothesis that, besides playing a role in swimming, the spinal CPGs are used both for the C-start and for spawning by salmon. Here the same sets of neurons in spinal CPGs would thus be activated by descending signals coming from different regions in the brain during swimming, escape, and spawning. I have tentatively termed this the "multifunction spinal CPGs" theory. My hypothesis is based on the insight that animal bodies must as far as possible be designed rationally and reasonably. Similarly, it has been reported that swimming and struggling by *Xenopus* tadpoles are driven by common CPG neurons in the cord (Soffe 1993).

3 Identification of a Swimming Initiation Region in the Fish Brain

3.1 Discovery of a Swimming Region

Kashin et al. (1974) first discovered a brain area which probably included neuronal structures initiating swimming in the carp midbrain. Oka et al. (1986) further suggested that the candidate neuronal structure concerned with swimming initiation was either the nucleus of medial longitudinal fasciculus (Nflm) or the nucleus ruber (red nucleus) located in the midbrain tegmentum (Oka et al.1986). At nearly the same time, it was demonstrated in various teleosts that some Nflm neurons project axons to the spinal cord (Kimmel et al. 1982, Prasada Rao et al. 1987, Becker et al. 1997). Experiments involving exhaustive electric and chemical stimulation of the tegmentum of the midbrain and the medulla oblongata in carp have provided evidence that the Nflm is the brain center initiating swimming (Uematsu and Ikeda 1993, Uematsu and Todo 1997). As already reported in mammals (Grillner et al. 1998), if the strength of stimulation of the Nflm is increased, the fish increases its tail beat frequency (Uematsu and Ikeda 1993).

3.2 Electrical Activities and Functions of Nflm Neurons

In the tadpole and lamprey, it has been suggested that tonic excitatory drive from the brainstem is responsible for initiating and maintaining locomotion (Roberts et al. 1998, Grillner et al. 1998, Li et al., 2006). If the Nflm is a brain region controlling swimming in fish, we would expect electrode recording to reveal neurons active during swimming in the Nflm . Two types of neuronal activities were recorded from the Nflm in both immobilized and swimming carp (Baba et al. 2003, Uematsu et al. 2007). One type of neuron, encountered more often, was characterized by tonic, continuous firing that occurred just before swimming and was maintained at high levels throughout a swimming episode. In contrast, the other type of Nflm neuron had a markedly decreased firing rate during swimming. It was proposed that the former type of Nflm neuron probably drives spinal CPGs tonically during swimming in fish (Baba et al. 2003, Uematsu et al. 2007). It was also suggested that some tonic Nflm neurons may remain inactive during slower swimming and be recruited to strengthen the drive to spinal CPGs for faster swimming (Uematsu et al. 2007). The functions of the other type of Nflm neurons remain unclear.

3.3 Cytoarchitecture of Nflm and its Connections to Spinal Neurons

3.3.1 Cytoarchitecture of Nflm neurons

The Nflm is located in the midbrain tegmentum just caudal to the posterior commissure and beneath the midbrain ventricle. As a whole the Nflm has a shape like a gull wing around the rim of the floor of the midbrain ventricle when viewed rostrally (Uematsu et al. 2007). The caudal end of the nucleus is more difficult to determine, since it is continuous with the oculomotor nucleus (NIII) and the trochlear nucleus (NIV). However, the overall appearance of the Nflm and other nuclei became apparent in three-dimensional reconstructions of Nissl-stained sections. The total number of neurons in this region was estimated to be about 600 in the carp (Uematsu et al. 2007).

The difficulty in determining the boundaries of this region could be overcome using a molecular approach, with *in situ* hybridization targeting mRNA coding choline acetyltransferase (ChAT), the acetylcholine-synthesizing enzyme. Neurons in NIII and NIV should contain ChAT mRNA, since they are somatic motor nuclei consisting of neurons containing acetylcholine as the transmitter (Moon et al., 2005). Neurons in Nflm were labeled retrogradely by applying Biotinylated Dextran Amines (BDA) in the rostral cord. On alternating serial sections, there were 164 Nflm neurons labeled retrogradely and 254 ChAT mRNA-positive neurons at maximum among three carp examined (Uematsu et al. 2007). The Nflm and NIII overlap rostrocaudally over a range of 180-200 μm in the carp. Nflm neurons have been classified into two types according to the location and the projection patterns of their dendrites: one type is located medially with dendrites extending into both ipsilateral and contralateral sides of the brain, the other is located laterally with dendrites confined ipsilaterally in the zebrafish (Kimmel et al. 1982, Lee and Eaton 1991, Lee et al. 1993). Gahtan et al. (2005) have recently demonstrated using a laser microdissection technique that they share functions in controlling prey capture behavior in the zebrafish.

If Nflm neurons drive spinal CPG neurons, their transmitter may well be glutamate, since glutamate is the usual excitatory amino acid transmitter in vertebrates. Immunocytochemistry using a specific anti-glutamate monoclonal antibody enabled us to demonstrate the existence of approximately 400 glutamate-immunoreactive neurons in Nflm (Uematsu et al. 2007). Many labeled axons projecting to the cord were also observed in the medial longitudinal fasciculus (flm).

In a very important report Gahtan and O'Malley (2003) emphasised two features of the patterns of projection of Nflm neurons in larval zebrafish. First, collaterals of all three types of Nflm neurons axon give rise to their most extensive arbors in the caudal medulla and rostral cord. They suggested that these arbors might influence other neurons involved in swimming. Second, axon collaterals of one type of neuron have contralateral in addition to ipsilateral projections. This discovery was quite new, and suggests that the mechanisms controlling fish swimming may be complex.

In a similar experiment using rather classical methods, Uematsu et al. (2007) demonstrated that 70 to 460 Nflm neurons were retrogradely labeled when BDA-soaked gels were placed in incisions made at several segmental levels in the cords of nineteen adult carp. Unexpectedly, a clear linear correlation was found between the number of labeled neurons and the segment level in which BDA was applied, indicating that rostral portions of cord receive projections from larger numbers of Nflm neurons (Uematsu et al. 2007). This finding is consistent with the observation that more rostral levels of the cord receive stronger excitatory input in *Xenopus* tadpoles (Tunstall and Roberts, 1994).

On the other hand, Prasada Rao *et al.* (1987) reported that approximately 80 Nflm neurons in the goldfish were retrogradely labeled by tracer injections into the caudal level of the spinal cord. The authors argued that every Nflm neuron projected throughout the spinal cord, in contrast to the findings we have obtained.

3.3.2 Spinal Projections of Nflm neurons and Their Contacts with Spinal Neurons

The connections of Nflm neurons with spinal neurons also require clarification. It has already been clearly demonstrated that many individually identifiable spinal interneurons make contact with the axon collaterals of Nflm neurons (Gahtan and O'Malley 2003). In our study, it was demonstrated using two different labeling procedures that the axon collaterals labeled anterogradely made contact not only with glycinergic and GABAergic interneurons labeled immunocytochemically but also with motor neurons labeled retrogradely (Uematsu et al. 2007). We have also observed connections with GABAergic and glycinergic spinal interneurons labeled with specific antibodies (Uematsu et al. 2007). Both of these studies strongly suggest that some Nflm neurons project to the cord and directly connect with spinal neurons. It can also be concluded that as many as 400 glutamatergic Nflm neurons project onto spinal neurons included in the CPGs and exert excitatory drive on them to initiate swimming (Uematsu et al. 2007)..

3.3.3 Remote Control of Fish Swimming; Wired and Unwired Methods

The demonstration that the Nflm is a brain region essential for the initiation and maintenance of swimming movement by fish is of considerable interest. Microstimulation of the mecencephalic area including Nflm was attempted through an electrode implanted in the brain in free-moving goldfish under wired and wireless conditions. For this purpose, we developed a two-channel wireless microstimulator in which the stimulating currents, channels used, and durations of stimulation were controlled via infrared command signals emitted from a control box equipped with an infrared LED (Kobayashi et al. in preparation). Using this system the artificial control of goldfish swimming in the horizontal plane was successfully performed. When the Nflm on one side was stimulated, the fish continued turning to that side, but when Nflm was stimulated on both sides simultaneously the fish swam forward (Kobayashi et al. in preparation). These findings provide strong evidence that the Nflm is a center controlling swimming in fish.

4 The Cerebellum as a Regulatory Center of Fish Swimming

4.1 Efferent Connections of the Cerebellum and Other Brain Regions Concerned with Swimming

The cerebellum of teleosts has certain features unique among vertebrates (Finger 1983). One is that the Purkinje cells lack efferent axons extending out of the cerebellum; the cerebellum instead has unique efferent neurons, the eurydendroid cells (Pouwels 1978, Finger 1978 and 1983). In addition, unlike other animals, the cerebellum in teleosts receives various types of sensory input directly from the entire body. Overall, the afferent pathways of cerebellum in teleosts have been studied thoroughly (Finger 1983). In contrast, findings on the efferent system, especially of the valvula cerebelli, are quite limited and feature many uncertainties, as pointed out by Ikenaga et al. (2002). The valvula cerebelli is another unique part of the teleost cerebellum. The brain areas targeted by cerebellar efferent fibers were demonstrated by an anterograde tract-tracing method using a precision microinjection into the corpus and valvula cerebelli in goldfish (Ikenaga et al. 2002 and 2006). Fibers arising from the corpus terminated in many nuclei in the diencephalon, Nflm, optic tectum, and nucleus ruber in the midbrain, and the medial and lateral lobes of the valvula cerebelli. These findings suggest that the cerebellum is probably involved in motor control in fishes. It has also been proposed that the cerebellum of teleost fishes has higher functions already known to exist in mammalian cerebellum (Ikenaga et al. 2002).

4.2 Efferent Neurons of the Teleost Cerebellum

Until quite recently, no real picture of the cerebellar efferent neurons, the eurydendroid cells, was available, with only old and limited information, most of which had been obtained by Golgi or earlier HRP techniques. For this reason, a tract-tracing experiment using tracers and antibodies was performed to demonstrate the cytoarchitecture of the cerebellar efferent neurons. Neurons assuming spindle or polygonal shape with a few primary dendrites of the corpus cerebelli were back-filled by BDA injection at the ventral base of the cerebellum (Ikenaga et al. 2005). While most of them were located in the Purkinje cell layer (ganglion cell layer), some were also found in the molecular and granular cell layers. Based on their appearance and location, it was concluded that they were so-called eurydendroid cells (Ikenaga et al. 2005). In addition, it was suggested that approximately half of eurydendroid cells were aspartate-immunopositive, and that all cells retrogradely labeled were negative for staining with Zebrin II antiserum (Ikenaga et al. 2005). Zebrin II is a specific marker for Purkinje cells of vertebrates (Lannoo et al. 1991; Meek et al. 1992). One half of the deep cerebellar nucleus (DCN) neurons in tetrapods contains aspartate as a transmitter (Ottersen 1993). The large cells with GABA-like-immunoreactivity found in the Purkinje cell layer are probably Purkinje cells (Ottersen 1993). In addition, we observed that retrogradely labeled neurons were entangled with Zebrin II-positive fibers, as has been reported in a mormyrid (Lannoo et al. 1991), suggesting that the eurydendroid cells are innervated by Purkinje cells. Based on these findings, it was concluded that the eurydendroid cells are cerebellar efferent neurons in cyprinids and are homologous with DCN neurons.

4.3 Effects of Ablations of the Cerebellum on Swimming by Fish

Although the cerebellum of mammals is an important center for regulation of forms of locomotion such as walking, running, stepping, and scratching (Arshavsky and Orlovsky 1986, Orlovsky et al. 1999), removal of a part of the corpus cerebelli in the goldfish and the rainbow trout *Oncorhynchus mykiss* appeared to affect neither swimming nor maneuverability (Roberts et al. 1992). To clarify the functions of the corpus cerebelli in the control of sustained swimming or cruising in the goldfish, Matsumoto et al. (2007a) carried out experiments to test the effects of partial ablation of the corpus cerebelli on swimming against running water. Most of the ablated fish could perform sustained swimming, indicating that the corpus cerebelli is not required for generation of the simple swimming pattern. However, they found that both tail-beat amplitude and frequency in the ablated goldfish tended to be greater than those in the control fish at the same swimming speed. The thrust index employed was significantly larger in the ablated fish than in the control fish at higher swimming speeds (faster than 2.0 body lengths per second). This finding may suggest that the ablated goldfish needed more thrust generated by tail beats than the control fish to maintain sustained swimming at higher speeds. They therefore concluded that, in the goldfish, the corpus cerebelli plays no major role in posture control or generation of simple forward swimming movement, although the integrity of the corpus cerebelli is important for normal swimming gait. This conclusion is consistent in part with the finding that the cerebellum is involved in reflex movement of the pectoral fins in the dogfish *Scyliorhinus canicula* (Paul and Roberts, 1979).

4.4 Electrical Activities of Cerebellar Neurons in Fish

Identification and classification of neuronal activities in the fish cerebellum have been carried out (Peterson 1972, Kotchabhakdi 1976, Lee and Bullock 1990). Following these studies, Matsumoto et al. (2007b) examined neuronal activities extracellularly in the cerebellum of immobilized and swimming goldfish to determine which types of neuronal activities could be discriminated. By analyzing the waveforms of action potentials of cerebellar neurons and the depths of recording, 5 types of neural activities could be estimated to originate from 3 identified types of neurons, including Purkinje cells, eurydendroid cells, and granule cells, in both *in situ* and *in vivo* experiments (Matsumoto et al. 2007b). Among these neurons, Purkinje cells are the easiest to identify in animals from fishes to mammals, because they all discharge with two typical spikes, simple and complex (Kotchabhakdi 1976). In addition, it was observed for the first time that firing patterns of Purkinje cells and the eurydendroid neurons, both of which were individually identified, changed in relation to turning movements (Matsumoto et al. 2007b).

These studies have also indicated that disturbance of fish movement is to some extent inevitable in the "wired" condition. We have therefore recently developed a prototype of a wireless neural-signal sensing LSI (Yoshida et al. 2004ab). Using such recent techniques and devices, we believe that we will be able to comprehensively determine the neuronal basis of various fish swimming patterns in the near future.

5 Conclusion

Probably the majority of Nflm neurons are glutamatergic and innervate spinal CPG neurons to initiate swimming. However, many issues remain to be addressed regarding the Nflm and swimming, including the following: 1. The specific functions of morphologically different Nflm neurons projecting in the cord. 2. The afferent and efferent connections of Nflm neurons with other brain regions, especially the cerebellum, the optic tectum, and the telencephalon. Simultaneous intracellular recordings from Nflm neurons and others structures will be needed to clarify these. 3. The types of brainstem neurons involved in different fish behaviors, especially spawning by salmon. 4. Determination whether the same neuron groups in the spinal cord are shared for different behaviors. Studies of this type will also provide important clues to deepening understanding of brain mechanisms.

The cerebellum projects to the Nflm as well as to the nucleus ruber and hence appears to be an important center for the control of motor function in fish. Cerebellar efferents originate from eurydendroid cells which are distinct from Purkinje cells. However, the functions of the cerebellum and particularly the valvula cerebelliin in teleosts remain largely unclear. Electrophysiological and micro-lesion studies of the cerebellum will prove fruitful in gaining further insight into this brain region.

Acknowledgement

I should like to thank Masayuki Yoshida and Alan Roberts for comments on the manuscript.

References

Arshavsky TI and GN Orlovsky (1986) Role of the cerebellum in the control of rhythmic movements. In: Grillner S, Stein PSG, Stuart DG, Forssberg H, Herman RM (Eds)Neurobilogy of Vertebrate Locomotion. Macmillan, Houndmills , London, pp.677-689

Baba Y, Kake Y, Yoshida M, Uematsu K (2003) Activities of mesencephalic nucleus neurons during fictive swimming of the carp Cyprinus carpio. Fish Sci 69: 581-588

Becker T, Wullimann MF, Becker CG, Bernhardt RR, Schachner M (1997) Axonal regrowth after spinal cord transection in adult zebrafish. J Comp Neurol 377: 577-595

Biewener AA (2003) Animal Locomotion. Oxford University Press, New York, pp.281

Bone Q (1978) Locomotor Muscle. In: Hoar WS, Randall DJ (Eds) Fish Physiology Vol. 7. Academic Press, New York San Francisco London, pp 361-424

D'Août K, Aerts P (1999) A kinematic comparison of forward and backward swimming in the eel *Anguilla anguilla*. J Exp Biol 11: 1511-1521

Day MH (1981) Vertebrate Locomotion. Academic Press, London, New York, pp. 471

Faber DS, Fetcho JR, Korn H (1989) Neuronal networks underlying the escape responses in goldfish. General implications for motor control. Ann N Y Acad Sci 563: 11-33

Fetcho JR (1992) The spinal motor system in early vertebrates and some of its evolutionary changes. Brain Behav Evol 2-3: 82-97

Fetcho JR, Faber DS (1988) Identification of motoneurons and interneurons in the spinal network for escapes initiated by the Mauthner cell in goldfish. J Neurosci 8: 4192-4213

Fetcho JR, Liu KS (1998) Zebrafish as a model system for studying neuronal circuits and behavior. Ann NY Acad Sci 860: 333-345

Finger TE (1978) Efferent neurons of the teleost cerebellum. Brain Res 153: 608-614

Finger TE (1983) Organization of the teleost cerebellum. In: Northcutt RG, Davis RE (Eds) Fish Neurobiology Vol. 1. University of Michigan Press, Ann Arbor, pp. 261-284

Gahtan E, O'Malley DM (2003) Visually guided injection of identified reticulospinal neurons in zebrafish: a survey of spinal arborization patterns. J Comp Neurol 459:186-200

Gahtan E, Bier H (2003) Of lasers, mutants, and see-through brains: functional neuroanatomy in zebrafish. J Neurobiol 59:147-161

Gahtan E, Tanger PH, Bier H (2005) Visual prey capture in larval zebrafish is controlled by identified reticulospinal neurons downstream of the tectum. J Neurosci 25: 9294-9303

Gahtan E, Sankrithi N, Campos JB, O'Malley DM (2002) Evidence for a widespread brain stem escape network in larval zebrafish. J Neurophysiol 87: 608 - 614

Gibb AC, Jayne BC, Lauder GV (1994) Kinematics of pectoral fin locomotion in the bluegill sunfish *Lepomis macrochirus*. J Exp Biol 189: 133-161

Gray J (1936) Studies in animal locomotion. IV. The neuromuscular mechanism of swimming in the eel. J exp Biol 203: 170-180

Gray J (1968) Animal Locomotion. Weidenfeld and Nicolson, London, pp. 479

Grillner S, Parker D, El Manira A (1998) Vertebrate Locomotion – A lamprey perspective. In: Kiehn O, Harris-Warrick RM, Jordan LM, Hultborn H, Kudo N (Eds) Neuronal Mechanisms for Generating Locomotor Activities. The New York Academy of Sciences, New York, pp. 1-18

Grillner S, Buchanan JT, Wallén P, Brodin L. (1988) Neural control of locomotion in lower vertebrates: from behavior to ionic mechanisms. In: Cohen AH, Rossignol S, Grillner S (Eds) Neural Control of Rhythmic Movements in Vertebrates. John Wiley and Sons, New York, pp 1-40

Grillner S, Deliagina T, Ekeberg Ö, El Manira A, Hill RH, Lansner A, Orlovsky GN, Wallén P (1995) Neural networks that co-ordinate locomotion and body orientation in lamprey. Trends Neurosci 18: 270-279

Higashijima S, Mandel G, Fetcho JR (2004a) Distribution of prospective glutamatergic, glycinergic, and GABAergic neurons in embryonic and larval zebrafish. J Comp Neurol 480: 1-18

Higashijima S, Schaefer M, Fetcho JR (2004b) Neurotransmitter properties of spinal interneurons in embryonic and larval zebrafish. J Comp Neurol 480: 19-37

Ikenaga T, Yoshida M, Uematsu K (2002) Efferent connections of the cerebellum of the goldfish, *Carassius auratus*. Brain Behav Evol 60: 36-51

Ikenaga T, Yoshida M, Uematsu K (2005) Morphology and immunohistochemistry of efferent neurons of the goldfish corpus cerebelli. J Comp Neurol 487: 300-311

Ikenaga T, Yoshida M, Uematsu K (2006) Cerebellar efferent neurons in teleost fish. The Cerebellum 5: 268-274

Kashin SM, Feldman AG, Orlovsky GN (1974) Locomotion of fish evoked by electrical stimulation of the brain. Brain Res 82: 41-47

Kimmel CB, Powell SL, Metcalfe WK (1982) Brain neurons which project to the spinal cord in young larvae of the zebrafish. J. Comp Neurol 205: 112-127

Kobayashi N, Yoshida M, Matsumoto N, Uematsu K (2007) Artificial control of swimming in goldfish by brain stimulations. Comp Biochem Physiol (in preparation)

Kotchabhakdi N (1976) Functional circuitry of the goldfish cerebellum. J Comp Physiol 112: 47-73

Lannoo MJ, Brochu G, Maler L, Hawkes R (1991) Zebrin-II immunoreactivity in the rat and in the weakly electric teleost *Eigenmannia* (Gymonotiformes) reveals three modes of Purkinje cell development. J Comp Neurol 310: 215-233

Lee LT, Bullock TH (1990) Cerebellar units show several types of early responses to telencephalic stimulation in catfish. Brain Behav Evol 35: 278-290

Lee RKK, Eaton RC (1991) Identifiable reticulospinal neurons of the adult zebrafish, *Brachydanio rerio*. J Comp Neurol 304:34-52.

Lee RKK, Eaton RC, Zottoli SJ (1993) Segmental arrangement of reticulospinal neurons in the goldfish hindbrain. J Comp Neurol 329: 539-556

Li W-C, Soffe SR, Roberts A (2004) Glutamate and acetyl choline co-release by spinal neurons. PNAS 101 (43), 15488-15493

Li W-C, Soffe SR, Wolf E, Roberts A (2006) Persistent Responses to Brief Stimuli: Feedback Excitation among Brainstem Neurons. J Neurosci 26:4026-4035

Lindsey CC (1978) Form, Function, and Locomotory Habitds in Fish. In: Hoar WS, Randall DJ (Eds) Fish Physiology Vol. 7. Academic Press, New York, San Francisco, London, pp 1-100

Liu DW, Westerfield M (1990) The formation of terminal fields in the absence of competitive interactions among primary motoneurons in the zebrafish. J Neurosci 10: 3947-59

Matsumoto N, Yoshida M, Uematsu K (2007a) Effects of partial ablation of the cerebellum on sustained swimming in goldfish. Brain Behav Evol (in printing)

Matsumoto N, Yoshida M, Uematsu K (2007b) Recording cerebellar neuron activities in swimming goldfish. Fish Sci (in printing)

Matsushima T, Kitamura S, Takei K, Satou M, Okumoto N, Ueda K (1986) An electromyographic study of sexual behavior in himé salmon (landlocked socheye salmon, *Oncorhynchus nerka*). Zool Sci 3: 563-567.

Meek J (1992) Comparative aspects of cerebellar organization. From mormyrids to mammals. J Morph 30:37-51

Meek J, Hafmans TGM, Maler L, Hawkes R (1992) Distribution of zebrin-II in the gigantocerebellum of the mormyrid fish *Gnathonemus petersii* compared with other teleosts. J Comp Neurol 316:17-31

Middleton FA and Strick PL (1998) The cerebellum: an overview. Trends Neurosci 21:367-369

Moon SJ, Fujikawa Y, Nishihara T, Kono S, Kozono K, Ikenaga T, Esaka M, Iijima N, Nagamatsu Y, Yoshida M, Uematsu K (2005) Partial cloning and expression of mRNA coding choline acetyltransferase in the spinal cord of the goldfish, *Carassius auratus*. Comp Biochem Physiol B 141: 253-260

Mori S, Matsui T, Kuze B, Asanome M, Nakajima K, Matsuyama K (1998) Cerebellar-induced locomotion: Reticulospinal control of spinal rhythm generating mechanism in cats. Ann NY Acad Sci 860: 94-105

Myers PZ (1985) Spinal motoneurons of the larval zebrafish. J Comp Neurol 236: 555-561

Nursall JR (1955) The lateral musculature and the swimming of fish. Proc Zool Soc Lond 126: 127-143

Oka Y, Satou M, Ueda K (1986) Descending pathways to the spinal cord in the himé salmon (landlocked red salmon, *Oncorhynchus nerka*). J Comp Neurol 254: 91-103

Orlovsky GN, Deliagina TG, Grillner S (1999) Neural Control of Locomotion. Oxford University Press, Oxford, NewYork, pp.175-204

Ottersen OP (1993) Neurotransmitters in the cerebellum. Rev Neurol 149: 629-636

Paul DH, Roberts BL (1979) The significance of cerebellar function for a reflex movement of the dogfish. J Comp Physiol A134: 69-74

Peterson RH (1972) Electrical responses of goldfish cerebellum, responses to parallel fibre and peduncle stimulation. Brain Res 41: 67-79

Paul DH, Roberts BL (1979) The significance of cerebellar function for a reflex movement of the dogfish. J Comp Physiol. A134: 69-74

Pouwels E (1978) On the development of the cerebellum of the trout, *Salmo gairdneri*. III. Development of neuronal elements. Anat Embryol 153: 37-54

Prasada Rao PD, Jadhao AG, Sharma SC (1987) Descending projection neurons to the spinal cord of the goldfish, *Carassius auratus*. J Comp Neurol 265: 96-108

Rayner MD, Keenan MJ (1967) Role of red and white muscles in the swimming of the skipjack tuna. Nature 214: 392-393

Reidy SP, Kerr SR, Nelson JA (2000) Aerobic and anaerobic swimming performance of individual Atlantic cod. J Exp Biol 203: 347-357

Roberts BL, van Rossem A, de Jager S (1992) The influence of cerebellar lesions on the swimming performance of the trout. J Exp Biol 167:171-178.

Roberts A, Dale N, Soffe SRW (1986) Sustained responses to brief stimuli: swimming in *Xenopus* embryos. J Exp Biol 112: 321-335

Roberts A, Soffe SR, Dale N (1984) Spinal interneurons and swimming in frog embryos. In: Grillner S, Stein PSG, Stuart DG, Forssberg H, Herman RM (Eds) Neurobiology of Vertebrate Locomotion. Macmillan Press, Hampshire London, pp. 279-306

Roberts A, Soffe SR, Wolf ES, Yoshida M, Zhao, F-Y (1998) Central circuits controlling locomotion in young frog tadpoles. In: Kiehn O, Harris-Warrick, Jordan LM, Hultborn H, Kudo N (Eds) Neuronal Mechanisms for Generating Locomotor Activities. The New York Academy of Sciences, New York, pp. 19-34

Romer AS, Parsons TS (1977) The Vertebrate Body, 5th ed,, W. B. Saunders Co.Piladelphia,USA,

Rosenberger LJ (2001) Pectoral fin locomotion in batoid fishes: Undulation versus oscillation. J Exp Biol 204: 379-394

Soffe SR (1993) Two distinct rhythmic motor patterns are driven by common premotor and motor neurons in a simple vertebrate spinal cord. J Neurosci 13: 4456-4469
Tsukamoto,K. (1984a) The role of the red and white muscles during swimming of the yellow tail. Nippon Suisan Gakkaishi 50: 2025-2030
Tsukamoto,K. (1984b) Contribution of the red and white muscles to the power output required for swimming of the yellow tail. Nippon Suisan Gakkaishi 50: 2031-2042
Tunstall MJ, Roberts A (1994) A longitudinal gradient of synaptic drive in the spinal cord of *Xenopus* embryos and its role in co-ordination of swimming. J. Physiol (Lond) 474: 393-405
Uematsu K, Yamamori K, Hanyu I, Hibiya T (1980) Role of the trunk musculatures in oviposition of chum salmon, *Oncorhynchus keta.* Nippon Suisan Gakkaishi 46: 395-400
Uematsu K, Yamamori K (1982) Body vibration as a timing cue for spawning in chum salmon. Comp Biochem Physiol 72A: 591-594
Uematsu K, Ikeda T (1993) The midbrain locomotor region and induced swimming in the carp, *Cyprinus carpio.* Nippon Suisan Gakkaishi 59: 783-788
Uematsu K, Todo T (1997) Identification of the midbrain locomotor nuclei and their descending pathways in the teleost carp, *Cyprinus carpio.* Brain Res 773:1-7
Uematsu K, Shirasaki M, Storm-Mathisen J (1993) GABA- and glycine-immunoreactive neurons in the spinal cord of the carp, Cyprinus carpio. J Comp Neurol 332: 59-68
Uematsu K, Baba Y, Kake Y, Ikenaga T, Moon S-J, Miyai Y, Yoshida M (2007) Central mechanisms underlying fish swimming. Brain Behav Evol 69: 142-150
Williams T, Grillner S, Smoljaninov VV, Wallén P, Kashin S, Rossignol S (1989) Locomotion in lamprey and trout: The relative timing of activation and movement. J Exp Biol 143: 559-566
Wallén P, Grillner S, Feldman JL, Berglt S (1985) Dorsal and ventral myotome metoneurons and their input during fictive locomotion in lamprey. J Neurosci 5: 654-661
Webb PW (1978) Hydrodynamics: Nonscombroid fish. In: Hoar WS, Randall DJ (Eds) Fish Physiology Vol. 7. Academic Press, New York, San Francisco, London, pp 190-237
Westerfield M, McMurray JV, Eisen JS (1986) Identified motoneurons and their innervation of axial muscles in the zebrafish. J Neurosci 6: 2267-2277
Yoshida M, Fudoji M, Sakamoto H, Uematsu K (1999) Posthatching development of spinal motoneurons in the angelfish Pterophyllum scalare. Brain Behav Evol 53: 180-186
Yoshida T, Mashimo T, Akagi M, Iwata A, Yoshida M, Uematsu K (2004a) A design of neural signal sensing LSI with multi-input-channels. IEICE Trans FUNDAMENTALS E87A: 376-383.
Yoshida T, Akagi M, Mashimo T, Iwata A, Yoshida M, Uematsu K (2004b) Design of a wireless neural-sensing LSI. IEICE Trans Electron E87C: 996-1002

Chapter 10

Underwater Acoustical Sensing Behavior of Porpoises

Tomonari Akamatsu[1], Ding Wang[2], Kexiong Wang[2], and Yasuhiko Naito[3]

[1] Fisheries Research Agency of Japan, 7620-7 Kamisu, Ibaraki 314-0408, Japan
[2] Institute of Hydrobiology, Chinese Academy of Sciences, Wuhan 430072, People's Republic of China
[3] National Institute of Polar Research, 9-10, Kaga 1-chome, Itabashi-ku, Tokyo 173-8515, Japan

Summary. Detecting objects in their paths is a fundamental perceptional function of moving organisms. Potential risks and rewards, such as prey, predators, conspecifics or non-biological obstacles, must be detected so that an animal can modify its behavior accordingly. However, to date few studies have considered how animals in the wild focus their attention. Dolphins and porpoises are known to actively use sonar or echolocation. A newly developed miniature data logger attached to a porpoise allows for individual recording of acoustical search efforts and inspection distance based on echolocation. In this study, we analyzed the biosonar behavior of eight free-ranging finless porpoises and demonstrated that these animals inspect the area ahead of them before swimming silently into it as depicted in the figure. Before to get out of the inspected area, the porpoise use sonar. The porpoises inspected distances up to 77 m, whereas their swimming distance without using sonar was less than 20 m. The inspection distance was long enough to ensure a wide safety margin before facing real risks or rewards. Once a potential prey item was detected, porpoises adjusted their inspection distance from the remote target throughout their approach. Application of biosonar for biomimetic sonar is discussed.

Key words. echolocation, cetacean, finless porpoise, sonar, dolphin

1 Introduction

Sonar abilities of dolphins and porpoises have been extensively studied [1]. Not only can these animals detect a remote target using sonar, but they can also precisely discriminate its size, thickness and material composition [1]. In addition to controlled captive experiments, sonar signals of wild dolphins and porpoises have been examined [2, 3, 4] Sonar signals of dolphins and porpoises consist of high intensity and high frequency ultrasonic pulse trains that stand out well against ambient noises even in the wild [5]. The number of ultrasonic pulses emitted is considered an appropriate indicator of the search effort made by an animal. Moreover, the target range of the sonar used by dolphins can be estimated by the inter-click interval [1, 6]. These signals can easily be recorded and can provide information about the timing and the distance over which the echolocating animal uses its biosonar. This makes dolphins and porpoises some of the most appropriate model organisms in which to observe biosonar behavior in the wild.

However, the problem of recording the sonar signals of free-ranging dolphins and porpoises simultaneously with underwater behavior has remained largely unsolved. Hydrophone systems can observe individual vocalizing cetaceans for only a limited time and they record little behavioral information [7, 2, 8]. A mobile video camera carried by a human diver can record behavior and vocalizations only during brief encounters with the animals [9]. Although underwater sound recording systems have recently been attached to large marine mammals [10, 11, 12, 13], few studies have succeeded at recording ultrasonic sonar signals concurrently with underwater behavior of individual dolphins and porpoises [14, 15, 16]. Furthermore, none of the data logger systems attached to these animals can exclude vocalizations from other nearby individuals. These data logger systems to record vocalizations are therefore inappropriate for studies of animals that swim in groups, such as dolphins and porpoises.

To investigate the biosonar behavior of aquatic animals, we developed a miniature stereo acoustic data logger and used it on eight free-ranging porpoises in an oxbow of the Yangtze River, China. Time-sequence data of sonar events and underwater behavior were recorded for individuals. In addition, the direction of the sound source was stored in the data logger in order to exclude sonar signals from other individuals. Here we examine the acoustic activity and behavior of free-ranging echolocating finless porpoises in a large open water system.

2 Methods

Until 1972, the experimental site was a tributary of the Yangtze River, and water still enters from the main stream during every flood season [17]. This oxbow lake, part of Tian-e-Zhou Baiji National Natural Reserve of the Yangtze River, Hubei, China (29.30'-29.37'N, 112.13'-112.48'E), is approximately 21 km long and 1 to 2 km wide. It was established by the Chinese government in 1992 as a reserve for baiji (*Lipotes vexillifer*) and finless porpoises. Since 1990, 49 finless porpoises have been introduced from the main population in the river. Finless porpoises in the lake survive without supplemental food and are observed to reproduce annually. The environment of the lake is considered to be similar to the natural habitat of this species [18].

A miniature stereo pulse event recorder (W20-AS, Little Leonardo, Tokyo, Japan, diameter: 22 mm; length: 122 mm; weight: 77 g) was developed for this research. The data logger consisted of two hydrophones (System Giken Co. Ltd., -210dB/V sensitivity), an analogue-digital converter, a band pass filter (70 kHz to 300 kHz) to eliminate noise and 256 MB flash memory and a CPU (PIC18F6620, Microchip, China) inside a pressure-resistant aluminum cylinder. The storage capacity of this system allows recording of the intensity, timing and source direction of ultrasonic pulses for up to 60 hours with a sampling rate of 2,000 events per second. The two miniature hydrophones were situated at the front and the back ends of the device. The sound source direction is calculated from inter-hydrophone differences in sound arrival time stored in the data logger. The source direction was used to exclude vocalizations coming from nearby individuals. The dynamic range was between 129 dB peak to peak (reference pressure 1 uPa at 1 m) to 157 dB. The resolution of time difference was 542 nanoseconds. Simultaneously, swim speed, dive depth, heaving and surging body acceleration were recorded by a behavior data logger (diameter: 21 mm; length: 114 mm; weight: 60 g, PD2GT, Little Leonardo, Tokyo, Japan). Swimming speed was essential for calculating the swimming distance of the animals; the acceleration parameters were not used in analyses thereafter.

Nine finless porpoises were captured safely in the reserve using round net in October 2003 and secured safely in an enclosure. One animal carried two acoustic data loggers on both sides for calibration purposes to compare the received sound pressure levels on each side. The data obtained from this individual were not used for this paper. Each data logger was equipped with a suction cup (Canadian Tire Corp.), a VHF transmitter (MM130, Advanced Telemetry Systems, USA) and floatation material (expanded polyvinyl chloride, Klegecell #55, Kaneka Co. Ltd., Japan) to ensure positive buoyancy for easier retrieval after spontaneous release from the animals. The antenna of the transmitter was designed to be out of the water after detachment and to remain approximately perpendicular to the surface to detect strong radio transmission. The drag force of the whole data logger system was measured in a fluid dynamical experiment tank at the National

Research Institute of Fisheries Engineering, Japan, and was found to be less than 60 g at the animals' average 0.89 m/s cruising speed. The weights of the complete acoustic and behavior data logger systems in air were 216 g and 197 g, respectively. The weights of the data logger systems were adjusted to 20 g positive buoyancy in water. The data logger was fixed to the side of the body in the upper posterior direction from the pectoral fin; this area was the least affected by body movements and therefore ensured a long attachment time. Recording the sonar signals of finless porpoises was possible at this position even for off-axis sound detection [15].

To retrieve the data logger systems, the radio signals were monitored by two antennae (RX-155M7/W, Radix, Japan) from the top of the three-story field station building beside the oxbow. When a continuous radio signal was received, a data logger was considered to be floating. Retrieval operations were started six or more hours after release to avoid any disturbance to the animals. This experiment was conducted under a permit issued by the Fisheries Management Department of Hubei Province.

The time-sequence data from the acoustic and behavior data loggers were synchronized by the time stamp of the data file. Low-intensity signals below 136 dB, such as surface reflection, were excluded by off-line analyses using new custom software on MATLAB (The MathWorks, Massachusetts, USA) and Igor (WaveMetrics, Oregon, USA). Splash noises during respiration were easily identified and excluded at dive depths less than 0.3 m, which is six times the resolution of the depth meter (5 cm) on the behavior data logger. The sonar signals of dolphins and porpoises have a pulse train structure [1]. Inter-click train interval was defined as 200 ms or longer. We calculated the swimming speed in meters per second from the rotation of the propeller deployed in the behavior data logger using the methods of Fletcher et al. [19].

3 Results

We recorded the biosonar behavior as well as the underwater body movements of eight finless porpoises. The simultaneous recording time of the acoustic and behavioral parameters varied from 1 hour to 35 hours for each individual (average 8.75 hours). In total, we identified 2,425,095 pulses; 49,470 pulse trains and 5,113 dive events. The porpoises produced pulse trains on average every 5.1 seconds.

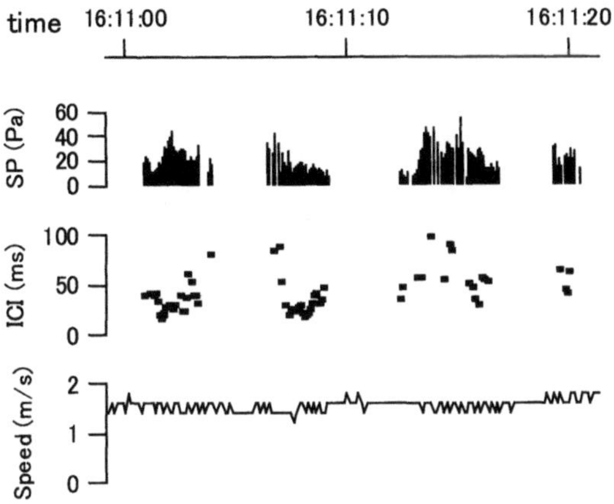

Fig. 1. Periods of silence in sonar signal records for a finless porpoise. The time series of the sound pressure (SP) shows periods of silence between pulse trains indicated by horizontal bars. The distance inspected acoustically was calculated by the inter-click interval (ICI) of a train prior to the silence.

Many periods of silence between pulse trains were recorded (figure 1). Silent periods sometimes lasted more than 10 seconds, whereas the swim speed during these periods ranged up to 3 m/s with an average cruising speed of 0.89 m/s. Because the underwater visibility in the lake was less than 1 m due to turbidity, the porpoises could not detect any remote obstacles in front of them without using sonar. The distance over which an animal traveled silently was calculated by multiplying the duration of every period of silence by the simultaneously recorded swimming speed.

We also calculated the distance inspected acoustically by the animals just prior to a period of silence. The acoustically inspected distance is given by the two-way distance traveled by sound within an inter-click interval. However, in cetacean bio-sonar systems, an additional lag time is required within each inter-click interval for the animal to process returning echoes (Au 1993). This lag time should be less than the shortest inter-click interval. The minimum accumulated distribution of the inter-click interval in a pulse train was 1 ms (figure 2a). The lag time in our study was

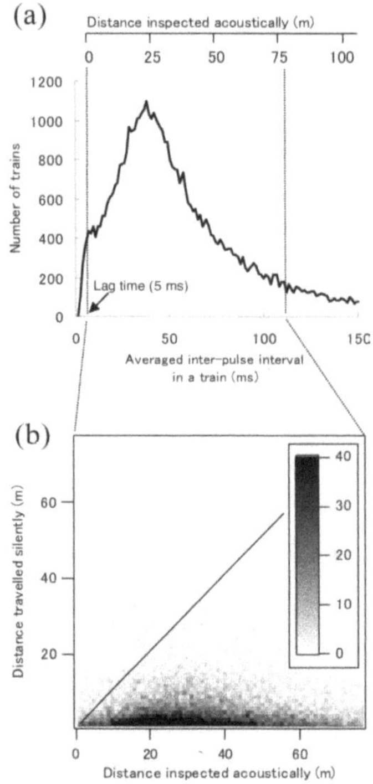

Fig. 2. Comparison of the distance travelled in silence and the distance previously inspected acoustically. (a) Distribution of the average inter-click intervals in pulse trains of eight finless porpoises. The lag time (echo signal processing duration) was assumed to be 5 ms. In 90% of cases, the acoustically inspected distance was less than 77 m. (b) The distance inspected acoustically was far greater than the distance subsequently travelled in silence. The gray scale shows the number of observed pulse trains in each bin. Most click trains fell under the diagonal line, indicating that porpoises inspected an area ahead for long enough before silently swimming into it.

conservatively assumed to be no more than 5 ms, since frequency of occurrence dropped sharply below this interval. The upper inset of figure 2a shows the scale of the calculated distance inspected acoustically. The distance inspected acoustically in advance was far greater than the distance traveled silently (figure 2b, accumulated data from all monitored individuals). This relationship was consistently observed for all individuals. In 90% of all cases, the acoustically inspected distance was less than 77 m and the modal focal distance was 24 m. Once a porpoise detected a potential target and approached it, the animal often emitted clicks with a linearly decreasing inter-click interval, hereafter termed the approach phase (figure 3a). Changes in the inspected distance from the beginning to the end of the

approach phase were calculated from changes in inter-click interval. Additionally, the distance traveled during the approach phase was calculated by multiplying the duration of the approach phase by swim speed (figure 3b). Finless porpoises changed the inspected distance by 11.0 m on average (maximum 42.7 of the 117 approach phases analyzed) and swam 8.6 m on average (maximum 43.0) during the approach phase. These two independent measures-the distance traveled and the distance inspected during the approach phase-were well correlated (figure 3b; $r = 0.80$, $p < 0.005$).
At the end of the approach phase, the minimum inter-click interval was near the assumed lag time (5 ms). A few seconds after the end of the approach phase, the swim speed frequently dropped to zero (see swimming speed profile in figure 3a).

4 Discussion

The finless porpoises observed in this study produced sonar signals nearly constantly, on average every 5.1 seconds. This result suggests that the finless porpoises strongly rely on their acoustic sensory systems for navigation and traveling.

When moving through an environment in which vision is limited, prior acoustic inspection of the area ahead is essential for these animals. This strategy offers substantial advantages both in terms of risk avoidance and prey capture. In particular, animals must detect potentially lethal risks as early as possible. For example, harbour seals have been shown to avoid the long-range vocalizations of killer whales, their primary natural predators [20]. Potential risks for finless porpoises in a freshwater habitat include injuries from or entanglement in underwater debris and entrapment in shallow waters, as well as rough water surfaces or floating materials that prevent safe respiration. Taking such risks can be costly; thus, the porpoises must detect risks with sufficient time to avoid them by changing behavior. As shown in figure 2b, finless porpoises almost always inspect upcoming areas for long enough before swimming into them silently. The porpoises seemed to employ a large safety margin in their sonar range when inspecting the area ahead relative to the distance they swam silently.

The inspection distance of the porpoises was as high as 77 m (figure 2a). This long-range sonar ability in dolphins and porpoises has also been reported in previous studies. In one study, the detection distance of a metal sphere target by a trained bottlenose dolphin (*Tursiops truncatus*) was up to 113 m [1]. Estimated inspection distances of three different odontocetes in the wild reached tens of meters, which was significantly greater than the distances inspected by captive individuals [4]. This suggests that the large

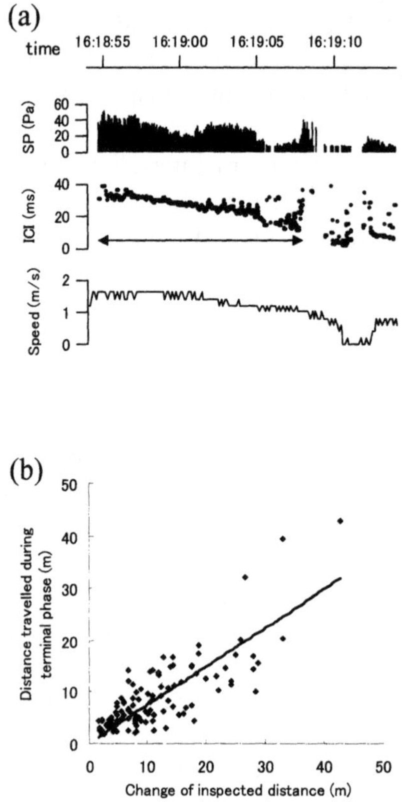

Fig. 3. Comparison of the distance travelled during the approach phase and the change in the distance inspected acoustically. **(a)** The approach phase (*arrow*) is characterised by a linear decrease in the inter-click intervals (ICI) of the sonar signals. The swimming speed dropped to nearly zero after the approach phase. **(b)** The distance travelled during the approach phase was closely correlated with the estimated change in the distance inspected.

safety margins observed in this study may be used by other odontocetes as well. The inter-click interval fluctuated and mean interval was larger than the two-way sound travel time to the target range in the target detection experiment by a harbour porpoise (Phocoena phocoena) [21]. The inter-click intervals may be an index for maximum target range, otherwise the harbour porpoise might scan the longer distance behind the target and adjusted the inter-click interval before the certain detection of the target.

On the other hand, for the detection of prey, the porpoises are acutely aware of the relative distances to a target that is still out of the visible range. The distance traveled during the approach phase and the change in the distance that was inspected acoustically were similar, suggesting that the porpoises recognized a remote target and focused their sonar on it at a distance sometimes exceeding 30 m (figure 3b). A harbour porpoise which is in the same family of a finless porpoise is known to detect a water-filled stainless steel sphere with outer diameter 5.08 cm up to 14 m by its sonar [21]. In the case above, the present finless porpoise might detect larger target such as fish school from a long distance.

During the approach phase, the porpoises seemed to employ smaller safety margins, since the inspection distance was adjusted to the distance from a potential prey item second by second. Once a porpoise detects an available resource ahead, it focuses the inspection distance on this target. In exchange for the small safety margin, a predator gains the advantage of locating the position of the moving remote prey well in advance.

Increasing repetition rate of sonar signals during approach to a target have been previously described in bat sonar [22, 23] before the capture of an insect. At the end of the approach phase in porpoises, the swimming speed dropped. This speed profile is considered the turning behavior of the animal [24]. The precise adjustment of sonar during the approach phase and the quick body movement strongly suggest that prey capture occurred.

On average, the approach phase of the animals in this study lasted over 11 m in swimming distance, whereas the sonar vocalizations at a higher repetition rate of a bat (*Eptesicus fuscus*) started 1 m to 2 m from a target in a large flight room [22]. The body lengths of these two species also differ by approximately ten times (140 cm for finless porpoises, 12 cm for big brown bats). Therefore, both species locked their sonar on a target at a distance of eight times their body length.

Because animals face limitations in detection ability and because their search effort is time constrained, they must use their sensory systems appropriately. The finless porpoises in this study usually searched no more than 77 m ahead, a distance that could be reached within a couple of minutes at their typical cruising speed. Objects farther away are not important to the animal, regardless of whether they are lethal or beneficial. Inspecting the immediate area through which a porpoise is going to travel allows time for decision-making before the animal faces real risks or rewards.

Biosonar of dolphins and porpoises is an appropriate model of biomimetic sonar. Precise target positioning as well as the wide frequency range response to discriminate targets are the potential advantages of the broadband sonar. The way to use sonar by porpoises is possibly applied for the development of signal transmission and processing system of future biomimetic sonar.

We thank Q. Zhao, Z. Wei, X. Zhang, X. Wang, B. Yu, J. Xiao, S. Li, J. Zheng, J. Xia, K. Lucke, K. Sato, A. Takahashi, V. Deecke, A. Kato, H. Tanaka, M. Suzuki, H. Muramoto, N. Sugiyama, K. Oda, A. Matsuda, S. Suzuki, T. Ishida, M. Ota, M. Nakamura, H. Hiruda, S. Numata, T. Sakai, T. Tobayama, M. Soichi and H. Katsumata for their help in conducting this research. This work was funded by the Program for Promotion of Basic Research Activities for Innovative Biosciences of Japan, the National Natural Science Foundation of China (30170142), the Chinese Academy of Sciences (CAS), and the Institute of Hydrobiology, CAS (No. KSCX2-SW-118 and 220103). This paper is originally appeared in Proc. R. Soc. Lond. B, 272, 797-801 (2005) and modified for ISABMEC 2006.

References

[1] W. W. L. Au, *The sonar of dolphins*, Springer, New York, 1993

[2] W. W. L. Au, and K. J. Benoit-Bird, Automatic gain control in the echolocation system of dolphins, *Nature,* vol.423, pp.861-863, 2003

[3] M. O. Lammers, W. W. L. Au, and D. L. Herzing, The broadband social acoustic signaling behavior of spinner and spotted dolphins, *J. Acoust. Soc. Am.,* vol.114, pp.1629-1639, 2003

[4] T. Akamatsu, D. Wang, K. Nakamura, and K. Wang, Echolocation range of captive and free-ranging baiji (*Lipotes vexillifer*), finless porpoise (*Neophocaena phocaenoides*), and bottlenose dolphin (*Tursiops truncatus*)., *J. Acoust. Soc. Am.*, vol.104, pp.2511-2516, 1998

[5] W. J. Richardson, C. R. Greene, Jr., C. I. Malme, and D. H Thomson, *Marine mammals and noise.,* Academic Press, New York, 1995

[6] J. A. Thomas, and C. W. Turl, Echolocation characteristics and range detection threshold of a false killer whale (*Pseudorca crassidens*). In Sensory abilities of cetaceans: laboratory and field evidence (ed. J.A. Thomas & R.A. Kastelein), pp.321-334, Plenum Press, New York, 1990

[7] C. W. Clark, and P. J. Clapham, Acoustic monitoring on a humpback whale (*Megaptera novaeangliae*) feeding ground shows continual singing into late spring. *Proc. R. Soc. Lond.,* vol. B 271, pp. 1051-1057, 2004

[8] V. M. Janik, S. M. van Parijs, and P. M. Thompson, A two-dimensional acoustic localization system for marine mammals. *Mar. Mamm. Sci.,* vol.16, pp.437-447, 2000

[9] D.L. Herzing, Vocalizations and associated underwater behavior of free-ranging Atlantic spotted dolphins, *Stenella frontalis*, and bottlenose dolphins, *Tursiops truncatus*. *Aquatic Mammals,* vol.22, pp.61-79, 1996

[10] P. L. Tyack, M. Johonson, P. T. Madsen, and W. M. Zimmer, Echolocation in wild toothed whales. *J. Acoust. Soc. Am.,* vol.115, pp.2373, 2004

[11] D. P. Nowacek, M. P. Johnson, and P. L. Tyack, North Atlantic right whales (*Eubalaena glacialis*) ignore ships but respond to alerting stimuli. *Proc. R. Soc. Lond.,* Vol. B 271, pp. 227-231, 2004

[12] P. T. Madsen, R. Payne, N. U. Kristiansen, M. Wahlberg, I. Kerr, and B. Mohl, Sperm whale sound production studied with ultrasound time/depth-recording tags. *J. Exp. Biol.,* vol, 205, pp.1899-1906, 2002

[13] W. C. Burgess, P. L. Tyack, B. J. Le Boeuf, and D. P. Costa, A programmable acoustic recording tag and first results from free-ranging northern elephant seals. *Deep-Sea Research II,* vol. 45, pp.1327-1351, 1998

[14] R. E. Thomas, K. M. Fristrup, and P. L. Tyack, Linking the sounds of dolphins to their locations and behavior using video and multichannel acoustic recordings. *J. Acoust. Soc. Am.,* vol. 112, pp.1692-1701, 2002

[15] T. Akamatsu, D. Wang, K. Wang, and Y. Naito, A method for individual identification of echolocation signals in free-ranging finless porpoises carrying data loggers. *J. Acoust. Soc. Am.*, vol.108, pp.1353-1356, 2000

[16] P. Tyack, and C. A. Recchia, A data logger to identify vocalizing dolphins. *J. Acoust. Soc. Am.*, vol. 90, pp.1668-1671, 1991

[17] Z. Wei, D. Wang, X. Kuang, K. Wang, X. Wang, J. Xiao, Q. Zhao, and X. Zhang, Observations on behavior and ecology of the Yangtze finless porpoise (*Neophocaena phocaenoides asiaeorientalis*) group at Tian-e-Zhou Oxbow of the Yangtze River. *The Raffles Bulletin of Zoology, Supplement,* vol.10, pp.97-103, 2002

[18] X. Zhang, Z. Wei, X. Wang, J. Yang, and P. Chen, Studies on the feasibility of establishment of a semi-natural reserve at Tian-e-zhou (swan) oxbow for baiji, *Lipotes vexillifer*. *Acta Hydrobiologica Sinica,* vol.19, pp.110-123, 1995

[19] S. Fletcher, B. J. Le Boeuf, and D. P. Costa, Onboard acoustic recording from diving northern elephant seals. *J. Acoust. Soc. Am.*, vol.100, pp.2531-2539, 1996

[20] V. B. Deecke, P. J. B. Slater, and J. K. B. Ford, Selective habituation shapes acoustic predator recognition in harbour seals. *Nature,* vol.420, pp.171-173, 2002

[21] J. Teilmann, L.A. Miller, T. Kirketerp, R.A. Kastelein, P.T. Madsen, B.K. Nielsen, and W.W.L. Au, Characteristics of echolocation signals used by a harbour porpoise (*Phocoena phocoena*) in a target detection experiment. *Aquatic Mammals,* vol.28, pp.275-284, 2002

[22] C. F. Moss, and A. Surlykke, Auditory scene analysis by echolocation in bats. *J. Acoust. Soc. Am.*, vol.110, pp.2207-2226, 2001

[23] B. Tian, and H. Schnitzler, Echolocation signals of the greater horseshoe bat (*Rhinolophus ferrumequinum*) in transfer flight and during landing. *J. Acoust. Soc. Am.*, vol.101, pp.2347-2364, 1996

[24] T. Akamatsu, D. Wang, K. Wang, Z. Wei, Q. Zhao, and Y. Naito, Diving behavior of freshwater finless porpoises (*Neophocaena phocaenoides*) in an oxbow of the Yangtze River, China. *ICES J. Mar. Sci.*,vol. 59, pp.438-443, 2002

Chapter 11

Micro-structural Approach to Developing the Resonance Model of the Indirect Flight Mechanism

Hitoshi Miyake

Physical Education and Health Care Center, Nagaoka University of Technology, 1603-1 Kamitomioka-cho, Nagaoka, Japan

Summary. Insects (Phylum Arthropoda Class Insecta) are divided into two groups. One has a direct flight mechanism (wing driven by the "direct" muscles) and the other has an indirect one. The latter is known as "constant wing oscillation", where is different from the asynchronous flight muscle contraction-relaxation frequency, although the precise mechanism is unknown. We have already proposed a mechanically reconstructed resonance model. It works as if the "indirect" muscles cause the chest wall to vibrate of its proper frequency and as if the vibration then causes the higher wing oscillation. According to this, there is no "amplifier" apparatus. To clarify the indirect flight mechanism and the "amplifier" using the micro-structural approach, honeybees (*Apis mellifera*) were investigated and the micro-XCT was employed. Ordinary cross-section tissue specimens were also investigated. The results were; 1) No "amplifier" apparatus was found in the area adjacent to the wing joint. Contrary to this, our resonance model does not have any inconsistency. It needs not higher muscle contraction-relaxation frequency. 2) In the joint of wing, some motion limitation apparatus of wing were found, but it was to be thought to act as a passive shock absorber, not as an active one. 3) Consequently, our resonance model is the only significant explanatory theory without inconsistency for the indirect flight mechanism.

Key words. micro-structure, indirect flight, insect, resonance, model

1 Introduction

1.1 Two Types of Wing Driven Mechanism in Insect Flight

Insects (Phylum Arthropoda Class Insecta) are divided into two groups according to the wing driven mechanism for flying. One, like birds, has a direct flight mechanism (wing driven by the "direct" muscles), that is, butterflies, dragonflies, etc., the other has an indirect flight mechanism (wing driven by the "indirect" muscles) (Nachtigall (1989), Chapman (1982)). The latter group which includes smaller insects, such as flies, bees, mosquitoes, is known as "constant wing oscillation", where is different from the asynchronous muscle contraction-relaxation frequency of flight, although the precise mechanism is still unknown.

1.2 New Mechanical Model and Resonance Model of the Indirect Flight Mechanism

We reported a new mechanism at the 1st International Symposium on Aqua Bio-Mechanisms (Miyake 2000a) and proposed a mechanically reconstructed resonance model at the 2nd symposium (Miyake 2004), because almost earlier literature seemed to include an inconsistency (Chapman (1982)). The resonance model works as if the "indirect" muscles cause the chest wall to vibrate of its proper frequency and as if the vibration then causes the wing oscillation. According to this theory, there is no "amplifier" apparatus. Some earlier literature suggested it was "the crick mechanism" (Nachigall (1989)).

1.3 Where is the "amplifier" located?

The objective of this study was to discover the micro-structural mechanism of the "amplifier" apparatus for converting chest wall vibration to wing oscillation using our resonance model, because the oscillation frequency was much higher than the muscle contraction-relaxation frequency (Miyake H (2004)).

Fig.1. Two types of posture, up- and down-wing stroke of honey bees were prepared.

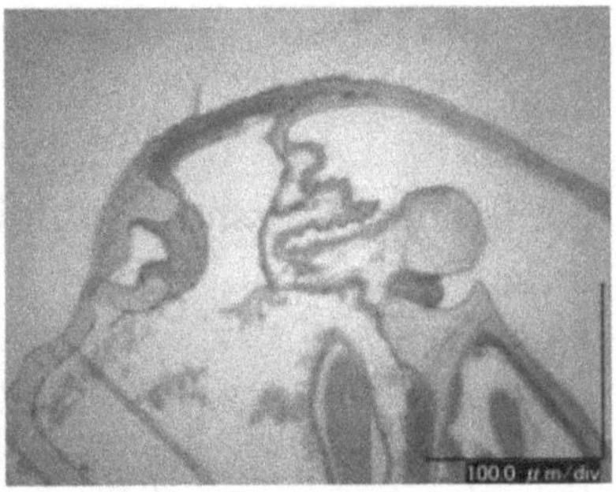

Fig.2. An example of tissue specimen (minimum resolution; 1.4 X 10^{-6} m, less than the micro-XCT's)

2 Methods

2.1 Materials

To clarify the indirect flight mechanism using the micro-structural approach, honeybees (*Apis mellifera*) were investigated. Just before observation, the chest part of each sample was separated sharply with a knife from the head and the abdomen. Also wings and legs were cut at the root. Two posture types, that is, up- and down-wing stroke's posture, were prepared as shown Fig.1.

2.2 Micro-XCT and Tissue Specimens

The micro-XCT (Shimadzu SMX-130CT-SV), which can show micro structure without cutting the specimen, was employed. Ordinary cross-section tissue specimens were also investigated. The slice width was 7.4 X 10^{-6} m and the slice pitch was 7.4 X 10^{-6} m. This means the minimum resolution was 7.4 X 10^{-6} m. Three-dimensional (3D) images were obtained easily using the micro-XCT results and examined carefully for the micro "amplifier" structure.
The tissue specimen's minimum resolution was 1.4 X 10^{-6} m (Fig. 2).

3 Results

3.1 Standard Sample (outside view)

Firstly standard samples were obtained to reconstruct 3D images from the micro-XCT images. Fig. 3 and Fig. 4 show typical 3D images of the left half portion of the chest near the root of wing. Fig. 5 also shows another typical outside view of the left chest near the root of wing. Some semi-lunar elevations were observed.

3.2 Cross-Sectional Observations (muscle layout)

From the cross-sectional observations, the layout of the two main wing drive muscles, which indicates the force transmission direction, were determined. Both muscles were attached to the tergum. They were not crossed at right angles (Fig. 6).

3.3 Cross-Sectional Observations of the Up- and Down-Stroke Wing Posture

The cross-sectional images of the two main muscles and the middle portion are shown in Fig. 7. The images of the top row show the slice direction of each CT image. The images in the second row show the Dorso-Ventral Muscle's cross-section. The third row shows the middle of the two muscles' cross-section and the images in the bottom row show the Dorso-Longitudinal Muscle's cross-section. The two figures on the left show on the up-stroke wing's posture. The two figures on the right show on the down-stroke wing's posture. The difference between the tow postures is most obvious at the root of wing. Fig. 8 shows the enlarged images of the up-stroke posture at the root. And Fig. 9 shows the same for the down-stroke posture. Comparing these images, the formed angles where the line "a" and the line "b" cross each other were different. That is, this line was the root of wing and this was not a point but a line formed by a membranous plain. In this area, super-elastic resirin was also found.

To summarize these observations, the following were clear;

1) On the side of the chest wall, the root of wing and its joint were found. (Fig. 3, Fig.4)

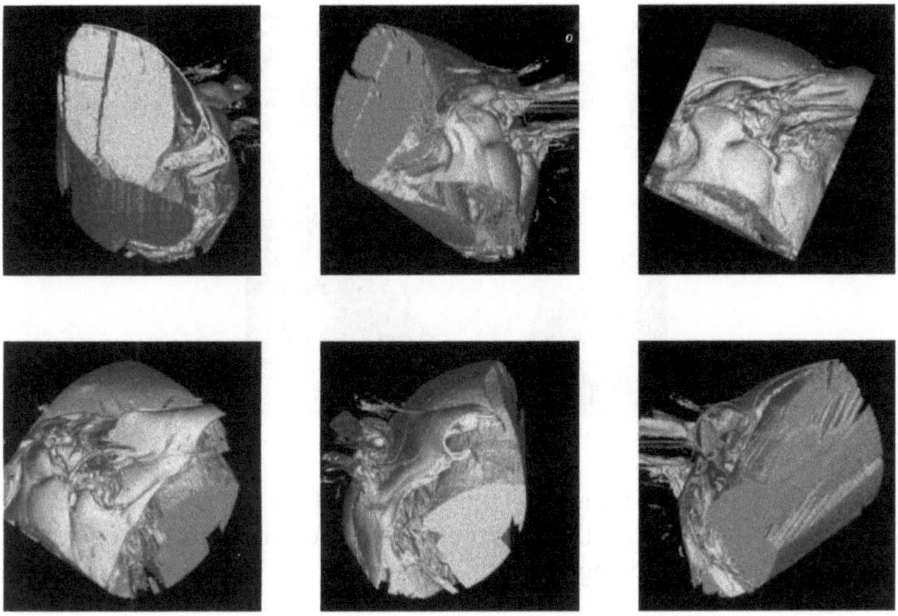

Fig.3. Examples of typical 3D images of the left half portion of the chest near the wing root shown as rotating from upper left to lower right from the CT images

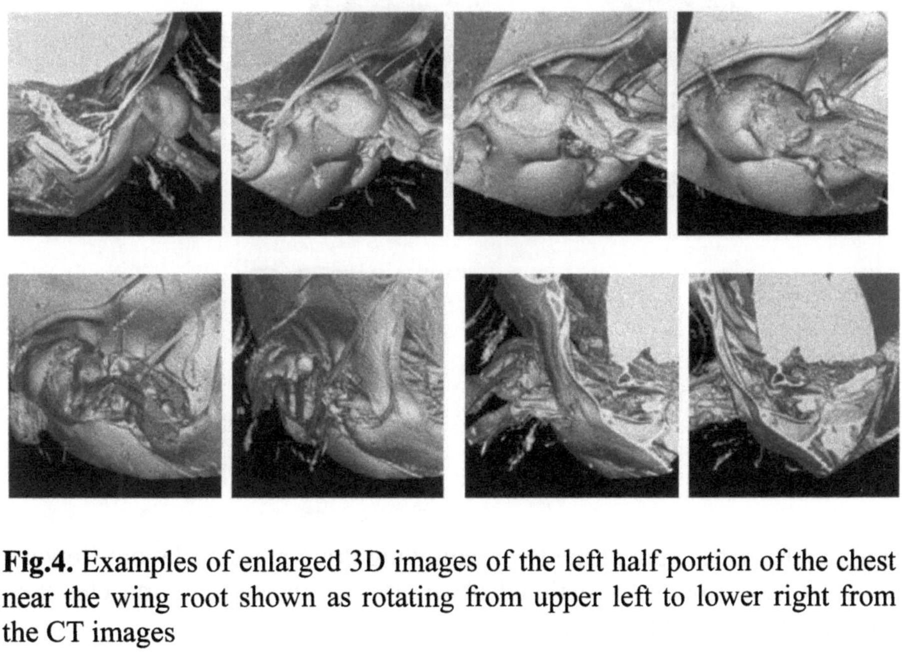

Fig.4. Examples of enlarged 3D images of the left half portion of the chest near the wing root shown as rotating from upper left to lower right from the CT images

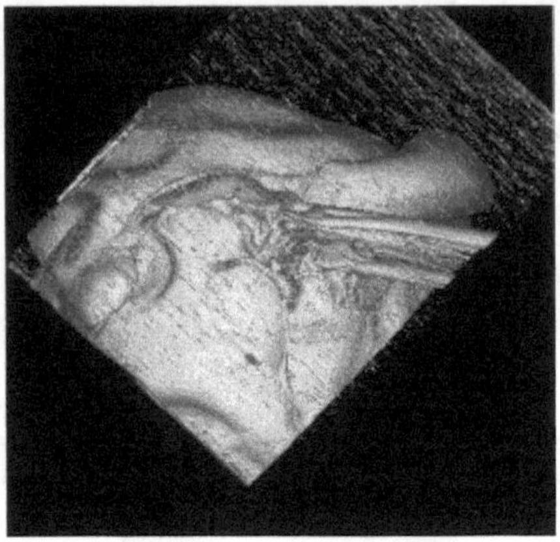

Fig.5. An example of typical 3D outside view of the left chest near the root of wing (the head is on the left, the abdomen on the right) reconstructed from the micro-XCT images

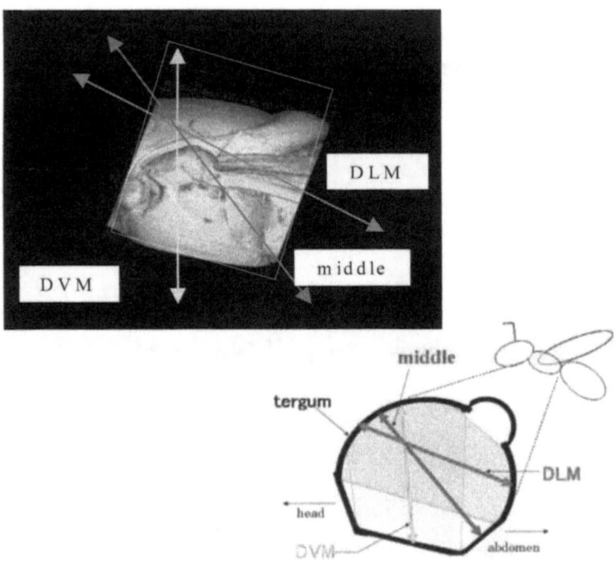

Fig.6. Schematic view of the two main wing drive muscles' layout (lower) ; derived from the cross-sectional observations (upper) (The muscles were not in the rectangle layout.)

2) A semi-lunar shaped pipe was found around the chest wall. (Fig.5)
3) At the root or joint of the wing, a membranous part was found. This was thought to act as passive shock absorber made of super-elastic resirin, not as an active one. (Fig. 8)
4) Also, at the joint, three semi-circular elevations were found. (Fig.7)
5) The pivot of the joint was found under the membranous part. (Fig. 8, Fig.9)
6) When the wing was on the up-stroke, the joint was in the curved area. And when it was on the down-stroke, this part was outside the area. That is, the structure of joint part spread out. (Fig. 8, Fig.9)
7) Both muscles (Dorsoventral muscle (DVM) and Dorsolongitudinal muscle (DLM)) were attached to the tergum. They did not cross at right angles. That is, the muscles only move the tergum when they work cooperatively. (Fig. 6)
8) Most of the chest wall cavity was occupied by the muscles. (Fig. 3)
9) The chest wall was made of three parts, that is, a hard upper and lower part and a softer middle part. (Fig. 6, Fig.7, Fig.8)

4 Discussions

4.1 Where is the crick mechanism?

Many theses suggest the existence of the crick mechanism (Nachigall (1989), Chapman (1982), Brodsky (1994)). We only observed the pivot of the wing joint and found no special mechanism there. But it may be that the crick mechanism has the function of so-called jack-knife movement. This suggests that it only works to strengthen the up-stroke and not the down-stroke, which needs a twisting movement to reduce air current resistance.

4.2 Does our resonance model have a weak point?

Our resonance model may have a weak point. That is, the driving force does not transmit in contact with the string. So the model was revised (Miyake 2000b). But even the revised model does not represent the actual muscle movement.

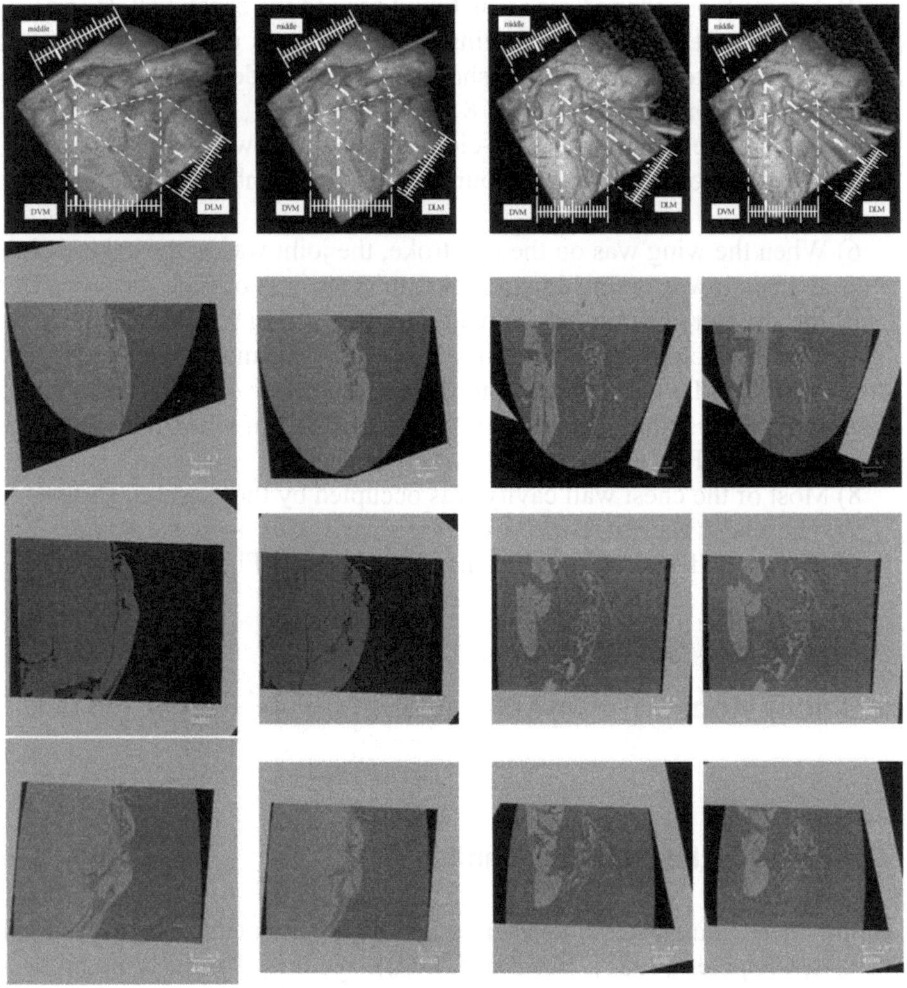

Fig.7. Images of the slice direction of each CT image (top row). Images of the Dorso-Ventral Muscle's cross-section (second row). Images of the middle of the two muscles' cross-section (third row). Images of the Dorso-Longitudinal Muscle's cross-section (bottom row). Left two file show on the up-stroke wing's posture. Right two file show on the down-stroke wing's posture. The scale in the top row shows arbitrary cross-sectional position of 20 cross-sections. The difference between the tow postures is most obvious at the root of wing. Following two figures (Fig.8 and Fig.9) are more explainable about this difference.

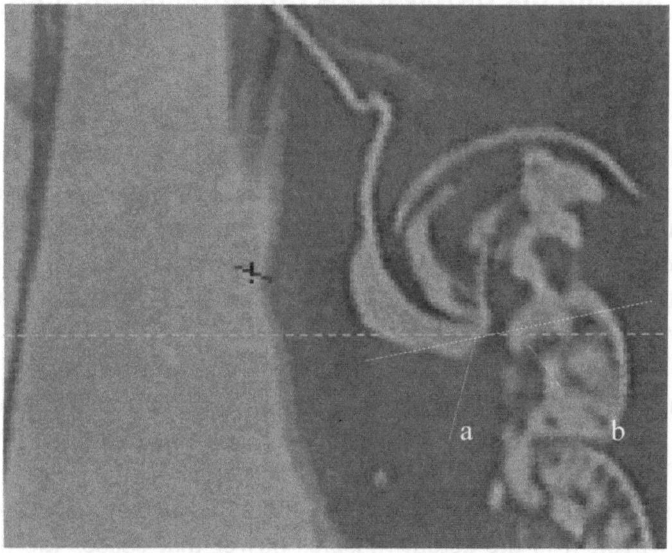

Fig.8. Enlarged images of the up-stroke posture at the root of wing, a short dotted line shows the wing posture. Note the position of "a" and "b", and thick material around "a", which may be made of super-elastic resirin.

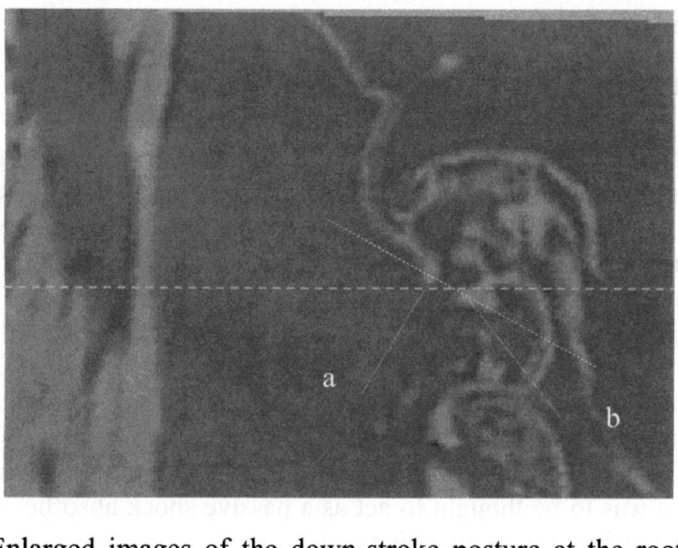

Fig.9. Enlarged images of the down-stroke posture at the root of wing, thick material around "a" is disappear.

4.3 How should we understand the muscles movement as the true model?

Argonne National Laboratory detected insect muscle micro-structure movement using the X-ray movies (Rgonne 2005). From this report, indirect type muscles move with the frequency of the wing beat, for example, 1k Hz. But the fact that vertebrata animals' muscles move at most of 100 Hz has been certified by EMG. That is, we must remind that muscle movement is not equal to EMG, and that muscles produce power only when they are stimulated by nerve pulses. MARCUS J. KING et al., 1996 described the DVM excitation frequency was approximately 35Hz. So we can assume that the indirect flight muscles move in two manners, that is, in the active phase and in the passive phase. So, even if the maximum frequency of the muscles contraction and relaxation is 100 Hz, the muscles themselves might be able to "move" at 1k Hz. According to our resonance model, however, it is the actual muscles power that causes the proper vibration of the tergum, which is much higher than the highest/maximum muscle contraction and relaxation frequency. As a phenomenon, the muscles themselves can be observed to "move" or to vibrate at a hither frequency. This resolves the apparent inconsistency.

5 Conclusions

This paper dealt with the indirect flight mechanism in insects. In order to discover the micro-structural mechanism of the "amplifier" apparatus for converting chest wall vibration to wing oscillation using our resonance model, honeybees (*Apis mellifera*) were investigated using the micro-XCT. The results were;

1) No "amplifier" apparatus was found in the area adjacent to the wing joint from the micro-XCT images. Contrary to this, our resonance model does not have any inconsistency. It needs not higher muscle contraction-relaxation frequency.

2) In the joint of wing, some motion limitation apparatus of wing were found, but it was to be thought to act as a passive shock absorber made of super-elastic resirin, not as an active one.

3) Consequently, our resonance model is the only significant explanatory theory without inconsistency for the indirect flight mechanism.

Acknowledgement

The author is extremely thankful to Prof. Hara of Niigata University, Department of Engineering, and also to the students of the Dept of Engineering, NUT.

References

Chapman R F (1982): THE INSECTS 3rd ed., Harvard Univ. Press, US

Miyake H (2000a): Approaches to the Indirect Flight Mechanism, Proceedings of the 1st International Symposium on Aqua Bio-Mechanisms, 239-244, August 27-30

Miyake H (2000b): Basic Studies of the Artificial Wing Driven by an Indirect Wing Movement Mechanism, Japanese Journal of Artificial Organs, 29(1), 271/275, (in Japanese)

Miyake H (2004): Resonance Model of the Indirect Flight Mechanism, in Biomechanisms of Swimming and Flying (N.Kato et al Eds.) 39/50 Springer, Tokyo

Nachtigall W (1989): INSECT FLIGHT, Graham J. Goldsworthy and Colin H. Wheeler Eds., CRC Press, US

RGONNE, Ill. (2005): http://www.anl.gov/ Media Center/ News/ 2005/ news050121.html

MARCUS J. KING et al., (1996): ACTIVITY OF ASYNCHRONOUS FLIGHT MUSCLE FROM TWO BEE FAMILIES DURING SONICATION (BUZZING), The Journal of Experimental Biology 199, 2317–2321

Part II

Hydrodynamics of Swimming and Flying

Chapter 12

Studies of Hydrodynamics in Fishlike Swimming Propulsion

Xi-Yun Lu[1], Xie-Zhen Yin[1], and Bing-Gang Tong[1,2]

[1]Department of Modern Mechanics, University of Science and Technology of China, 96 Jinzhai Rd., Hefei, Anhui 230026, China
[2]Department of Physics, Graduate University of the Chinese Academy of Sciences, 19A Yuquan Rd., Beijing 100049, China

Summary. We will attempt to provide an overview on the hydrodynamics of fishlike swimming propulsion based on our recent work performed experimentally, numerically and theoretically. In the present paper, we mainly review some typical work, including measurement on kinematics of free-swimming fish and prediction of dynamics acting on an arbitrarily deformable body, numerical and experimental simulations of flow over flapping and traveling wavy bodies, and theoretical modeling and analysis on flow around a waving plate and a flapping plate.

Key words. Fishlike swimming, Free-swimming fish, Animal propulsion, Hydrodynamics, Vortex dynamics

1 Introduction

Aquatic and aerial animals in fluid media, i.e. fish swimming in water and bird and insects flying through air, have developed their superior and complete performance of flying and swimming, which any modern man-made air and underwater vehicles cannot possess. These animals may be regarded as living machines, which act under their neural control with their muscle contraction as a motor to transform the biochemical energy to the mechanical energy and then implement their wing-flapping or body-undulating propulsion. We are interested in understanding the design concepts of these living machines, which include the neural science, muscle mechanics, morphology diversity, locomotion pattern (kinematics), propulsion and control (dynamics), energy costs and efficiency (energetics), biological material properties, etc. All these are involved in a discipline of flying and swimming biomechanics.

During recent years, some studies on biofluiddynamics of animal flight and swimming and the relevant biomimetic technology have been taken in the authors' team. In the present paper, we mainly review some recent work on the hydrodynamics of fishlike swimming propulsion.

2 Measurement of Free-Swimming Fish and Prediction of Dynamics on a Deformable Body

Measurement of free-swimming fish is a challenging task. Usually, the kinematic behaviors of free-swimming fish may be obtained by particle image velocimetry (PIV) (e.g. Li et al. 2003). However, how to measure the dynamics of free-swimming animals still is an unsolved problem. Thus, it is desirable to develop reliable and feasible method to predict the dynamics based on the kinematic data obtained in a certain region around an arbitrarily deformable body. For neatness, we mainly show a measurement of the kinematics of fish swimming in a fast-start and a theoretical method for predicting the dynamics on a deformable body, respectively.

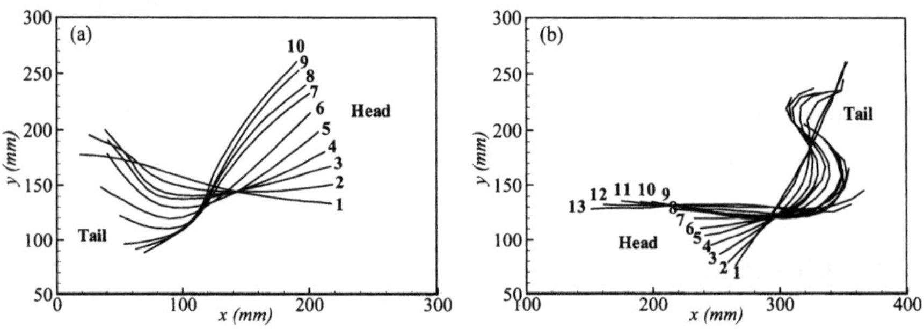

Fig. 1. The tracings of the fish in single-bend (a) and double-bend (b) C-starts for Crucian Carp and Yellow Catfish. Times between tracings are in milliseconds.

A fast-start is a high-energy swimming burst starting either from rest or steady swimming. The fast-starts are classified into two main types, C- and S-starts, in which the fish is bent into a 'C' or 'S' shape at the end of the first contraction of the lateral musculature. Here, the kinematics of C-starts of Crucian Carp (*Carassius auratus*) and Yellow Catfish (*Pelteobagrus fulvidraco*) from rest are investigated by Jing et al. (2004, 2005). Based on whether a contralateral bend appears, as shown in Fig. 1 for the tracings of the fish in the escape response, we have identified that the escape responses of Crucian Carp and Yellow Catfish exhibit single-bend and double-bend types, respectively.

Based on the backbone curves in Fig. 1, the strain of muscle along the fish body using a simple model and the dynamical performance using the Weihs' model during the escape responses are qualitatively analyzed (Jing et al. 2004, 2005). Further, experimental measurements of hydrodynamic force and moment acting on several fishtail shape models with a motion of simulating C-start of fish swimming are also performed. We have found that the force and moment acting on the fishtail models in the C-start motion depend on mainly the fishtail shape as its second and third area moments (Zhao et al. 2006).

Now, with the rapid development of noninterference instantaneous flow-field survey techniques, such as PIV, it is in principle possible to obtain the kinematic data in a certain region around a body and their time dependence. To predict the dynamics, we propose two innovative formulas for the total force acting on any solid body that moves and deforms arbitrarily in a viscous incompressible fluid, solely in terms of control-surface integrals, which are derived based on derivative-moment transformations (DMT) and described in detail by Wu et al. (2005). The formulas provide a theoretical basis of obtaining the dynamics on a deformable body.

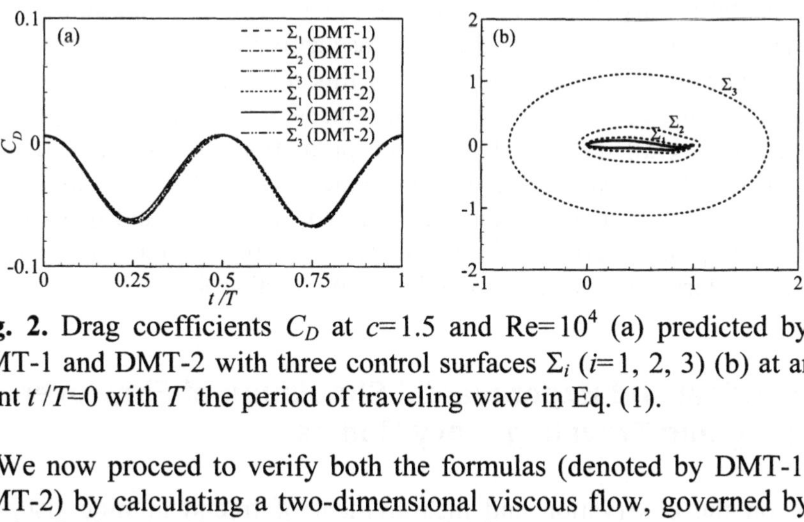

Fig. 2. Drag coefficients C_D at c=1.5 and Re=10^4 (a) predicted by the DMT-1 and DMT-2 with three control surfaces Σ_i (i=1, 2, 3) (b) at an instant t/T=0 with T the period of traveling wave in Eq. (1).

We now proceed to verify both the formulas (denoted by DMT-1 and DMT-2) by calculating a two-dimensional viscous flow, governed by the Navier-Stokes (N-S) equations, over a fishlike swimming body. In dimensionless form, the Reynolds number is defined as Re=UL/ν, where U is the free-stream velocity, L the length of body, and ν the kinematic viscosity. A NACA0012 airfoil is used as the contour of the body at an equilibrium position of undulating motion. The midline of the body is making a transversal oscillation in the form of a wave traveling in the streamwise direction, described by

$$y_m = A_m(x)\cos[2\pi(x-ct)],\ 0 \le x \le 1 \tag{1}$$

where A_m and c are the amplitude and phase speed of the traveling wave normalized by the length of swimming body. To model reasonably the lateral motion of the backbone undulation of fish swimming, $A_m(x)$ is approximated by a quadratic polynomial

$$A_m(x) = C_0 + C_1 x + C_2 x^2 \tag{2}$$

where the coefficients C_0, C_1, and C_2 are solved from the kinematic data of a steadily swimming saithe (Videler 1993), which gives $A_m(0)=0.02$, $A_m(0.2)=0.01$, and $A_m(1.0)=0.10$. To set the kinematic conditions on the deformable body, we assume that the body length is unchanged during swimming and its undulation is purely a lateral compressive motion.

Figure 2(a) shows that the drag coefficients C_D at $c=1.5$ and $\mathrm{Re}=10^4$, predicted by the DMT-1 and DMT-2 with three control surfaces Σ_i ($i=1, 2, 3$) in Fig. 2(b), agree with each other quite well. The independence of computed forces of the control domain is well confirmed. The above results suggest that, with high-accuracy flow data gained by the PIV in a neighborhood of Σ, both the formulas may easily lead to a force estimate with an accuracy higher than those estimates based on any conventional experimental means (Wu et al. 2005). In this regard, an important issue is that the formulas should be robust, i.e. insensitive to the inevitable measurement inaccuracy. A numerical test of this robustness has been carried out to mimic the effect of measurement inaccuracy in PIV data (typically 5%) on the overall quality of the force prediction. In the test, artificial noise of 5% free stream velocity is introduced. The error response is observed to be less than 1% or smaller. Further, a detailed and quantitative understanding of the effects of flow structures on the instantaneous overall force characteristics has been studied (e.g. Wu et al. 2007).

3 Numerical and Experimental Simulation of Flow over Flapping and Traveling Wavy Bodies

Based on the observation and measurement of free-swimming fish, the flapping and traveling wavy foils can be used as essential models to deal with the propulsion of fish swimming. Then, numerical and experimental simulations on viscous flow over the flapping and traveling wavy foils have been carried out. Here, we mainly discuss some typical cases, including flow around a flapping foil, flow past a traveling wavy foil and traveling wavy foils in a side-by-side arrangement, and flapping coupling between two filaments.

To understand the nature of flapping locomotion and the relevant dynamic response mechanisms, an up-and-down flapping foil is considered and its vertical position is given by

$$h(t) = A_d \cos(2\pi f_d t) \quad (3)$$

where A_d and f_d are the flapping amplitude and frequency, respectively. Then, a flapping Reynolds number is defined as Re=$\rho_o f_d A_d L/\mu$, where L is the chord of foil, ρ_o the fluid density, and μ the dynamic viscosity. When a foil takes only flapping without the horizontal movement, two typical parameters involved in this problem are the frequency parameter, defined as $\beta=\rho_o f_d L^2/\mu$, and the flapping amplitude $A=A_d/L$, with a relation to the flapping Reynolds number Re= βA.

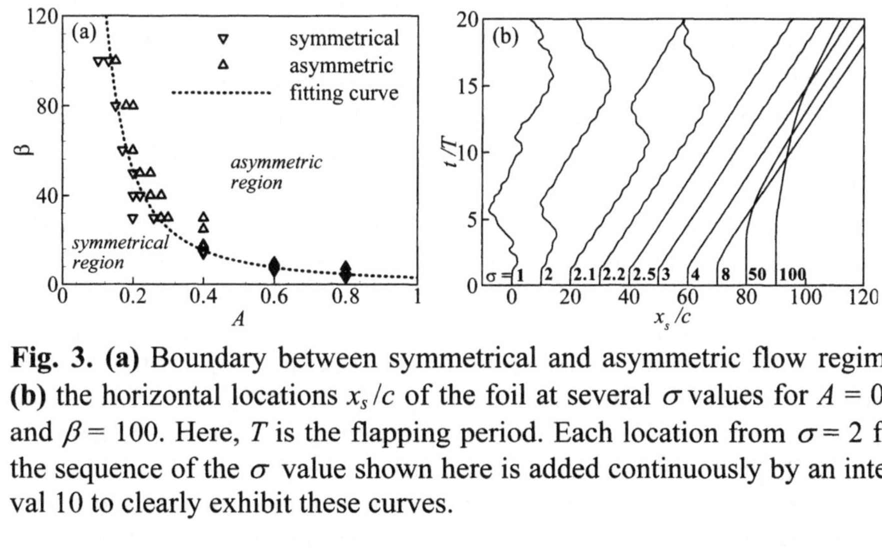

Fig. 3. (a) Boundary between symmetrical and asymmetric flow regime; **(b)** the horizontal locations x_s/c of the foil at several σ values for A = 0.5 and β = 100. Here, T is the flapping period. Each location from σ = 2 for the sequence of the σ value shown here is added continuously by an interval 10 to clearly exhibit these curves.

The foil exerts forces on the fluid through its own inertia response and is likewise acted on by the fluid flow pressure and viscous friction. Thus, there exists a critical flapping Reynolds number; as the critical Reynolds number exceeds, the flapping foil begins to move spontaneously in the horizontal direction (e.g. Lu and Liao 2006). To examine the critical flapping Reynolds number, the N-S equations are numerically solved for different pairs of β and A. For each pair of β and A under examination, the horizontal force component F_x is calculated. When F_x vanishes through the cycles, the flow is symmetric with the flow structure being left-right symmetric. When F_x is nonzero, the flow is characterized as asymmetric, the foil will move spontaneously in the horizontal direction. After extensive tests, the boundary between symmetrical and asymmetric flow regime on the $\beta - A$ plane shown in Fig. 3(a) is determined in the form of a fitting curve, $A = 1.9\beta^{-0.57}$. Then, by Re=βA, the critical flapping Reynolds number can be expressed as $\mathrm{Re}_c = 3.1A^{-0.75}$ approximately.

To analyze the dynamic response of spontaneously moving foil, the flapping frequency β and amplitude A are chosen in asymmetric regime in Fig. 3(a). In this case, the horizontal force component F_x, due to the pressure and friction forces exerting on the foil from the surrounding fluid, propels the foil to move horizontally through Newton's second law,

$$m_s \frac{d^2 x_s}{dt^2} = F_x \tag{4}$$

where x_s is the horizontal location of the foil, and m_s is the foil mass per unit spanwise length, or $\rho_s S$, with ρ_s the foil density and S the foil area. To normalize Eq. (4) by $\rho_0 f_d^2 L^3$, a typical parameter, the density ratio $\sigma = \rho_s/\rho_0$, is introduced. Here, we deal with the dynamics of the foil with different density ratios. The horizontal locations of the foil at several σ values are shown in Fig. 3(b). As $\sigma \leq 2.2$, the foil undergoes spontaneous oscillations, while as $\sigma \geq 2.5$, the foil moves to a stable state and reaches a steady movement. It is verified that there exist a threshold value of the density ratio with $\sigma_c \sim 2-4$ for all the cases in the (β, A) parameter space considered here. Thus, the horizontal movements of the foil demonstrate the existence of two dynamic responses. One is an oscillatory state when $\sigma < \sigma_c$ and the other a steady movement state when $\sigma > \sigma_c$.

The steady movement response is related to the forward flapping motion. The Strouhal number that governs a vortex shedding for the forward flapping foil has been calculated in a range of 0.2 – 0.4 (Lu and Liao 2006) and lies in the range where flying and swimming animals will be likely to tune for high propulsive efficiency. Further, the propulsive performance and vortex shedding of a foil in pitching and plunging motions have been investigated by Lu et al. (2003). Some different types of the leading-edge vortex shedding are identified numerically and also confirmed experimentally. Usually, the evolution of the leading-edge vortex has an effective influence on the vortex structures in the wake of the foil (Wu et al. 1998). Therefore, we find that the vortex structures in the wake (e.g. Karman or reverse Karman vortex-street) may be closely related to these types of leading-edge vortex shedding. In addition, the effects of some typical factors, such as the oscillating frequency and amplitude, and the phase difference between the pitching and plunging motions, on the vortex shedding and the unsteady forces have been analyzed (Lu et al. 2003).

To reveal the mechanisms of traveling locomotion, viscous flows over a traveling wavy foil undergoing lateral motion in the form of a streamwise traveling wave in Eqs. (1) and (2) are investigated. The hydrodynamic features of flow structure and vortex shedding near the traveling wavy foil are examined to get into physical insights to the understanding of fish-like swimming mechanisms in terms of drag reduction and optimal propulsive performance. The drag force acting on the wavy foil and the power needed for it to be propelled are directly relevant to the study of fish locomotion. The total drag force (C_D) on the wavy foil consists of a friction drag (C_{DF}) and a form drag due to pressure distribution (C_{DP}), where these drag force coefficients, i.e. C_D, C_{DF} and C_{DP}, have been described (e.g. Dong and Lu 2005, Lu and Yin 2005). Similarly, the total power P_T required for the propulsive motion of the foil consists of two parts. One is the swimming power P_S, required to produce the lateral oscillation of traveling wave motion, and the other is the power P_D, needed to overcome the drag force. Then, the total power $P_T = P_S + P_D$ is obtained.

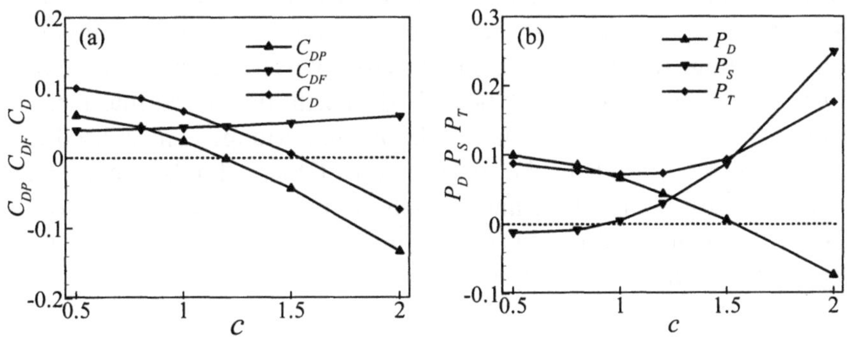

Fig. 4. Time-averaged drag force and power coefficients: **(a)** drag force; **(b)** power.

Time-averaged drag force coefficients and powers P_T, P_S and P_D are shown in Fig. 4 for Re=5×10^3 based on the uniform velocity and the chord length of the foil. With the increase of c, the time-averaged form drag coefficient C_{DP} and total drag coefficient C_D decrease, and the friction drag coefficient C_{DF} increases somewhat in Fig. 4(a). It is noted that C_{DP} becomes negative and acts as thrust force when $c > 1.2$ approximately, and C_D gets negative when $c > 1.5$ approximately. In Fig. 4(b), as c increases, P_S increases and becomes positive for $c > 1$ approximately. The negative value of P_S means that the wavy foil motion can be actuated by the flow and no power input is needed. The power to overcome the drag force P_D decreases monotonically with c, because of the decrease of C_D shown in Fig. 4(a). When P_D is negative, it means that the wavy foil is propelled by the thrust; however, the thrust is at the expense of the swimming power P_S. Both P_S and P_D present the competing mechanisms. The distribution of total power P_T versus c is concave upwards with a minimum around $c = 1.2$, which is consistent with the value used for traveling wave-like swimming motion of live fish in nature (Videler 1993).

Considering the undulating body subject to a net thrust (i.e. $C_D < 0$), we introduce a propeller efficiency (or Froude efficiency), defined in a time-averaged manner as the ratio between $-P_D$ and P_S, or $\eta = -P_D/P_S$. Note that for a self-propelled motion, the time-averaged net thrust force is equal to zero, and the propeller efficiency is also zero. As shown in Fig. 4, the total drag C_D is near zero at c=1.5, thus η approaches zero, while a net thrust occurs at c=2.0, η is around 30%.

The hydrodynamic aspect of fish schooling is a challenging topic and involves complex flow phenomena. Since the hydrodynamic behaviors of fish schooling are closely associated with the complex interferences among the fish, e.g. streamwise and lateral interference, thus a reasonable way is to study them separately for revealing some typical mechanisms involved in fish schooling. Relatively, the streamwise interference has been extensively studied (e.g. Liao et al. 2004). To deal with the lateral interference, flow over traveling wavy foils in a side-by-side arrangement is used as a simplified model. Based on the observation of fish schools, we consider two typical cases, i.e. in-phase and anti-phase traveling wave motion of the adjacent foils (Dong and Lu 2007). The midline of the each foil is making a lateral oscillation described by Eqs. (1) and (2).

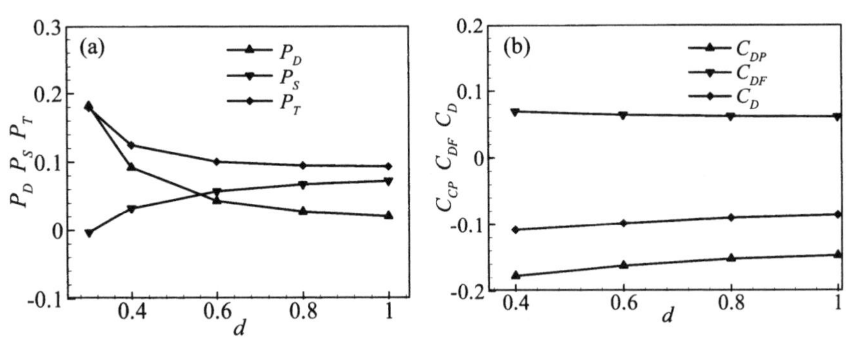

Fig. 5. Distributions of time-averaged powers versus the lateral distance d in in-phase case at c = 1.5 **(a)** and time-averaged drag force coefficients in anti-phase case at c = 2.0 **(b)**.

The distributions of time-averaged powers P_T, P_S and P_D versus the lateral distance d in the in-phase case are shown in Fig. 5(a) for c = 1.5 and Re=5×10^3, where d represents the lateral position of the midline of foil at an equilibrium position normalized by the length of swimming body. It is interesting to find that the swimming power P_S decreases as d decreases, and even becomes negative at d = 0.3. It means that the in-phase traveling wavy motion can be actuated by the flow without power input when P_S is negative. In addition, the time-averaged drag force coefficients in the anti-phase case are shown in Fig. 5(b) at $c = 2.0$. Due to the lateral interference, a thrust (i.e. $-C_D$) becomes larger as d decreases. It is noted that the increase of the drag or thrust is mainly contributed by the form drag C_{DP}. Thus, it is revealed that the lateral interference is of benefit to saving the swimming power in the in-phase case and enhancing the forces in the anti-phase case. In addition, some typical vortex structures, e.g. vortex-pair row, single vortex row, in-phase and anti-phase synchronized vortex-street, are observed in the wake of the traveling wavy foils (Dong and Lu 2007). The results obtained provide physical insight into the understanding of hydrodynamics and flow structures for flow over the traveling-wavy foils and swimming mechanisms relevant to fish schooling.

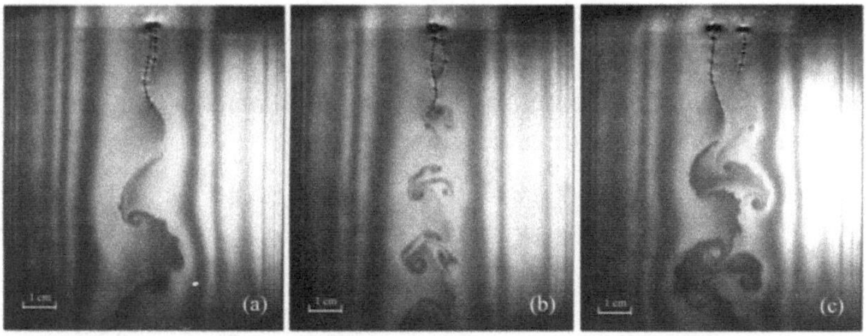

Fig. 6 Experiment photos of the coupling between two different length filaments with left filament length L=30*mm* and right filament L=15*mm*: (a) lateral distance between the upper fixed locations d=1.5*mm*, (b) d=2*mm* and (c) d=10*mm*.

Fish swimming in water includes complicated interactions between the body and the surrounding fluid flow as well as among bodies in fish schooling. Such fundamental fluid-body interactions are investigated experimentally by use of two filaments in a flowing soap film (Jia et al. 2007). Fig. 6 shows the experimental photos of both the filaments with different lengths. We have found three typical coupling modes, i.e. the anti-symmetrical in-phase mode, the symmetrical out-of-phase mode, and the indefinite mode. The values of the Strouhal number relevant to the vortex shedding are approximately 0.2. To understand the mechanisms involved in this problem, the temporal linear instability analysis is carried out based on a simplified hydrodynamic model (Jia et al. 2007).

Natural selection helps animals to obtain the highest propulsive efficiency. It means that animals can use the minimal energy to propel themselves over a longer distance in a preferred Strouhal number. The Strouhal number of filament flapping indicates that the fundamental flapping mode may correspond to the minimal energy state. As shown in Fig. 6, the out-of-phase mode has a higher frequency than the in-phase one when the Strouhal number keeps around 0.2, implying that the flapping amplitude of the out-of-phase mode is smaller than that of the in-phase one. The smaller amplitude can help the filament keep itself at a lower potential energy level. The results obtained by Jia et al. (2007) are of helpful in understanding the mechanism of fluid-body interaction and natural selection of animal locomotion for high propulsive efficiency.

4 Theoretical Modeling and Analysis on Flow around a Waving Plate and a Flapping Plate

To understand the fundamentals of hydrodynamic characteristics relevant to animal propulsion, theoretical modeling and analysis are performed for flow around a waving plate and a flapping plate.

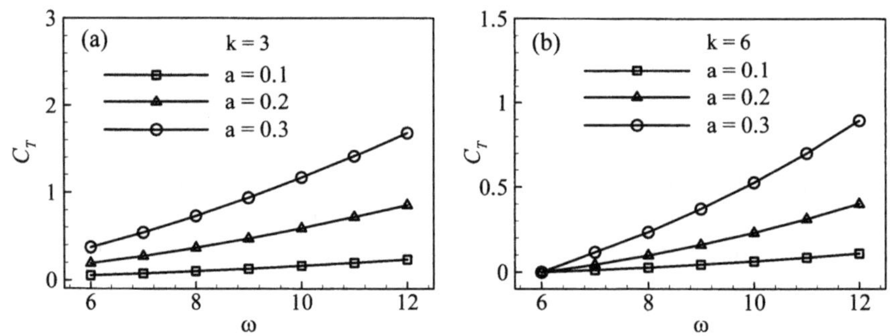

Fig. 7. Time-averaged thrust coefficient versus the frequency for the different amplitudes and wavenumbers: **(a)** k=3 and **(b)** k=6.

The hydrodynamic characteristics of two-dimensional unsteady flow around a waving plate are examined theoretically. Under large Reynolds number approximation, the flow is assumed to be a combination of the outer potential flow and a thin vortex layer, which consists of a boundary layer and a shed free shear layer. The viscous effect is assumed to be confined to the thin boundary layer. Then a nonlinear mathematical formulation for describing the outer unsteady potential flow coupled with an unsteady boundary layer equation for the inner viscous flow adjacent to the waving plate is developed (Zhang et al. 2007). Considering flow around a waving plate, the motion of the plate is taken as

$$y = ax^2 \sin(kx - \omega t) \tag{5}$$

where a is the constant amplitude, ω the frequency, and k the wavenumber. We suppose that no stretch occurs during the course of deformation of the waving plate. The thrust and viscous drag coefficients, propulsive efficiency, and the pattern of wake vortex sheet are analyzed for different amplitudes, frequencies, and wavenumbers (Zhang et al. 2007). The thrust coefficient C_T is shown in Fig. 7, which indicates that the thrust increases with the frequency ω and the amplitude a at constant wavenumber k. It is seen that the thrust decreases with the increase of the wavenumber.

Further, an innovative theoretical modeling approach is established for highly unsteady and viscous flow excited by a flapping plate (Yu et al. 2003, Yu and Tong 2005). Based on the reasonable assumption that the flapping Strouhal number is a dominant parameter in the flow, or the Strouhal number St >>1/Re, we propose a simplified dynamic model. The outside flow field of the flapping plate is regarded as the unsteady potential flow, where Laplace equation is applicable, and Kutta conditions should be satisfied at both the leading- and trailing-edge of the plate, where the vortex motions are described by discrete vortices. The present method has been validated to prove that it can give qualitatively correct and quantitatively reasonable results (Yu et al. 2003). Furthermore, the total fluid-dynamic force based on this approach can be decomposed into three parts: one due to the added inertial (or mass) effect, the other and the third due to the induction of vortices shed from the leading- and trailing-edge and their images, respectively, and this decomposition is of helpful in analyzing the control mechanisms in animal locomotion.

Acknowledgements

This work was supported by the National Natural Science Foundation of China (No. 10332040), the Innovation Project of the Chinese Academy of Sciences (No. KJCX-SW-L04), the Program for Changjiang Scholars and Innovative Research Team in University, and the Fund for Foreign Scholars in University Research and Teaching Programs. The authors sincerely thank Professors L. X. Zhuang, J. Z. Wu, J. M. Yang and Y. L. Yu for their generous collaboration.

References

Dong GJ, Lu XY (2005) Numerical analysis on the propulsive performance and vortex shedding of fish-like travelling wavy plate. Int J Numer Methods Fluids 48: 1351-1373

Dong GJ, Lu XY (2007) Characteristics of flow over traveling-wavy foils in a side-by-side arrangement. Phys Fluids (in press)

Jia LB, Li F, Yin XY, Yin X Z (2007) Coupling modes between two flapping filaments. J Fluid Mech 581: 199-220

Jing J, Yin XZ, Lu XY (2004) The hydrodynamic analysis of C-start in Crucian Carp. J Bionics Engin 1: 102-107

Jing J, Yin XZ, Lu XY (2005) Experimental and theoretical investigation on fast-start of yellow catfish (Pelteobagrus fulvidraco). Prog Nat Sci 15: 34-40

Li XM, Wu YF, Lu XY, Yin XZ (2003) Measurements of fish's wake by PIV. Proc SPIE – Int Soc Opt Eng 5058: 139-145

Liao Q, Dong GJ, Lu XY (2004) Vortex formation and force characteristics of a foil in the wake of a circular cylinder. J Fluids Struct 19: 491-510

Lu X Y, Liao Q (2006) Dynamic responses of a two-dimensional flapping foil motion. Phys Fluids 18: 098104

Lu XY, Yin XZ (2005) Propulsive performance of fish-like travelling wavy wall. Acta Mech 175: 197-215

Lu XY, Yang JM, Yin XZ (2003) Propulsive performance and vortex shedding of a foil in flapping flight. Acta Mech 165: 189-206

Videler JJ (1993) Fish Swimming. Chapman & Hall, London

Wu JZ, Lu XY, Zhuang LX (2007) Integral force acting on a body due to local flow structures. J Fluid Mech 576: 265-286

Wu JZ, Pan ZL, Lu XY (2005) Unsteady fluid-dynamic force solely in terms of control-surface integral. Phys Fluids 17: 098102

Wu JZ, Lu XY, Denny AG, Fan M, Wu JM (1998) Post-stall flow control on an airfoil by local unsteady forcing. J Fluid Mech 371: 21-58

Yu YL, Tong BG (2005) A flow control mechanism in wing flapping with stroke asymmetry during insect forward flight. Acta Mech Sinica 21: 218-227

Yu YL, Tong BG, Ma HY (2003) An analytic approach to theoretical modeling of highly unsteady viscous flow excited by wing flapping in small insects. Acta Mech Sinica 19: 508-516

Zhang C, Zhuang LX, Lu XY (2007) Analysis of two-dimensional unsteady flow around waving plates. J Engin Math (submitted)

Zhao L, Jing J, Lu XY, Yin XZ (2006) Measurements and analysis of force and moment of caudal fin model in C-start. Prog Nat Sci 16: 796-802

Chapter 13

Optimisation of Fish Shape and Swim Mode in Fully Resolved 2-D Flow Field by Genetic Algorithm with the Least Square Prediction Method

Shintaro TAKEUCHI[1,2], Sho KUSUDA[1] and Takeo KAJISHIMA[1]

[1] Department of Mechanical Engineering, Osaka University, 2-1 Yamada-oka, Suita-city, Osaka 565-0871 Japan

Summary. Optimisation of fish shape and swim mode is attempted in fully-resolved flow fields by genetic algorithm (GA) facilitated by efficient prediction method of target functions. Arbitrary Lagrangian Eulerian method is employed for solving the flow field around the moving 2-D fish swimming by undulatory motion propagating from head to tail tip. The fish performance is optimised with characteristic five parameters for targeting maximum cruising speed. For identifying the average cruising speed (in steady swimming state), a least square prediction (LSP) method is developed. The LSP is found to effectively reduce the computation time spent for searching the optimal solutions in the above GA process. The results demonstrate the capability of our method for efficiently solving this type of optimisation problem including the surrounding fully resolved flow field in a feasible computational time.

Keywords. Fish Shape and Swim Mode, Cruising Velocity, Genetic Algorithm, Least Square Prediction, Random-Search and Mathematical-Programming

1 Introduction

Finned aquatic animals, like fish and dolphin, often exhibit extraordinary physical capabilities for swimming speed, efficiency and steep turn. Meanwhile, ship and undersea vessel have been using rotary propeller to get thrust. If the above fish-like propulsion system is applied to the man-made vessels, it may be able to control noise or improve efficiency. In fact, several prototypes of robot fish have been launched already, and in the future the competition in the development and improvement of robot fish is expected to be further accelerated. However, it is not realistic to test all the possible types of robot, because one robot fish has many adjustable parameters to identify shape and swim mode. Therefore, a sophisticated method is needed to seek a best-performing solution, i.e. optimisation, efficiently.

A number of optimisation algorithms are known, and those can be mainly classified into (1) recursive approach based on mathematical-programming and (2) heuristic approach based on random-search algorithm. Examples of the first approach include linear/dynamic programming, variation method, simplex method and penalty method. These methods commonly assume the gradient(s) of the target function to

[2]Present Address: Dept. Mechanical Engineering, The University of Tokyo, Tokyo 113-8656 Japan

be zero in a continuous function space. Therefore the methods offer efficient optimisation when the system is known or easily assumed. However, for large systems, it tends to be computationally expensive to obtain the gradients, and a unique solution is not always guaranteed when the solution space has multiple minima. The second approach is, on the other hand, suitable for problems with multi-minimum solutions in a multi-dimensional parameter space, as it does not assume gradient of the target function within the optimisation theories, but retrieve a better solution with the high-performance candidates through some heuristic mechanisms. Therefore this approach is suitable for optimisation of unknown systems. Among the heuristic approaches, there is one particular technique, genetic algorithm (GA), used for optimisation of machine learning or function design under Pareto-efficiency. GA got its basic ideas from the natural evolution and selection systems of individuals in a group of families; i.e. *reproduction*, *crossover* and *mutation*, which are essentially non-linear stochastic mixing processes. Due to the controlled non-linear processes over long generations, GA works effectively with a non-linear fitness (target) function when the problem exhibits strong non-linearity (Michalewicz 1992). Therefore, with a digital selection mechanism, GA is expected to solve the non-linear problems, inaccessible by the mathematical-programming methods.

GA has gained popularity for a variety of optimisation problems. For robot fish, for example, Barrett (2002) conducted some experiments to optimise fish locomotion by directly incorporating the experimental results into GA, and reported that GA works effectively for achieving desired robot-fish functions. Also, Kuo and Grierson (2003) employed GA to search efficient prospective gait based on computational results. The number of parameters optimised in their work was five; each of them is a major characteristics determining carangiform locomotion. They obtained fully evolved samples by allowing 50 samples to cross-bread for 120 generations. However, the model for fluid-object interaction by Kuo and Grierson was rather simplified and restricted by employing an empirical fluid drag model. Generally, object's movements are affected by fluid motion passing by it and vortical structures in the wake (Smith and Wright 2004; Takeuchi et al. 2006). Also, some pioneering researchers (Liu et al. 1996; Carling et al. 1998) attempted full-scale simulations past undulatory object and analysed the interaction between the flow structures and object behaviours in detail. It is of importance to include the effect of fluid motions with the full scales into the GA process.

The present authors have recently demonstrated the usefulness of the GA for optimising fish's performance (travelling distance in a certain period of time and travelling efficiency) in a fully resolved flow field with a parallel computation library (Kusuda 2007). However, even with the aid of parallel processing, optimisation on steady travelling speed is not an easy task, as time-averaged cruising speed is an expensive target (as explained later). In addition, the elapsed times until the cruising speeds are achieved are different from individuals, which causes computational inefficiency as some processors (of superior samples) have to wait for the poorer

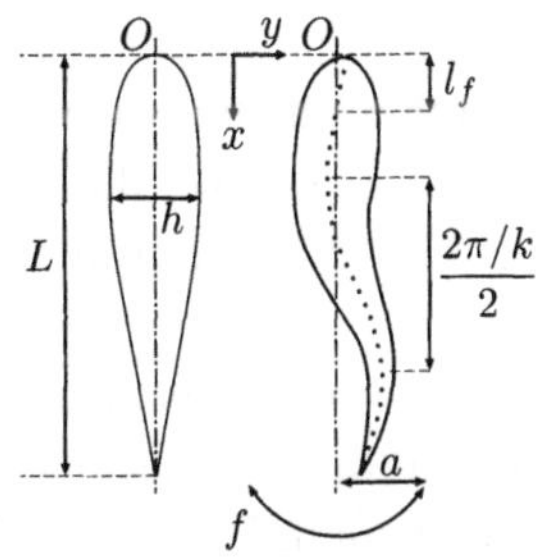

Fig. 1. Stationary form (left) and typical deformed form (right)

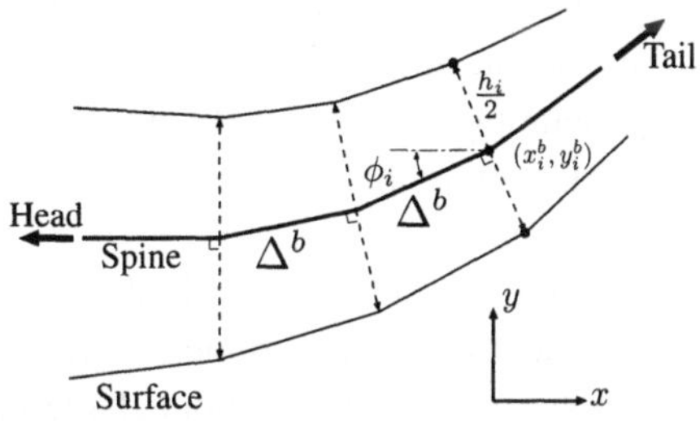

Fig. 2. Schematic of fish's deformation

samples until the next cross-breeding event takes place. In this study, we look at the possibility of reduction in the computational time for optimised performances on cruising speed through an original method of least square prediction (LSP).

2 Governing Equations and Numerical Methods

2.1 Governing Equations

A body-fitted curvilinear coordinate system is employed for fluid phase. In the present study, we apply altering coordinates to fit the object surface which changes the shape with time in a prescribed manner.

Governing equations of the fluid are the Navier-Stokes (N-S) equation and the equation of continuity for incompressible fluid. Arbitrary Lagrangian Eulerian (ALE) method (Hirt et al. 1974) is used to include the effect of the altering mesh.

Motion of the fish body is governed by preservation of linear and angular momentums for translating and rotating motions, respectively. For time-advancement, a time-development scheme based on Wilson's linearly-extrapolated acceleration method (Smith and Griffiths 2004) is applied.

2.2 Fish model

Figure 1 shows a schematic of the stationary and deformed positions of the fish. NACA00xx series hydrofoil is used as the fish model at the stationary position.

In the present study, the maximum amplitude a is varied in a range including $0.2L$, which is reported as a typical amplitude of a real fish's movement (Bainbridge 1958). From the thickness of the hydrofoil h, the hydrofoil model (NACA00xx) is determined as xx=$100h/L$. Fish is allowed undulatory motion from head to tail continuously. Wavenumber k of undulatory motion (Fig.1, right) is tested in the range between 0 and 0.4. Here, zero wavenumber means that every point on the backbone is set in phase (in-phase motion), and non-zero wavenumber causes some phase-lag between arbitrary two points on the deformable part (phase-lag motion).

Frequency f determines the speed of the undulatory motion propagating down towards the tail tip as $2\pi f/k$. The major deformable length of the body (responsible for generating thrust) is $L-l_f$ in the rear, and the position $x=l_f$ gives the minimum amplitude along the entire body as suggested by Yu et al. (2004) for robotic-fish's manoeuvre.

Deformation on the fish body is decided as follows. First, a displacement of the fish centre line (i.e. spine) is given in y-direction, and then displacement in x-direction is determined without changing the entire body length. Position of the i-th node of the backbone, $x_i^b(t)$ and $y_i^b(t)$, is prescribed by the following equations of travelling wave:

$$y_i^b(t) = R(x_i^b)W(t)\sin(2\pi f t - k\,x_i^b) \quad (1)$$

$$x_i^b(t) = x_{i-1}^b(t) + \Delta^b\cos\phi_i, \quad (2)$$

$$R(x) = \begin{cases} a(x-l_f)^2/(L-l_f)^2 & (0 \le x < l_f) \\ a(x-l_f)/(L-l_f) & (l_f \le x) \end{cases} \quad (3)$$

$$W(t) = \begin{cases} t-\sin(2\pi t)/(2\pi) & (0 \le t < 1) \\ 1.0 & (1 \le t) \end{cases} \quad (4)$$

where Δ^b is length of each backbone segment, ϕ_i angle between fish peduncle and the neutral line given as $\tan\phi_i = (y_i^b - y_{i-1}^b)/(x_i^b - x_{i-1}^b)$. The number of backbone segments is 60, and therefore Δ^b=$L/60$. The first weight function $R(x)$ determines the local amplitude at the position x. The second weight function $W(t)$ takes the advantage of zero gradients for $\dot{W}$ and $\ddot{W}$ at t=0 and 1, which enables a smooth start-up and connection to the steady undulatory motion at both times. And then, the fish surface is placed at a given distance ($h_i/2$) in both directions from the spine, where h_i is the initial body thickness at the backbone node i.

Figure 2 illustrates the above deformation process. For both extreme values of k (0 and 0.4), volume deviations during the swim are found to differ no more than 10^{-3} from the initial value.

2.3 Numerical conditions

In the present study, fish is set in an open space of stationary fluid. C-type mesh is used to represent the object surface and the fluid domain. Figure 3(a) shows a mesh near the object, and Fig.3(b) outlines the computational domain and boundaries. The computational domain is a semi-circle of diameter $8L$ in the front of the fish and $7L$ in length in the behind. Boundary condition of fluid velocity is the non-slip condition at the surface of the fish (N), and the instantaneous velocities of the object are assigned for the grid velocities at the inlet boundary (A). The convective boundary condition is used at the out-flowing boundary (C), and a traction-free boundary condition is employed at the upper and lower straight sections of the C-type mesh (T). At the object surface, the same pressure boundary condition of Nakamura et al. (1993) is applied.

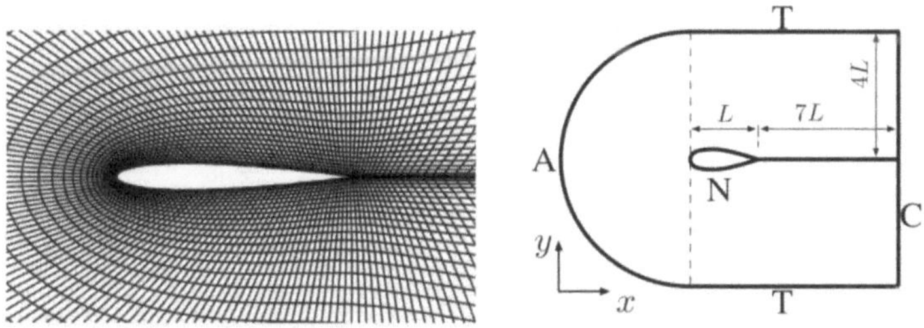

(a) Grids near the object (b) Domain size and boundaries

Fig. 3. Grid system and boundary conditions.

The number of grid points is 240×40. There are 120 grid points along the fish surface. With an increased spatial resolution, 280×50, we have confirmed that the simulation results (velocity profiles of the fish) differed by no more than 1%. Computational mesh fitted to a deformed fish is given every time step by an elliptic mesh generation algorithm. Residuals in the grid generation procedure were reduced to 10^{-6} within 50 iterations at steady state. The simulation results were found to agree very well with those employing a stricter convergence threshold of 10^{-8}.

The flow field obtained by the ALE was validated through comparison with the simulation results conducted independently in a body-fixed (non-inertial) frame of reference by the present authors (Takeuchi et al. 2006). We confirmed good agreements for time histories of drag and lift coefficients as well as instantaneous pressure distributions.

3 Specific Study of the Propulsive Stage

Some specific cases of typical fish characteristics are studied to see the fundamentals of the interaction between the fluid and the deformable hydrofoil. In this section, a, h, l_f and f are set to $0.2L$, $0.12L$, $0.1L$ and 1, respectively. Two wavenumbers k=0 and 0.4 are tested to observe the fluid behaviours past the object. Hereafter, all the velocity values are normalised by the reference velocity travelling one body length L in a unit-time.

Figures 4(a) and 4(b) compare the results of fish velocities and instantaneous flow fields in the transitional propulsive stage for different k values. Negative velocity value means that the object is moving in $-x$ direction. From the velocity diagrams, the fishes experience large accelerations in both x and y directions after t=1. Then, for both k cases, x component of the velocity develops exponentially with fluctuations, and it is expected that the velocities would eventually attain stable oscillations around the constant cruising velocities. Within the time range tested,

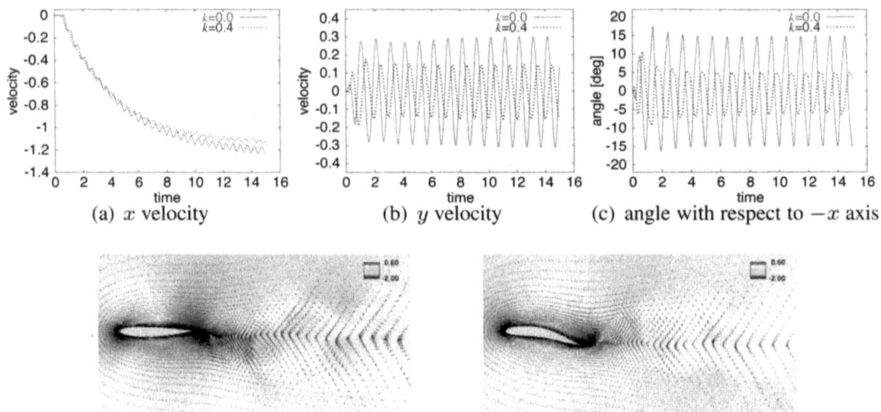

(a) x velocity (b) y velocity (c) angle with respect to $-x$ axis

(d) k=0 (In-phase motion) (e) k=0.4 (Phase-lag motion)

Fig. 4. Fish velocities and flow fields. (a)(b)(c) time evolutions of fish velocities and angles; (d)(e) instantaneous pressure and velocity fields in slip streams.

the terminal velocities are not attained, and the diagrams suggest that it could generally take a relatively long time to estimate steady cruising speed. On the other hand, stable oscillations are observed for y components of the velocity and angles of the body with respect to $-x$ axis for both cases. The time-averaged velocities in y direction are -3.6×10^{-3} and -3.0×10^{-4} for k=0 and 0.4, respectively. Also, the time-averaged angles with respect to $-x$ axis are 0.32[deg] and -0.11[deg] for k=0 and 0.4, respectively. Therefore, the fishes travel almost straight ahead for both cases. In our previous study (Kusuda et al. 2006) the fish exhibited specific travelling orientations (-3.0[deg] and 18.4[deg] for the respective k values). However, those orientating tendencies have been cancelled by employing the present undulatory mode which covers from head to tail. The result suggests that the study in fully resolved flow field allows analysis of sensitive directivity (or controllability) of fish.

Figures 4(d) and 4(e) compare instantaneous pressure and velocity fields in the slip streams of k=0 and 0.4, respectively. Flow fields similar to our previous results (Kusuda et al. 2006, 2007) are obtained, as the present fish employs the same undulatory motion for the deformable rear part as Kusuda et al. (2006, 2007). The case with the smaller wavenumber (k=0) exhibits stronger vortices in the slip stream than the larger k case. Generally, fish's total gain in momentum is dependent on the momentum held by the discharged vortices. This may explain the larger fluctuations in the velocities and the angle observed in Fig.4(a) through to 4(c). The major outcome from the analysis of the flow fields of Kusuda et al. (2006) is briefly summarised for the convenience of discussion in the subsequent sections. The fish with a larger wavenumber (phase-lag motion) is more efficient in converting the hydrodynamic force into propulsion, while the fish with a smaller phase-lag (smaller k) effectively transfer kick power to fluid to get a stronger thrust by generating stronger vortices. The results show that only a fully-resolved flow field enables analysis and prediction of dynamic fish-fluid interaction at this level.

4 GA with Least Square Prediction (LSP)

It would be interesting to determine what fish shape and swim mode give optimal solution in terms of cruising speed in steady state. To address this issue, a genetic algorithm (GA) is employed. As observed in Fig. 4(a), fish's travelling velocity in the longitudinal direction increases exponentially until it reaches a steady state. However, time-averaged terminal velocity is only obtained by averaging the velocities over several periods after reaching steady state. This process is often computationally expensive. Therefore, an efficient method for predicting the steady speed is needed. In the present study, both the velocity evolution and the terminal velocity are predicted by a least square prediction (LSP) method.

4.1 LSP of asymptotic cruising speed of fish

A data sequence (t_i, v_i) (i=1,$\cdots$, N) obtained by simulation is fitted with the following exponential equation:

$$v(t) = V_t(1 - e^{-gt}) \tag{5}$$

where V_t and g are terminal velocity and temporal decrease rate, respectively. The above two parameters are determined so that the following sum of the squared errors from the simulation result is minimised:

$$L = \sum (v(t_i) - v_i)^2 = \sum \left\{ V_t(1 - e^{-gt_i}) - v_i \right\}^2 . \tag{6}$$

The minimum state is attained when $\frac{\partial L}{\partial V_t} = \frac{\partial L}{\partial g} = 0$ is satisfied. After eliminating V_t, the following equation is obtained for g:

$$\sum v_i(1-e^{-gt_i}) \sum t_i e^{-gt_i}(1-e^{-gt_i}) - \sum (1-e^{-gt_i})^2 \sum v_i t_i e^{-gt_i} = 0 . \tag{7}$$

The above equation is solved by Marquardt method (Marquardt 1963). The least-squared V_t is given as follows:

$$V_t = \sum v_i(1 - e^{-gt_i}) \Big/ \sum (1 - e^{-gt_i})^2 . \tag{8}$$

Figure 5 shows some typical results of travelling speed in x direction (dotted line), covering from initial propulsive stage to steady stage. The LSP is applied to the sections depicted by thick solid line ($1 \leq t \leq 6$), and the terminal velocities are predicted with N=100 data points sampled at 0.05 unit-time intervals. In the present

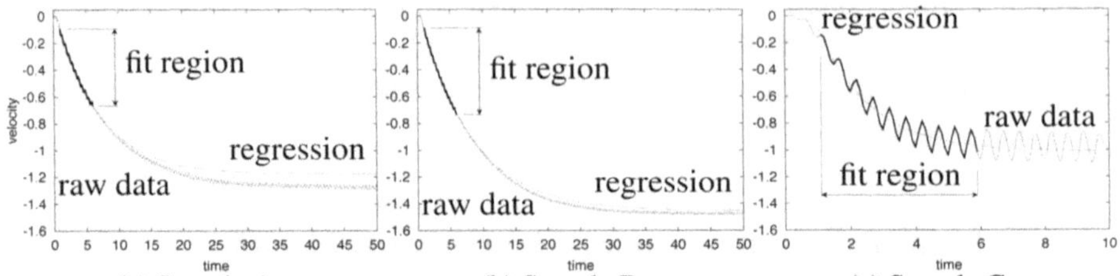

(a) Sample A (b) Sample B (c) Sample C

Fig. 5. Typical velocity diagrams and fitting results with regression equation (5). Dotted line: raw travelling speed in x direction, thick solid line: regression region of LSP, thin solid line: regression curve.

case, the data points are shifted by $(-t_1, -v_1)$ so that the fittings always start from the origin.

The regression curves are shown in the same figure by thin solid line. The maximum and minimum errors between the predicted and the real time-averaged terminal velocities are found to be 7.5% for Sample A and 1.2% for Sample B, respectively. Here the real terminal velocities are calculated by averaging the instantaneous velocities over 20 periods in the steady stages for the first two samples and over 5 periods for sample C. The terminal velocity for Sample A is clearly underestimated. However, our preliminary study shows that the errors in V_t are negligible as the asymptotic velocities and the decrease rate of the exponential curves become large. Therefore, the LSP provides sufficient accuracy for predicting time-averaged terminal velocities for high-performance samples. Also, it should be noted that the LSP enables significant reduction in the computation time (including the periods for averaging the velocities) for obtaining asymptotic terminal velocities: by 3/5 times for sample C and 1/8 times for samples A and B. This is an important aspect of the LSP as the GA can identify a potential fast swimmer from a small fraction of its swim record.

4.2 Numerical Conditions for GA

The following five parameters are varied in search of the optimised performance for the fish: a, l_f, h, f and k. Each parameter has 16 equally-divided quanta in the respective range including the values studied in the previous section. The ranges are shown in Table 1. Density ratio of the fish to fluid is set to 5. Fit regions of the LSP are restricted within 5 unit-times, and the fitting starts between t=0 and 1 depending on the onset of the exponential increase in the velocity.

Along with the attempt to optimise the time-average terminal velocity, the GA result obtaining the highest travelling speed in a given time is also compared. The

Table 1. Ranges of GA parameters

	a/L	l_f	h/L	f	k
Min.	0.08	0	0.08	0.60	0
Max.	0.23	$0.25L$	0.23	1.35	0.4

Table 2. GA conditions

Number of individuals	8
Number of generations	30
Probability of Crossover	0.3
Probability of Mutation	0.01

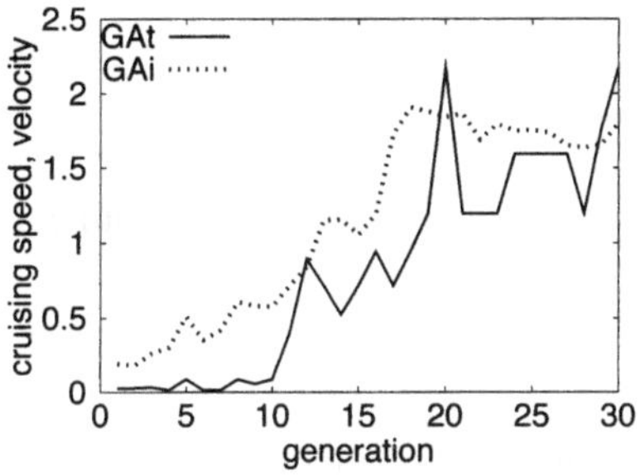

Fig. 6. Progress of travelling speed in GAt and GAi

following two fitness functions, f_t and f_s, are introduced:

GAt : $f_t = V_t$: average terminal cruising speed in $-x$ direction by LSP

GAi : $f_s =$ instantaneous speed in $-x$ direction at $t = 10$

where GAt and GAi are GA session names targeting optimisation of f_t and f_s, respectively. In both cases, a higher fitnesses value means a better performance. The major difference of the two GA sessions is that GAt does not consider the elapsed time to attain the terminal velocity, whereas GAi accounts for acceleration in the initial stage. The conditions used for both sessions are summarised in Table 2. Both GA sessions start with the individuals of the same set of parameters determined randomly. In a preliminary study, we confirmed similar trends of fitness for two different sets of initial random numbers (and in turn different sets of individuals) for two independent GA sessions.

In a GA session, the samples within one generation are evolved in parallel with a parallel computation library. As the LSP provides predictions from a fix-term record for each sample, the samples of better performance are less likely to be kept waiting for the others. This parallelisation policy promises an efficient use of computational resources.

4.3 Results and discussion

Figure 6 compares the progresses in fitness (f_t and f_s) in GAt and GAi. Both GAt and GAi acquire remarkable capabilities of travelling speed as generation goes, except in the generations between 20th to 30th. The reason for the decrease in

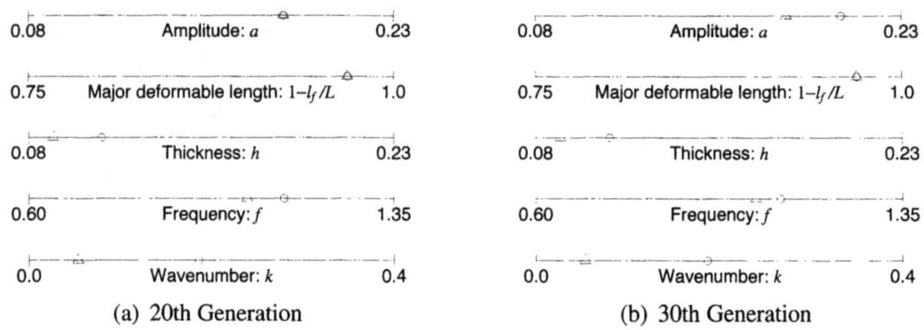

Fig. 7. Comparison of fish parameters at 20th and 30th generations in GAt (○) and GAi (△)

performance in this decade is that the random number generator of the GA failed to hand over the genes of the good performers. However, the GA finally bred again the characters showing roughly the same performance as the 20th generation through its selection mechanism.

GAt and GAi follow the similar growth curves, though GAt exhibits some negative data-shift against GAi. This may be because in the initial generations GAt underestimates the terminal velocity, as observed in Fig.5(a). However, the LSP is shown to improve the accuracy of prediction as the generation descends.

It should be noted that the total computation time for GAt is reduced to about 60% compared to GAi as terminal velocity of each individual is estimated between t=5 and 6 by the LSP.

Figure 7 compares the parameters for GAt and GAi at the 20th and 30th generations. Circular symbol represents GAt, and triangular symbol is GAi. Similar trends are observed for these two generations. Among the optimised parameters of GAt and GAi, a significant difference in wavenumber exists for these generations. As summarised in the previous section, a smaller wavenumber is suitable for obtaining a stronger thrust. Therefore the fishes optimised by GAi show a stronger acceleration in the initial propulsive stage and, so attains a faster travelling speed in a short time. The results of GAt and GAi lead to the following conclusions: to attain a faster travelling speed (whether or not within a finite time), the fish should have a thin body (h) and twist a large part of the entire body ($L - l_f$) at a moderate frequency (f) with a large amplitude (a). Detailed analysis of the effect of each parameter on travelling speed (with higher resolution for the parameters) is the subject of ongoing research by the present authors.

Figure 8 shows variation of the velocity of the fish optimised in the GAt session. The best performing fish attains a terminal velocity of about the same value as predicted by GA with LSP. The results suggest that the combination of heuristic and analytical approaches enables prediction of asymptotic terminal velocity at reduced

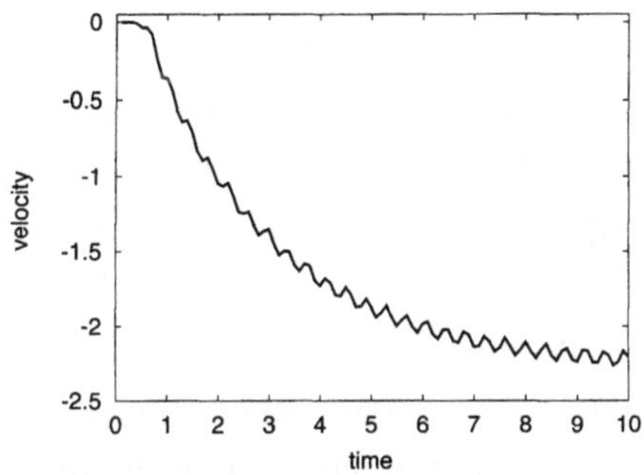

Fig. 8. Velocity evolution of the best performing fish. The parameters at 30th generation of GAt are employed.

computation load.

5 Conclusions

Two dimensional simulation of fully resolved flow field past a fish-like deformable hydrofoil was conducted. Optimised performance of the fish was studied by applying a genetic algorithm (GA) facilitated by a prediction method for the performance.

The results with the fixed parameters show that a fully resolved flow field enables prediction of the detailed flow structures and the effect of the swim mode on the swimming dynamics (including fishes' terminal velocities and orientations).

Fish shape and swim mode were optimised for higher cruising speed by GA in conjunction with a least square prediction (LSP) method. The fishes' velocity profiles were successfully reconstructed, as asymptotic profiles, by the LSP from a small fraction of the swim record of individual fishes. The GA results showed that wavenumber of the undulatory motion was found to have more influence on the fish's accelerations and terminal cruising velocities than the other parameters employed (amplitude, major deformable length, body thickness and frequency).

The snap prediction method from a fixed-term record (LSP) allowed improved compatibility with parallel-processing. In addition, it facilitated the GA greatly in selecting the samples of a higher cruising velocity. Further, the LSP was shown to effectively cut down the turn-around times for prediction of average terminal velocities and, hence, the total computation time for GA.

The present study demonstrated the capability and effectiveness of our combined heuristic and analytical approach, i.e. GA with LSP, in solving the optimisation problem by employing a fully-resolved flow field surrounding a deformable object.

References

Bainbridge R (1958) The speed of swimming of fish as related to the size and to the frequency and amplitude of the tail beat, J. Exp. Biol. 35:109-133

Barrett D (2002) Optimization of swimming locomotion by Genetic Algorithm, NeuroTechnology for Biomimetic Robots, MIT Press, pp.207-221

Carling J, Williams TL, Bowtell G (1998) Self-propelled anguilliform swimming: simultaneous solution of the two-dimensional Navier-Stokes equations and Newtons' laws of motion, J. Exp. Biol. 201:3143–3166

Kuo PD, Grierson D (2003) Genetic Algorithm optimization of escape and normal swimming gaits for a hydrodynamical model of carangiform locomotion, Genetic and Evolutionary Comp. Conf. Late Breaking Papers, Chicago, USA, pp.170-177

Hirt CW, Amsden AA, Cook JI (1974) An Arbitrary Lagrangian-Eulerian Computing Method for All Flow Speeds, J. Comp. Phys. 14:227-253

Kusuda S, Takeuchi S, Kajishima T (2006) 2-D Analysis of Fish Locomotion and Optimization by Genetic Algorithm, Proc. 3rd International Symposium on Aero Aqua Bio-Mechanics, Okinawa, Japan, Paper No.202

Kusuda S, Takeuchi S, Kajishima T (2007, *To Appear*) Genetic Algorithm Optimisation of Fish Shape and Swim Mode in Fully-Resolved Flow Field, Parallel Computational Fluid Dynamics, Elsevier Science

Liu H, Wassersug RJ, Kawachi K (1996) A computational fluid dynamics study of tadpole swimming, J. Exp. Biol. 199:1245-1260

Marquardt DW (1963) An Algorithm for Least-Squares Estimation of Nonlinear Parameters, J. Soc. Indust. Appl. Math. 11-2:431-441

Michalewicz Z (1992) Genetic Algorithm, Springer-Verlarg

Nakamura Y, Jia W, Mizuno D (1993) A Numerical Method of Incompressible Flow around Arbitrarily Moving Body, Trans. JSASS 41-469:40-47

Smith IM, Griffiths DV (2004) Programming the Finite Element Method (4th edn), Wiley

Smith RW, Wright JA (2004) Simulation of *Robo Tuna* Fluid Dynamics using a New Incompressible ALE method, 84th AIAA Fluid Dynamics Conference and Exhibit, Portland, US, Paper No.2004-2347

Takeuchi S, Yamazaki T, Kajishima T (2006) Study of Solid-Fluid Interaction in Body-Fixed Non-Inertial Frame of Reference, J. Fluid Sci. and Tech. 1-1:1-11

Yu J, Tan M, Wang S, Chen E (2004) Development of a Biomimetic Robotic Fish and Its Control Algorithm, IEEE Trans. Sys. Man Cybernetics B, 34-4:1798-1810

Chapter 14

Modeling, Simulation and Optimization of Anguilliform Swimmers

Stefan Kern, Philippe Chatelain, and Petros Koumoutsakos

Computational Science and Engineering Laboratory, ETH Zurich, Universitaetstrasse 6, CH-8092, Switzerland

Summary. We present a computational framework for the modeling, simulation and evolutionary optimization of anguilliform swimmers. The hydrodynamics of anguilliform swimming is investigated by solving the three-dimensional Navier-Stokes equations of incompressible flow past a self-propelled body. An inverse design procedure is employed in order to identify optimal swimming behavior. In this procedure, objectives such as swimming efficiency and burst swimming speed are pre-specified and the motion of the body is obtained through an evolutionary algorithm. The results support the hypothesis that anguilliform swimmers modify their kinematics according to different objectives.
The present simulations provide a detailed quantitative analysis of the swimming motion and the forces experienced by the body. The optimization of swimming speed is associated with large tail motions and a straight anterior part of the body. In contrast, efficient swimming is associated with significant lateral undulations along the entire length of the body. The wake of the investigated swimming modes, consists largely of a double row of vortex rings with an axis aligned with the swimming direction. The vortex rings are responsible for producing lateral jets of fluid, which has been documented in prior experimental studies. Comparison with a trout-like swimming pattern reveals that anguilliform swimming has superior efficiency for the same body geometry and flow regime.

Key words. Anguilliform swimming, hydrodynamics, fluid-body interaction, self-propelled, optimization, evolutionary algorithms.

1 Introduction

Anguilliform swimming is the primary mode of locomotion for numerous aquatic species across a range of diverse taxa. Anguilliform swimmers propel themselves forward by propagating waves of curvature towards the posterior of the body and this type of locomotion is widespread among species, ranging in scales from nematodes to eels (Müller et al., 2001). Starting from the pioneering work of Gray in 1933 (Gray, 1933), anguilliform swimming has attracted the attention of researchers from diverse scientific fields, ranging from neuroscience to hydrodynamics (Ekeberg, 1993; Tytell and Lauder, 2004). A number of experimental studies using Particle Image Velocimetry (Müller et al., 2001, Tytell, 2004; Tytell and Lauder, 2004) reported on swimming kinematics and flow characteristics. Computational studies of anguilliform swimming include simulations with prescribed motion patterns (Carling et al., 1998, Liu and Kawachi, 1999, Leroyer and Visonneau, 2005) and prescribed bending moments (Pedley and Hill, 1999) generating the undulating swimming motion. A comprehensive survey of the hydrodynamics of fishlike swimming can be found in (Triantafyllou et al., 2000).

The goal of this study is to investigate anguilliform swimming using three-dimensional simulations of a self-propelled eel-like body immersed in a viscous fluid. The present simulations are coupled with an optimization procedure in order to test the hypothesis that distinct motion kinematics and resulting wake patterns correspond to different swimming modes (Müller et al., 2001). In this work, the fish motion is not a-priori specified. Instead, the motion parameters are obtained by optimizing objective functions corresponding to fast and efficient swimming, using an evolutionary optimization algorithm. The simulations provide detailed information of the complete flowfield enabling the quantification of the vortex formation and shedding process and allow comparisons with related experimental works (Müller et al., 2001; Tytell, 2004; Tytell and Lauder, 2004). In addition, the simulations enable the identification of the force distribution along the self-propelled body and their link with the kinematics of the body and the vorticity dynamics of the wake. It is demonstrated that distinct swimming kinematics and body force distribution correspond to different swimming modes. A comparison with a trout-like swimming pattern reveals superior efficiency of the anguilliform swimming modes for the given body geometry and flow regime.

2 Methods

The computational model solves the three-dimensional Navier-Stokes equations for the incompressible, viscous flow past a deforming, anguilliform body. The time dependent motion of the body is defined by the two-dimensional deformation of its mid-line, allowing for yaw, but no pitch or roll motions. The fluid-body interaction includes forces in all three coordinate directions and the torque perpendicular to the plane of the deformation of the mid-line (yaw). The reader is referred to Kern and Koumoutsakos (2006) for a complete description of the computational setup.

2.1 Geometrical Model and Parameterization of the Motion

The three-dimensional geometry of the anguilliform swimmer with body length L is constructed from spatially varying ellipsoid cross sections. The length of the two half axis $w(s)$ and $h(s)$ of the cross sections are defined as analytical functions of the arc length s along the mid-line of the body:

$$w(s) = \begin{cases} \sqrt{2w_h s - s^2} & 0 \le s < s_b \\ w_h - (w_h - w_t)\left[(s - s_t)/(s_t - s_b)\right]^2 & s_b \le s < s_t \\ w_t\,(L - s)/(L - s_t) & s_t \le s \le L \end{cases} \tag{1}$$

$$h(s) = b\sqrt{1 - \left[(s - a)/a\right]^2} \tag{2}$$

The parameters are set to w_h=s_b=0.04L, s_t=0.95L, w_t=0.01L, a=0.51L, and b=0.08L.

The motion of the body is described by the time dependent curvature κ of the mid-line of the body defined as

$$\kappa(s,t) = K(s)\cdot\sin\left[2\pi\left(t/T - \tau(s)\right)\right], \tag{3}$$

where $K(s)$ is the cubic spline through the m interpolation points K_i, i=1,…,m, $\tau(s)$ is the phase shift along the body, and T is the cycle time. In the current study we choose the phase shift $\tau(s)$ to be linearly proportional to the position s along the body, i.e. $\tau(s) = \tau_{\text{tail}}\, s/L$.

We have tried to maintain a minimum number of parameters and found that m=4 interpolation points, evenly distributed along s combined with a linear phase shift suffice to allow a wide range of motion patterns.

2.2 Optimization

The 6 parameters $K_{1\ldots4}$, τ_{tail}, and T define a realization of a motion pattern according to Eqn 3 and 4 and are obtained through an optimization process. We implement an Evolution Strategy with Adaptation of the Covariance Matrix (CMA-ES) (Hansen and Kern, 2004; Hansen and Ostermeier, 2001). Three different objectives have been optimized that may correspond to biological functions such as hunting/escaping (for the burst velocity) or migrating (for the efficient swimming). The corresponding objective functions are described in the following.

2.2.1 Maximize swimming efficiency for prescribed undulation frequency

We relate optimizing swimming efficiency η to maximizing the ratio between the kinetic energy of the forward motion and the work per swimming cycle. We define

$$f_\eta = \frac{m\bar{U}^2}{2W_{\text{cycle}}}, \; W_{\text{cycle}} = \int_t^{t+T} \oint_S -\bar{\bar{\boldsymbol{\sigma}}} \cdot \mathbf{n} \cdot \mathbf{u} \, dS \, dt, \tag{4}$$

where m is the total mass of the body, $\bar{U}$ the mean asymptotic swimming velocity, $\bar{\bar{\boldsymbol{\sigma}}}$ the viscous stress tensor with $\sigma_{ij}=-p\delta_{ij}+\tau_{ij}$, $\mathbf{n}$ the outer surface normal vector, and $\mathbf{u}$ the velocity of the moving surface S. f_η is maximized and $f=1/T=1$ is kept constant. Note that classical Froude efficiency is not applicable to a self-propelled body as drag and thrust cancel out in the asymptotic swimming state.

2.2.2 Maximize swimming efficiency for prescribed swimming speed

The mean total input power $\bar{P}_{\text{tot}}$ needed to swim at a prescribed speed $\bar{U}_{\text{P}}$ is minimized. We define

$$f_P = \bar{P}_{\text{tot}} + C_U \cdot (\bar{U}_{\text{P}} - \bar{U})^2, \; \bar{P}_{\text{tot}} = \tfrac{1}{T} \int_t^{t+T} \oint_S -\bar{\bar{\boldsymbol{\sigma}}} \cdot \mathbf{n} \cdot \mathbf{u} \, dS \, dt, \tag{5}$$

where $C_U=10.0$ is a scaling factor for the quadratic penalty term; f_P is minimized. We set $\bar{U}_{\text{P}}$ to the swimming speed obtained in the optimization of swimming efficiency with prescribed undulation frequency.

2.2.3 Maximize swimming speed for prescribed undulation frequency and constrained input power

The objective function for burst swimming is defined as the sum of the mean swimming velocity $\bar{U}$ and two terms that penalize high mean and peak input power requirements motivated by natural physiological limits. We define

$$f_U = \bar{U} - C_1 \cdot \theta(\bar{P}_{tot} - \bar{P}_{\max})(\bar{P}_{tot} - \bar{P}_{\max})^2 - C_2 \cdot \theta(P_{tot} - P_{\max})(P_{tot} - P_{\max})^2, \tag{6}$$

where $\theta(.)$ is the Heaviside function, $C_1=10^6$, $C_2=10^8$, $\bar{P}_{\max} = 2 \cdot 10^{-3}$, and $P_{\max} = 4.8 \cdot 10^{-3}$.

The maximum curvature of the mid-line was constrained in all optimization cases, in the interval [0,2π] in order to prevent too large deformations of the body

2.3 Equations and numerical method

The system of the deforming body interacting with the surrounding fluid is described by the 3D incompressible Navier-Stokes (NS) equations:

$$\frac{D\mathbf{u}}{Dt} = -\frac{1}{\rho}\nabla p + \nu\nabla^2\mathbf{u}\,,\ \nabla\cdot\mathbf{u} = 0\,, \tag{7}$$

where $D/Dt=\partial/\partial t+\mathbf{u}\cdot\nabla$ and ν denotes the kinematic viscosity of the fluid. These equations are solved along with the no-slip boundary condition on the body surface S, $\mathbf{u}|_S = \dot{\mathbf{x}}_S$ coupled with Newton's equations of motion for the self-propelled body

$$m\ddot{\mathbf{x}}_c = \mathbf{F}\,,\ \dot{I}_z\dot{\phi}_c + I_z\ddot{\phi}_c = M_z\,. \tag{8}$$

where $\mathbf{F}$ and M_z are the fluid force and yaw torque acting on the body surface, $\mathbf{x}_c$ is the position of the center of mass of the body, $\dot{\phi}_c$ its global angular velocity, and I_z the (time dependent) inertial moment about the yaw axis. The feedback of the fluid torque is limited to the yaw direction to simplify computations. The fluid force $\mathbf{F}$ and the torque M_z are computed as follows:

$$\mathbf{F} = \oint_S \bar{\bar{\sigma}}\cdot\mathbf{n}\,dS\,,\ M_z = \oint_S \Big(\big(\bar{\bar{\sigma}}\cdot\mathbf{n}\big)\times(\mathbf{x}-\mathbf{x}_c)\Big)\mathbf{e}_z\,dS \tag{9}$$

The feedback from the body to the fluid is realized by imposing no slip boundary conditions on the moving body surface. The far field boundary condition is set to a constant static pressure, modeling the fish propelling itself through an infinite tank of still fluid. The NS equations were solved using the Finite Volume package STAR-CD v. 3.15 that computes flows using arbitrary Lagrangian-Eulerian grids. For the fluid domain, a deforming structured grid with 3 10^5 cells in a cylindrical domain with a length of of 8L and a radius of 2L is used. The domain moves with the center of mass of the body. A detailed description of the approach, validation, and convergence study is provided in Kern and Koumoutsakos (2006).

The simulations are conducted in non-dimensional units with body length L=1, viscosity μ=1.4 10^{-4}, and density of fluid and body ρ=1. The resulting Reynolds number Re = ρL[redacted]/μ is in the range of 2400-3900 depending on the swimming mode. The time step used is Δt=5·10^{-4}. All simulations are started with the body at rest, and the motion is initialized by gradually increasing the curvature amplitude from $K(s) \equiv 0$ to its designated value during the first cycle.

3 Results

We present swimming kinematics, hydrodynamic forces, and wake morphology for the swimming patterns obtained using the three different objective functions f_η, f_P, and f_U. In addition, results for a trout-like motion pattern are presented where the undulation frequency is adjusted to match the swimming speed of the motion optimized for fast swimming.

Table 1 summarizes the optimized motion parameters and the kinematics of the resulting motions. The evolution of the unsteady longitudinal and lateral velocities is shown in Fig. 1. The motion pattern obtained using f_η resulted in an asymptotic swimming velocity of $\bar{U}$=0.33 in turn used as desired swimming velocity in f_P. The characteristics of the different swimming patterns become evident by the amplitude envelopes of the body plotted in Fig. 2. Efficient swimming involves a coordinated undulation of the whole body while the fast swimming motion is characterized by an increased tail amplitude and a reduced bending in the anterior part of the body. The trout-like motion amounts to relatively small amplitude undulations of the tail.

Table 1. Optimized[1] motion parameters, number of objective function evaluations needed in the optimization process n_{eval}, asymptotic swimming velocity $\bar{U}$, wave velocity V, slip $\bar{U}/V$, tail amplitude A, Strouhal number $St=2fA/\bar{U}$, and input power $\bar{P}_{tot}$ and $\bar{C}_P = 2\bar{P}_{tot}/(\rho\bar{U}^3 S)$ for the four investigated swimming motions. V is computed as the mean velocity of the zero crossings of the body mid-line.

	Efficient (f_η)	Efficient (f_P)	Fast (f_U)	Trout-like
$K_{1..4}$	[4.37,2.22,6.07,3.07]	[2.06,1.49,6.28,6.18]	[1.29,0.52,5.43,4.28]	*[0.25,0.50,1.50,3.80]*
τ_{tail}, T	1.71, *1.0*	1.69, 1.06	1.52, *1.0*	*1.0*, 0.48
n_{evals}	460	1485	460	-
$\bar{U}$	0.33	0.32	0.47	0.46
V	0.60	0.59	0.74	1.61
$\bar{U}/V$	0.55	0.54	0.63	0.29
A	0.11	0.11	0.14	0.08
St	0.67	0.65	0.59	0.75
$\bar{P}_{tot}$	$8.7\cdot10^{-4}$	$8.1\cdot10^{-4}$	$2.1\cdot10^{-3}$	$2.8\cdot10^{-3}$
$\bar{C}_P$	0.153	0.162	0.124	0.189

[1] The parameters not included in the optimization are in *italics*.

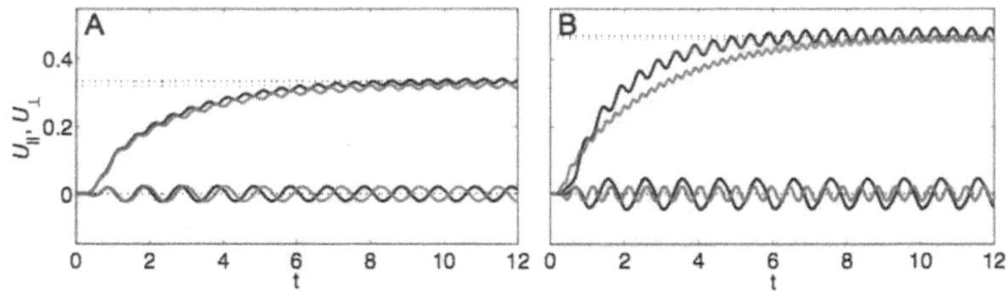

Fig. 1. Unsteady longitudinal and lateral velocities for A: the efficient swimming cases using f_η (black) and f_P (grey), and B: the fast swimming f_U (black), and the trout-like motion (grey).

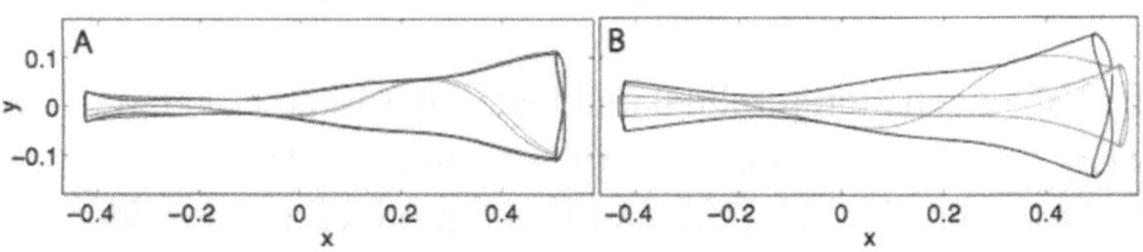

Fig. 2. Amplitude envelopes of the body mid-line including the feedback from the fluid forces of A: efficient swimming using f_η (black) and f_P (grey), and B: fast swimming (black), and trout-like motion (grey).

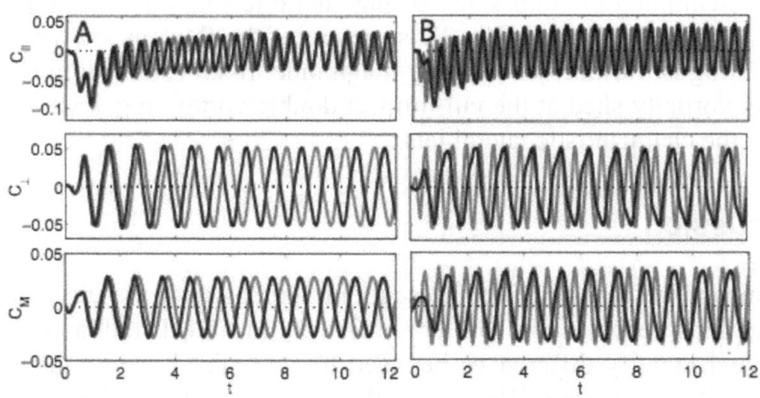

Fig. 3. (A) Unsteady longitudinal and lateral force and moment coefficients $C_{||}$, $C_{\perp}$, and C_M for the efficient swimming cases using f_η (black) and f_P (grey), and (B) $C_{||}$, $C_{\perp}$, and C_M for the fast swimming f_U (black), and the trout-like motion (grey).

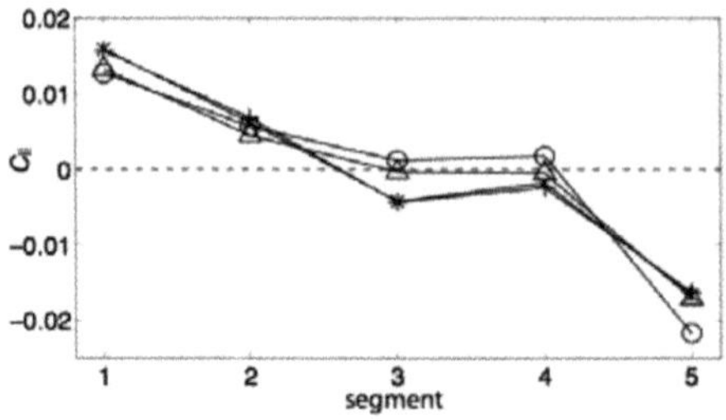

Fig. 4. Time averaged fluid forces in swimming direction acting on the fish surface for the five segments from head (1) to tail (5) for the efficient swimming motion using f_η (+), efficient swimming using f_P (×), fast swimming (Δ), and trout swimming motion (O). Positive values relate to drag, negative to thrust.

Figure 3 shows the longitudinal and lateral net force $C_{\parallel}=2F_{\parallel}/(\rho\bar{U}^2S)$ and $C_{\perp}=2F_{\perp}/(\rho\bar{U}^2S)$, and the moment on the body $C_M=2M_z/(\rho\bar{U}^2LS)$. Regions of the body responsible for drag and thrust production were identified by analyzing the longitudinal forces on five segments of equal length from head to tail (c.f. Fig. 4).

The flow fields of the different swimming motions are visualized in Figs. 5 and 6. Fig. 5 shows the vorticity normal to the image plane and the velocity components in plane for cross sections at mid-height of the body. All four swimming motions show strong lateral jets in the wake, separated by two same-sign vortices. The 3D structure of the flows is shown in Fig. 6 by plotting isosurfaces of vorticity magnitude. In all four swimming motions the vorticity shed at the tail forms a double vortex ring wake responsible for the characteristic lateral jets.

4 Discussion

We compare the results of the present simulations with the experimental investigations reported by Müller et al. (2001) and Tytell and Lauder (2004) and identify differences between the optimized swimming modes and trout-like swimming. We remark that the Reynolds number in the experimental studies is a factor of 5-30 higher than in our simulations. The good qualitative agreement of the wake structure obtained in our simulations with the reported flow patterns in these studies justifies, in our opinion, a comparison of the results.

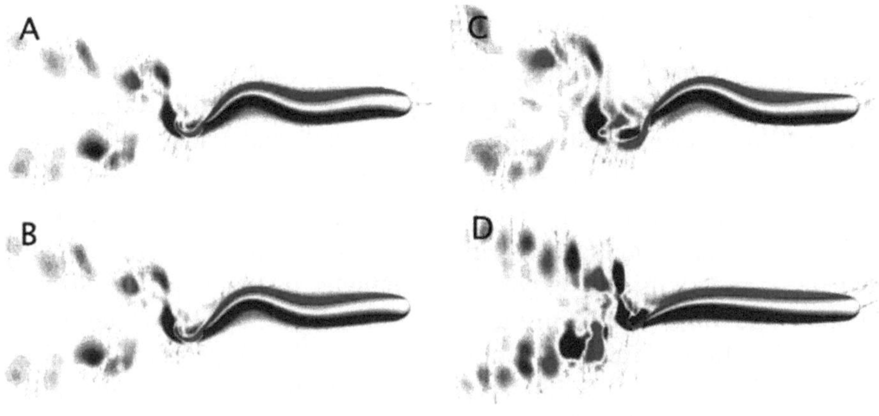

Fig. 5. Velocity field and vorticity normal to the image plane $\omega_z = \boldsymbol{\omega} \cdot \mathbf{e}_z$ for (A) efficient swimming using f_η, (B) efficient swimming using f_P, (C) fast swimming and (D) trout-like swimming.

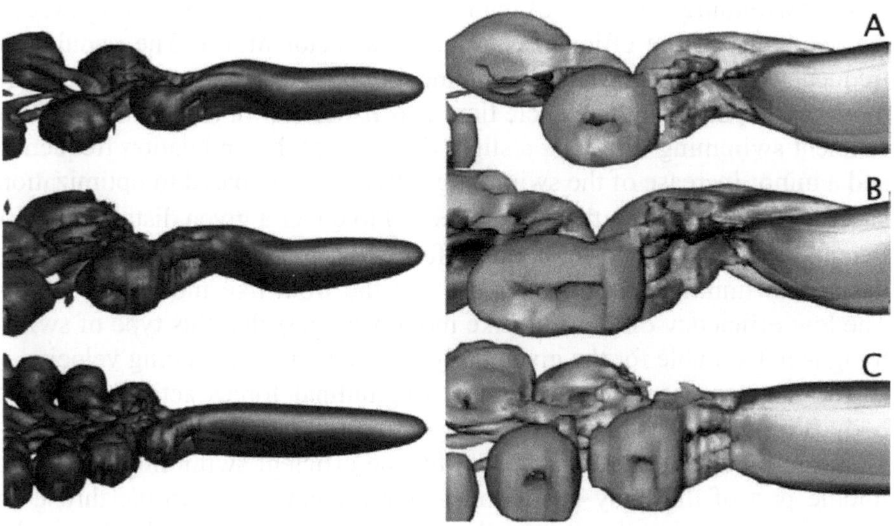

Fig. 6. Left: Snapshot of iso-surfaces of vorticity magnitude $||\boldsymbol{\omega}||=2$ for (A) efficient swimming using f_P, (B) fast swimming, and (C) trout-like motion after the body has reached its asymptotic mean swimming velocity. Right: Iso-surfaces of vorticity magnitude $||\boldsymbol{\omega}||=2$ colored by the local value $\omega_y = \boldsymbol{\omega} \cdot \mathbf{e}_y$ of for (A) efficient swimming using f_P, (B) fast swimming, and (C) trout-like motion.

4.1 Swimming kinematics and hydrodynamic forces

The mean swimming velocity obtained in the present simulations lies well within the range of 0.26-0.5 body length cycle^{-1} reported in experimental studies (Müller et al., 2001; Tytell, 2004; Tytell and Lauder, 2004). The values for the slip $\bar{U}/V$ of 0.55, 0.55, and 0.63 obtained for the two efficient and the fast swimming mode respectively, are slightly below the experimental values ranging between 0.6 and 0.73. The largest discrepancy between simulations and experimental studies is found in the Strouhal number. The values for St in our simulations lie clearly above the values reported in Müller et al. (2001) and Tytell (2004) and the range of 0.2-0.4 identified for efficient oscillatory propulsion of swimming and flying animals (Taylor et al., 2003). This discrepancy may be attributed to the reduced Reynolds number and will be addressed in future investigations.

Force and input power comparisons with experiments are difficult due to the limited data available. In Tytell (2004) a value of $\bar{C}_P = 0.038$ is computed based on two-dimensional PIV data, which is a factor 3-5 smaller than the values obtained in the present simulations.

The swimming velocities of the fast and the trout-like swimming exceed the velocities of the efficient patterns by a factor of 1.4. The undulation frequency for the trout-like swimming is doubled compared to the optimized cases. Including the cycle time T as parameter in the optimization of efficient swimming results in a slight decrease of the undulation frequency and a minor increase of the swimming efficiency compared to optimization based on f_η. Computing the energy needed to cover a given distance the efficient swimming based on f_P outperforms the motion based on f_η by 4%, the fast swimming motion by 77%, and the trout-like motion by 141%. The low efficiency of the trout-like motion suggest that this type of swimming is not suitable for the given body geometry and swimming velocity.

The segmentwise analysis of the longitudinal forces acting along the body (Fig. 5) reveals that for the fast swimming mode thrust is almost exclusively produced at the tail, while for the efficient swimming mode, the middle part of the body has a considerable contribution to the thrust. In trout-like swimming thrust is exclusively produced at the tail. The results for the efficient swimming pattern support the hypothesis (Blickhan et al., 1992; Müller et al., 2001), that anguilliform swimmers produce thrust not only with their tail but also with parts of the body anterior to the tail.

4.2 Wake morphology

The wake patterns obtained for our 3D simulations are in good qualitative agreement with 2D PIV measurements of real anguilliform swimmers

(Mueller et al., 2001, Fig 6, p. 2755, Tytell and Lauder, 2004, Fig. 5, p. 1831) and show the characteristic lateral jets reported in these experimental studies. The wake structure and the shedding mechanism are similar for all four simulated swimming patterns. The main differences in the vortical structures are related to the strength of the secondary vortices. The shedding of the double vortex rings is accompanied by complex secondary flow structures that are generated by the transversal motion of the body, convected downstream, and eventually shed in the wake. In the fast swimming case the secondary structures are much stronger than in the efficient cases and the vortex rings are stretched in the swimming direction, forming jets that are 1.4 times wider than the ones observed in efficient swimming. The short distances in swimming direction of the vortex rings for the trout-like swimming is also reflected in the very small value of $\bar{U}/V$=0.29.

We postulate that in addition to the motion pattern, the body geometry and the Reynolds number may have a considerable influence on the formation and shedding of the wake. Continuous caudal fins observed in many species of anguilliform swimmers presumably serve to control the formation of the secondary vortex structures. Simulations with increased Reynolds number and investigations of the influence of the body geometry (back and anal fins, etc.) and body motion will help to clarify this issue.

4.3 Concluding remarks

We have presented 3D simulations of optimized self-propelled anguilliform swimming in a viscous incompressible fluid. An inverse design procedure is employed in order to identify optimal swimming patterns: the swimming motions are not described a-priori but are obtained through an evolutionary optimization. This enables a systematic exploration of the motion parameter space and the identification of distinct swimming modes corresponding to different objective functions. In fast swimming, strong tail motions have been observed while efficient swimming involves a harmonious undulation of the whole body. The swimming kinematics and wake structures obtained in the present simulations agree well with experimental data and further elucidate the mechanisms of anguilliform swimming. The interplay of body shape and motion was further investigated by considering a trout-like motion applied to the eel-like body. The results indicated a largely reduced swimming efficiency compared to the optimized anguilliform motions.

We wish to acknowledge enlightening discussions with Ulrike Müller. Financial support of the Swiss National Science Foundation is acknowledged.

References

Blickhan R, Krick C, Zehren D, Nachtigall W and Breithaupt T (1992) Generation of a vortex chain in the wake of a subundulatory swimmer. Naturwissenschaften 79:220-221

Carling J, Williams TL, and Bowtell G (1998) Self-propelled anguilliform swimming: simultaneous solution of the two-dimensional Navier-Stokes equations and Newtons' laws of motion. JEB 201:3143-3166

Ekeberg O (1993) A combined neuronal and mechanical model of fish swimming. Biological Cybernetics 69:363-374

Gray J (1933) Studies in animal locomotion I. The movement of fish with special reference to the eel. JEB 10:88-104

Hansen N and Kern S (2004) Evaluating the CMA evolution strategy on multimodal test functions. In Parallel Problem Solving from Nature - PPSN VIII, vol. 3242, pp. 282-291

Hansen N and Ostermeier A (2001) Completely derandomized self-adaptation in Evolution Strategies. EC 9:59-195

Kern S and Koumoutsakos P (2006) Simulations of optimized anguilliform swimming. JEB 209:4841-4857

Lauder GV and Tytell ED (2006) Hydrodynamics of undulatory propulsion. Fish Biomechanics 23:425-468

Leroyer A and Visonneau M (2005) Numerical methods for RANSE simulations of a self-propelled fish-like body. J. Fluids Struct 20:975-991

Liu H and Kawachi K (1999) A numerical study of undulatory swimming. J. Comput. Physics 150:223-247

Müller UK, Smit J, Stamhuis EJ and Videler JJ (2001) How the body contributes to the wake in undulatory fish swimming: Flow fields of a swimming eel (Anguilla anguilla). JEB 204:2751-2762

Pedley TJ and Hill SJ (1999) Large-amplitude undulatory fish swimming: Fluid mechanics coupled to internal mechanics. JEB 202:3431-3438

Schmidt J (1923) Breeding places and migration of the eel. Nature 111:51-54

Taylor GK, Nudds RL, and Thomas ALR (2003) Flying and swimming animals cruise at a Strouhal number tuned for high power efficiency. Nature 425:707-711

Triantafyllou MS, Triantafyllou GS and Yue DKP (2000) Hydrodynamics of Fish-like Swimming, Annual Review of Fluid Mechanics 32:33-53

Tytell ED and Lauder GV (2004) The hydrodynamics of eel swimming I. Wake structure. JEB 207:1825-1841

Tytell ED (2004) The hydrodynamics of eel swimming II. Effect of swimming speed. JEB 207:3265-3279

van Ginneken V, Antonissen E, Müller UK, Booms R, Eding E, Verreth J, and van den Thillart G (2005) Eel migration to the Sargasso: Remarkably high swimming efficiency and low energy costs. JEB 208:1329-1335

Chapter 15

A Numerical Study of Hovering Aerodynamics in Flapping Insect Flight

Hikaru Aono[1] and Hao Liu[2]

[1] Graduate School of Science and Technology, Chiba University, 1-33 Yayoi-cho, Inage-ku, Chiba 263-8522, Japan
[2] Graduate School of Engineering, Chiba University, 1-33 Yayoi-cho, Inage-ku, Chiba 263-8522, Japan

Key words. Aerodynamics, CFD, Flapping-flying insect, Leading-edge vortex, Reynolds number

Summary. In this paper, the *Re* (Reynolds number)-dependence of the unsteady aerodynamics in a flapping-flying insect was investigated. To this end, we developed a multi-block- and overset-grid-based *in-house* CFD solver and constructed a wing-body geometric and kinematic model based on measurements of a hawkmoth. Insect-flight computations were carried out with the same insect model but under different *Re* conditions that characterize two insects, namely, fruit fly (*Re*=134) and hawkmoth (*Re*=6300), respectively. The computed flow fields at *Re*=6300 exhibited an evident leading-edge vortex (LEV) along with a strong spanwise flow inside it. At *Re*=134, an evident LEV is predicted as well but without the presence of a strong spanwise flow inside it. The main features of the flow fields computed at both *Re* conditions agreed well with previous experimental results. The computed mean lift coefficient at *Re*=134 was considerably smaller than that at *Re*=6300, which was found to result from the relative augment of viscous effect on force generation at lower *Re*.

1 Introduction

Aerodynamic mechanisms underlying insect flapping flights have been addressed both experimentally and numerically on numerous insects varying in size, shape and wing-kinematics, such as a hawkmoth (Ellington et al. 1996; van den Berg and Ellington 1997; Willmott and Ellington 1997; Liu and Kawachi 1998; Liu et al. 1998) and a fruit fly (Dickinson et al. 1999; Ramamurti and Sandberg 2002). These studies arrived at a general agreement that delayed stall, which was featured by a prolonged leading-edge vortex (LEV), was an essential aerodynamic mechanism responsible for augmenting lift production in insect flights. Since insects fly in a wide *Re* range (from ten to several thousands), there exists an interesting issue on how the aerodynamic performance of an insect depends on its wing-body size (or to say, *Re*). Concerning this issue research yet remains rare.

Ellington et al. (1996) visualized an LEV on the wing of a hawkmoth in tethered flight (*Re*=4000). Afterward, Dickinson et al. (1999) revealed a similar LEV in a fly-wing robotic experiment. More recently, the presence of the LEV on the wings of a butterfly in free flight was confirmed by Srygley and Thomas (2002). On the other hand, in recent years, CFD has been widely applied in insect flight studies. CFD-based insect flight simulations not only confirmed the existence of the LEV, but also deepened our understanding of the dynamic structure of the LEV, on a basis of detailed flow field information (Liu and Kawachi 1998; Liu et al. 1998; Wang 2000; Ramamurti and Sandberg 2002).

Recently, to investigate how the aerodynamic performance of a flapping wing changes with *Re*, several computational and experimental studies have been carried out (Wang 2000; Sun and Gang 2003; Birch et al. 2004). Wang (2000) elucidated that at *Re* larger than 150, lift generation was dominated by dynamic pressure rather than by viscous force. Sun and Gang (2003) concluded that the delayed stall was the main high-lift mechanism for insect flights operating in a fairly wide *Re* range (75~4000). Birch et al. (2004) demonstrated that the structure of the LEV at *Re*=120 differed significantly from that at *Re*=1400. Miller and Peskin (2004) proposed that the vortex dynamics around a flapping wing at low *Re* might become recognizably different from at high *Re*.

Stable attachment of the LEV on wings has generally been considered to benefit from the presence of a spanwise flow which drains energy from the vortex core into the tip vortex in large-insect flights (Ellington et al. 1996; Liu et al. 1998). Usherwood and Ellington (2002) confirmed stable attachment of the LEV at both high (5000) and low (120) *Re* but without a detailed analysis of the transient spanwise flow. It seems like that the spanwise flow only contributes to LEV stability at high *Re*.

The above-mentioned studies provided significant insights into aerodynamic mechanisms underlying insect flights. However, the majority of the numerical studies were limited to 2D (two-dimensional) computation, and the experiments were generally lacking in sufficient precision in visualizing flow field. For this reason, the detailed 3D (three-dimensional) dynamic structure of the LEV and the spanwise flow generated by flapping wings in insect flights is yet incompletely understood.

The purpose of this study is to investigate the dependency of aerodynamic performance of insect flight on insect size (*Re*), with special emphasis on the structure of the LEV and the spanwise flow. To this end, we develop a multi-block- and overset-grid based *in-house* CFD solver, and construct wing-body geometric and kinematic models based on measurements of a hawkmoth. Herein, we focus on fruit fly (fly at a characteristic *Re* of 134 with a wing length of 0.3cm) and hawkmoth (fly at a characteristic *Re* of 6300 with a wing length of 5.0cm), which represent two typical insects most popularly investigated in the field of insect flight. Some of the computed results are compared with the relevant experimental data or computed results reported previously. Based on the detailed flow field information offered by computations, we make an extensive discuss on the change in the LEV structure and the spanwise flow with *Re*. Moreover, in two *Re* conditions, we analyze quantitatively the relative contributions of inviscid and viscous terms to aerodynamic force as well.

2 Materials and Methods

2.1 Computational geometric model of a hawkmoth

The geometric modeling procedure consists of the *imaging* to digitally process images of an object; the *segmentation* to extract the object configuration in terms of wire frame and/or skeleton model; the *smoothing and curve/surface fitting* for the reconstructed geometric model; and the *modeling and domain discretization*, which involve surface and/or volume rendering of the reconstructed object as well as the computational domain decomposition in terms of grids. With these methods, a realistic geometric model of the hawkmoth, *Agrius convolvuli*, with a body length of about 5.0

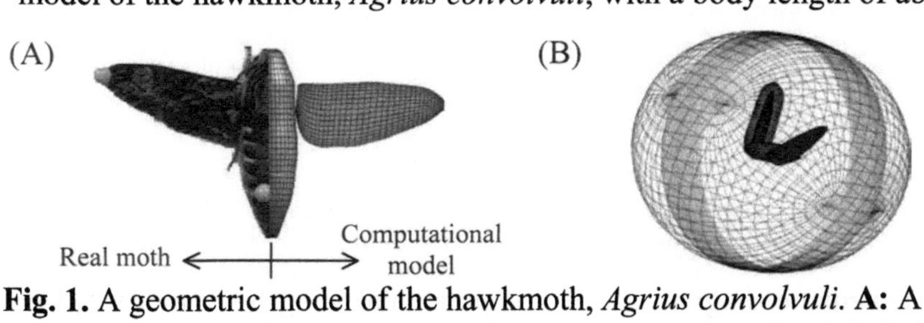

Fig. 1. A geometric model of the hawkmoth, *Agrius convolvuli*. **A:** A hawkmoth with a computational model superimposed on the right half. **B:** A multi-block grid system of the two wings and body of the hawkmoth (wing: $45\times45\times31$, body: $45\times47\times95$).

cm, and a wing length of 5.0 cm is constructed (figure 1A). Each cross section of the wing and the body is then defined to have an elliptic configuration by using the two cross-sectional radii as well as the wing- and body-axis. The thickness of the wing assumes a uniform, which is set to be 1.2% of the mean chord length as observed in most hawkmoths. More details on the methodology of geometrical modeling can be found in the reference (Liu and Kawachi 1998)

2.2 Computational wing-body kinematic model of a hawkmoth

The kinematic model of a hawkmoth-flapping wing was built on a basis of a Fourier series expression of the kinematics of a hovering hawkmoth, *Manduca sexta* (Willmott and Ellington 1997; Liu and Kawachi 1998). The body-fixed system (x, y, z) illustrated in figure 2A has its origin at the wing base, with the x-axis normal to the stroke plane, the y-axis vertical to

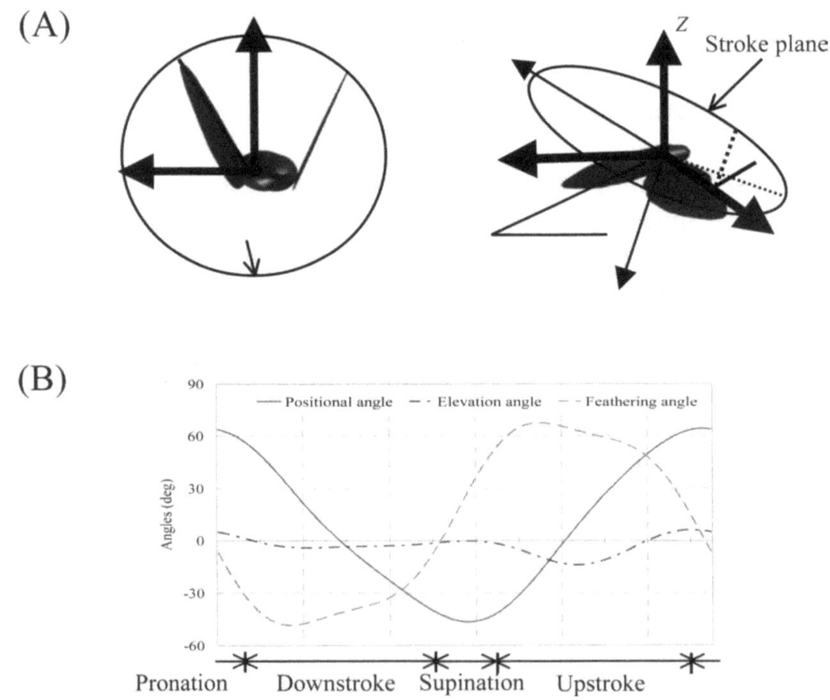

Fig. 2. Schematic diagram of the computational wing-body model. **(A)** Definition of the positional angle (ϕ), the elevation angle (θ), the feathering angle (α), stroke plane angle (β), and body angle (χ); and **(B)** Time courses of the three angles of the wing over a flapping cycle.

the body axis, and the z-axis parallel to the stroke plane. 3D movements of the wing consist of three basic motions (1) flapping about the x-axis in the stroke plane described by the positional angle; (2) rotation of wing, about the z-axis, out of the stroke plane on either side described by the elevation angle; (3) feathering of the wing (leading-edge) with respect to the stroke plane by varying the feathering angle.

Time courses of the three angles of the wing over a flapping cycle are plotted in figure 2B. The parameter K is the reduced frequency defined by $2\pi\, fc_m/2U_{ref}$, where f is wingbeat frequency; c_m is the mean wing chord length (reference length) and U_{ref} is a reference velocity at the wing tip. U_{ref} is defined as $2\Phi Rf$, where Φ is the wingbeat amplitude and R is the wing length. The Reynolds number(Re) is defined as, $Re=c_m U_{ref}/\nu$, where ν is the kinematic viscosity of air, 1.5×10^{-5} m^2sec^{-1}. For the hawkmoth, $c_m=$ 1.83 cm, $R=$ 5.0 cm, $\Phi=$ 2.0 rad (115 *degree*), $f=$ 26.1 sec^{-1}, thus Re is approximately 6300 and K is 0.298. The hawkmoth weight is 1.6 *grams*, body angle (β) is 40 *degree* as well as stroke plane angle (χ) is 15 *degree* (Willmott and Ellington 1997).

2.3 Computational fluid dynamics solver

2.3.1 Single block grid system

The governing equations are the 3D incompressible unsteady Navier-Stokes (NS) equations, written in strong conservative form for momentum and mass. The artificial compressibility method is used by adding a pseudo time derivative of pressure to the equation of continuity. More details of the CFD solver for single grid system can be found in the references (Liu and Kawachi 1998; Liu et al. 1998).

2.3.2 Multi-block grid system

The realistic geometric model of an insect with a body and two or four wings is often of geometric complexity. Since the wings are attached onto the body and flap rapidly, undergoing large-scale movements relative to the body. It is necessary to tackle a challenging CFD problem of highly unsteady vortical flows about multiple and moving bodies. A multi-block and overset-grid method is hereby introduced and incorporated into the *in-house* CFD solver (Liu 2005, in preparation), which utilizes three single, structured grid systems for the body and the two wings, respectively. Each single grid system is generated to fit the object (body or wing), moving and deforming with the object. A tri-linear interpolation technique is employed for communication of velocities and pressures among overlapping grids (Liu 2005, in preparation). As shown in figure 1B, three single structured grids are generated for the body and the two wings of the hawkmoth. The grid fitted to one wing has a dimension of 45×45×31 with its outer boundary 2.5 mean chord lengths away from the wing surface; and the grid for the other wing is simply copied using the relation of geometrical symmetry of the two wings about the body-axis. The grid fitted to the body is taken much larger because it is used as a background grid to envelope the two wing grids for the interpolation; and it has a dimension of 45×47×95 with a distance of approximately 20 mean chord lengths between the body surface and the outer boundary.

2.3.3 Boundary conditions

As shown in figure 1, the solutions to the NS equations with a multi-blocked, overset grid around a flapping-flying insect require appropriate boundary conditions for the overlapping zones among the different single grid block, the moving walls of the wing and body, and the far-field outside boundary. For each single grid block as for a single wing or body we use the fortified solutions to the NS equations by adding a forcing term with communication of a vector $\boldsymbol{q}^*$ to offer the boundary conditions for velocity and pressure in the overlapping zones of the two grids. Inside the computational domain except the holes and the single grid boundary the fortified governing equations are solved. Therefore, for the three single grid blocks as in the two-winged and body model of an insect, the fortified NS equations are solved three times at each time step.

On the body surface, the nonslip condition is used for the velocity components. To incorporate the dynamic effect due to the acceleration of the oscillating body (moving and/or deforming body surface), pressure divergence at the surface stencils is derived from the local momentum equation. For the background grid of the insect body we need to define appropriate boundary conditions at the outside boundary (figure 1B). Consider that, when an insect flies forward at a speed $\boldsymbol{V}_f$, the boundary conditions for the velocity and the pressure may be given such as: 1) at upstream $V(u, v, w)=\boldsymbol{V}_f$ while pressure p is set to zero; 2) at downstream zero-gradient condition is taken for both velocity and pressure.

3 Results and discussion

Computations carry out with the identical hawkmoth wing-body geometric and kinematic model under two *Re* conditions (*Re*=134 and *Re*=6300). The vortical flows around the hovering hawkmoth model are computationally visualized, with some of them compared with relevant experimental data for both *Re* conditions. In particular, the computed flow fields at *Re*=134 are analyzed in detail. Then, the transient and mean lift and drag coefficients are investigated in relation to the flow fields. Finally, the viscosity effect is evaluated in quantitative terms via decomposing aerodynamic forces into inviscid and viscous portions.

3.1 Flow fields around hovering hawkmoths

Figure 3A and B show the computed iso-vorticity surfaces during early and middle downstroke, at *Re*=6300 and *Re*=134, respectively. Correspondingly, figure 3C and D illustrate that the computed spanwise flow velocity contours and the velocity vector maps at the cross section at 60% of the wing length from the wing base during early and middle downstroke. Note that the iso-vorticity surfaces are colored by the absolute magnitude of vorticity.

3.1.1 *Re*=6300

A spiral vortex (LEV) attaches coherently on the leading-edge of the wing during early and middle downstroke (figure 3A). During middle downstroke, a strong LEV and an evident spanwise flow inside the LEV core are predicted (figure 3A and C), as well as a sudden LEV breakdown at approximately 70% of the wing length from the wing base (figure 3 A-2) are observed. The computationally visualized LEV agrees qualitatively with those described in previous works (Liu and Kawachi 1998; Liu et al. 1998; van den Berg and Ellington 1997). The readers are referred to Aono and Liu (2006) for more details on the time-variant LEV structure over a complete flapping cycle.

3.1.2 *Re*=134

Figure 3B and D show that an existence of a stable LEV but without an evident spanwise flow in the LEV core. And other than the LEV at *Re*=6300, the LEV at *Re*=134 remains unbroken during the down- and up-stroke (figure 3B). Moreover, a TV (wing tip vortex) at *Re*=134 seems more stable than that at *Re*=6300. The computed vortical structure in the vicinity of the leading edge is in reasonable agreement with that visualized in experiments (Birch and Dickinson 2001; Birch et al. 2004).

3.1.3 *Re*-dependence of the spanwise flow

As the wings flap considerable flows direct from the wing base to the wing tip above the wings are predicted (figure 3C and D). From an extensive point of view, these flows can be termed general spanwise flows. Herein, what we are really interested in is the spanwise flow present at the core of the LEV allowing for the potential role of the spanwise flow in stabilizing LEV (van den Berg and Ellington 1997; Liu et al. 1998). The computed results illustrate in figure 3C and D indicate that the general spanwise

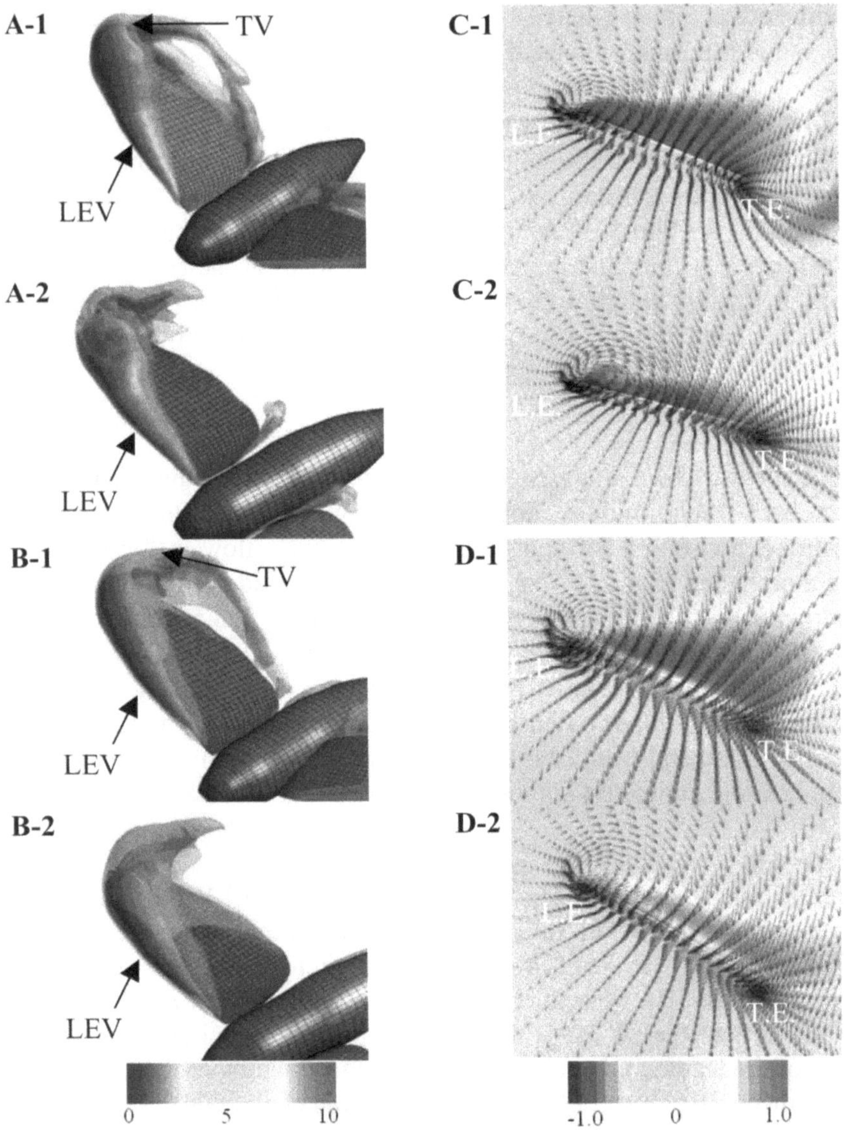

Fig. 3. Computed flow fields around a hovering hawkmoth during early and middle downstroke. (A, B) The LEV and the TV at *Re*=6300 and *Re*=134, respectively. (C, D) The spanwise flow velocity contours and the velocity vector maps at the cross section at 60% of wing length from wing base during early and middle downstroke at *Re*=6300 and at *Re*=134, respectively. Abbreviations: L.E., leading-edge; T.E., trailing-edge.

flows are not significantly different between *Re*=6300 and *Re*=134 in terms of flow intensity and distributing pattern; while the spanwise flows inside the LEVs differ considerably between the two *Re* conditions. At higher *Re* (6300), the spanwise flow emerges at the core of the LEV in the early stage of the downstroke and then enhances continuously following the growth of the LEV till middle downstroke. At lower *Re* (134), however, no identifiable spanwise flow is observed inside the LEV even over a complete flapping cycle (Birch and Dickinson 2001). It has been speculated that the spanwise flow, which drains energy from the LEV to the TV, might make for the stable attachment of the LEV on the wing thereby enhancing the delayed stall (Ellington et al. 1996; Liu et al. 1998). Given that the speculation is reasonable, according to our computations, the role of the spanwise flow in maintaining the stability of the LEV seems to operate only in the cases of higher *Re*. In the case of lower *Re*, despite of the absence of the spanwise flow, the LEV is quite stable without breaking down and shedding during the down- and up-stroke. Therefore, we consider that spanwise flow is not an indispensable factor for stabilizing LEV, it becomes necessary only when the operating *Re* of an insect is higher enough to yield strong LEVs that contain excessive energy. In fact, a comparison between the computed vorticities of the LEVs at the two *Re* conditions (*Re*=6300 and *Re*=134) indicates that the vorticity at *Re*=6300 is much stronger than that at *Re*=134 (figure 3).

3.2 Evaluation of aerodynamic force generation

3.2.1 Aerodynamic force

The computed lift and drag coefficients of a wing are plotted as functions of time in figure 4A-1 and B-1, respectively. Note that the lift and drag coefficients of the body are not included because their values are relatively small and hence negligible.

In both *Re* conditions, the waveforms of the time-variant lift and drag coefficients are basically similar over a flapping cycle. Two peaks of lift coefficient are predicted during each stroke. The lift peaks in a flapping cycle occur at almost the same timings (early down- and up-stroke) for the two *Re* conditions. Recalling the computed flow fields shown in figure 3A and B, we consider that the first lift peak is associated with the delayed stall. Such a viewpoint has also been held by other researches (Ellington et al. 1996; Liu et al 1998; Sun and Gang 2003). During the supination and the pronation, an evident valley of lift coefficient is predicted for both *Re* conditions (see figure 4A-1), which is led to by the shedding of the LEV.

The overall magnitude of the time-variant lift coefficient at *Re*=134 is slightly smaller than that at *Re*=6300, which is thought to result from the different vortical structures, especially the LEV, in the vicinity of the flapping wings at different *Re* (see figure 3A and B). By contrast, the overall magnitude of the time-variant drag coefficient at *Re*=134 is a bit larger than that at *Re*=6300, which is consider to be caused by the increase in friction force with the increased fluid viscosity (lower *Re*). Such an increase in drag with increased fluid viscosity is consistent with the previous experimental results (Brodsky 1994). Moreover, mean drag and lift coefficients are calculated by averaging the time-variant values over a flapping cycle. As a result, the mean lift coefficient is 0.445 at *Re*=134, and 0.533 at *Re*=6300; and the mean drag coefficient is 0.042 at *Re*=134, and 0.090 at *Re*=6300.

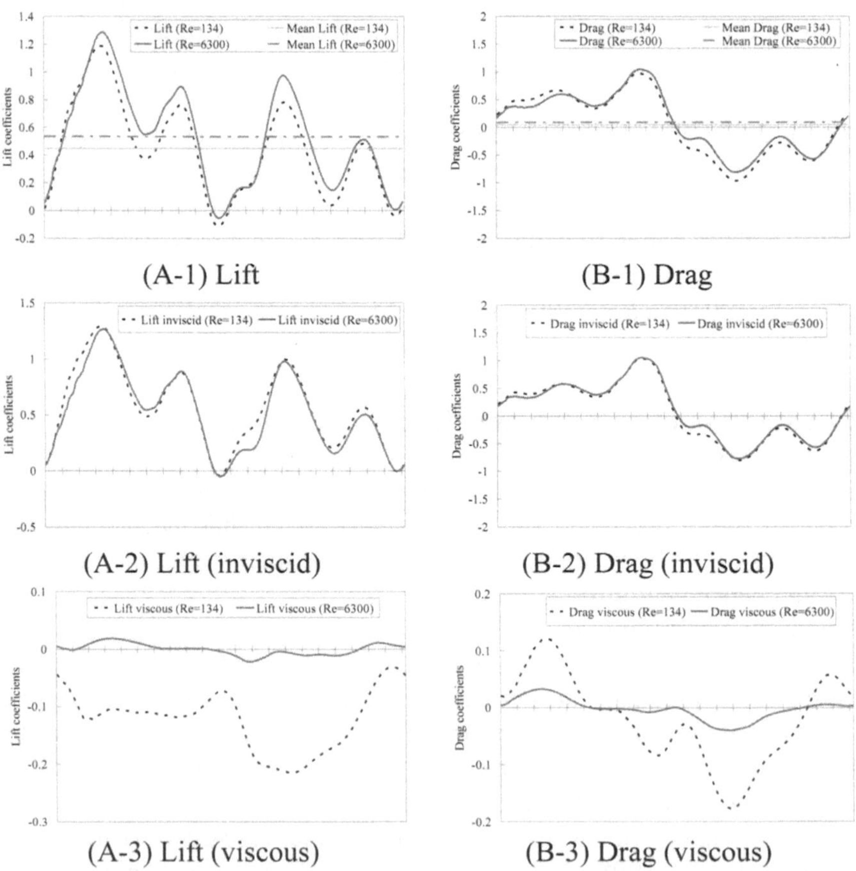

Fig. 4. Time courses of aerodynamic force coefficients over a flapping cycle. The curve of at *Re*=134 and at *Re*=6300 show blue-line and red-line, respectively. The mean aerodynamic force coefficient indicates green dot-line (*Re*=134) and orange dot-line (*Re*=6300), respectively.

3.2.2 Relative contributions of inviscid and viscous terms to aerodynamic force

Further quantitative analysis is conducted to investigate the relative contributions of inviscid and viscous terms to aerodynamic force. To this end, the transient lift and drag coefficients are decomposed into inviscid and viscous portions.

Inviscid aerodynamic force

The time-variant inviscid lift and drag coefficients are plotted in figure 4A-2 and B-2, respectively. The results for the two *Re* conditions are almost the same over a flapping cycle except during the early down- and up-stroke. The mean inviscid lift and drag coefficients are 0.568 (80.5% of total mean lift) and 0.064 (74.4% of total mean drag) at *Re*=134; and are 0.5326 (99.996% of total mean lift) and 0.093 (97.1% of total mean drag), respectively, at *Re*=6300.

Viscous aerodynamic force

Shown in figure 4A-3 and B-3 are the time-variant viscous lift and drag coefficients, respectively. It is evident that the time courses of both the viscous lift and drag differ significantly between *Re*=134 and *Re*=6300 in terms of the shape and the amplitude of waveform. For instance, the overall amplitudes of both the viscous lift and drag waveforms at *Re*=6300 are much smaller than those at *Re*=134. The mean viscous lift and drag coefficients are -0.123 (19.5% of total mean lift), and -0.022 (25.6% of total mean drag) at *Re*=134; and are 0.0002 (0.004% of total mean lift), and -0.0028 (2.9% of total mean drag), respectively, at *Re*=6300.

Quantitatively, the contribution of viscosity to aerodynamic force at *Re*=134 is more than 10 times that at *Re*=6300. Therefore, at lower *Re*, fluid viscosity is an important physical factor affecting the generation of lift and drag in flapping flight; while at higher *Re*, inviscid term (dynamic pressure) plays a dominant role in aerodynamic force generation. Similar conclusions have been drawn by Wang (2000) who carried out a series of computations on a flapping 2D single wing.

In this study, we elucidate the significant dependency of both the flow fields and the aerodynamic force on *Re* via computing the hovering hawkmoth under the two *Re* conditions (*Re*=134, 6300). *Re*=6300 represents the operating *Re* of a real hawkmoth, while *Re*=134 is in fact in the *Re* range of a real fruit fly. The reason we performed computations under different *Re* conditions but with an identical wing-body geometric and kinematic model of a hawkmoth is that we have intended to investigate *Re* effect only. However, as a matter of course, insects existing in the nature fly in a wide *Re* range and employ various wing-body kinematics. For instance, the flight of a fruit fly is quite different from that of a hawkmoth in terms of not only the range of *Re* but also the movement of wing. For a specific insect, the wing-body morphology, the size, and wing-body kinematics combine to determine the aerodynamic performance of flight.

Acknowledgement

We would like to acknowledge the anonymous referees for their valuable comments and suggestions on the manuscript. This work was partly supported by a PRESTO/JST (Japan Science and Technology Agency), entitled of A Biology-inspired Dynamic Flight Simulator, and the Grant-in-Aid for Scientific Research of No. 18656056 and No. 18100002. The computations were performed in a supercomputer (RSCC), RIKEN (The Institute of Physical and Chemical Research), Japan. Hikaru Aono thanks Dr. Fuyou Liang for his valuable comments on the manuscript and his help in the English writing.

References

Aono H, Liu H (2006) Vortical structure and aerodynamics of hawkmoth hovering. Journal of Biomechanical Science and Engineering 1:234-245

Birch JM, Dickinson MH (2001) Spanwise flow and the attachment of the leading-edge vortex on insect wings. Nature 412:729-732

Birch JM, Dickinson WB, Dickinson MH (2004) Force production and flow structure of the leading edge vortex on flapping wings at high and low Reynolds numbers. J. Exp. Biol. 207:1063-1072

Brodsky AK (1994) The evolution of insect flight, Oxford University Press, pp 49-52

Ellington CP, van den Berg C, Willmott AP, Thomas ALR (1996) Leading-edge vortices in insect flight. Nature 384:626-630

Dickinson MH, Lehmann F-O, Sane SP (1999) Wing rotation and the aerodynamic basis of insect flight. Science 284:1954-1960

Liu H (2005) Simulation-based biological fluid dynamics in animal locomotion. ASME, Applied Mechanics Review 58, 4:269-282.

Liu H Integrative modeling of insect flight: morphology, kinematics and aerodynamics. (in preparation)

Liu H, Kawachi K (1998) A numerical study of insect flight. J. Comput. Phys. 146:124-156

Liu H, Ellington CP, Kawachi K, van den Berg C, Willmott AP (1998) A computational fluid dynamic study of hawkmoth hovering. J. Exp. Biol. 201:461-477

Miller LA, Peskin CS (2004) When vortices stick an aerodynamic transition in tiny insect flight. J. Exp. Biol. 207:3073-3088

Ramamurti R and Sandberg WC (2002) A three-dimensional computational study of the aerodynamic mechanisms of insect flight. J. Exp. Biol. 205:1507-1518

Srygley RB, Thomas ALR (2002) Unconventional lift-generating mechanisms in free-flying butterflies. Nature 420:660-664

Sun M, Gang DU (2003) Lift and power requirements of hovering flight. Acta Mechanica Sinica. 19, 5:2413-2427.

Usherwood JR and Ellington CP (2002) The aerodynamics of revolving wings. I. Model hawkmoth wings. J. Exp. Biol. 205:1547-1564

van den Berg C, Ellington CP (1997) The three-dimensional leading-edge vortex of a ‘hovering’ model hawkmoth. Phil. Trans. R. Soc. Land B 352:329-340

Wang ZJ (2000) Two dimensional mechanism for insect hovering. Physical Review Letters 85, 10:2216-2219.

Willmott AP, Ellington CP (1997) The mechanics of flight in the hawkmoth *Manduca sexta* I. Kinematics of hovering and forward flight. J. Exp. Biol. 200:2705-2722

Chapter 16

Stabilization of Flapping-of-Wings Flight of a Butterfly, Considering Wakes

Kei Senda[1], Masakazu Sawamoto[1], Masahiko Kitamura[1], and

Tsuyoshi Tanaka[2]

[1] Graduate School of Natural Science and Technology, Kanazawa University, Kakuma-machi, Kanazawa, Ishikawa 920-1192, Japan

[2] Graduate School of Engineering, Osaka Prefecture University

Summary. This paper studies the flapping-of-wings flight of a butterfly, which is rhythmic and cyclic in motion. The objective is to clarify the principle of stabilization of the flapping-of-wings flight. For this purpose, an experimental system with a low-speed wind tunnel is constructed for fundamental data of flapping-of-wings motion, where the system measures the aerodynamic force and the motion simultaneously using a measure and an optical measurement system. A dynamics model of a butterfly is derived by Lagrange's method, where the butterfly is considered as a rigid body system. For the aerodynamic forces, a lumped-vortex method and a panel method are applied. Validity of the mathematical models is examined by the good agreement of the numerical results with the measured data. Then, periodic orbits of a flapping-of-wings flight are searched in order to fly the butterfly models. Almost periodic orbits are obtained, but both models in the flapping-of-wings flight are unstable. The unstable level of the panel method model is smaller by considering free-vortices in wakes. Meaning that the wake-induced flow has a type of feedback stabilization effect.

1 Introduction

Butterflies fly irregularly and their flapping-of-wings flights do not seem to be controlled. But, in reality their flights are well controlled because they do not crash to the ground and can perch on flowers for honey. Butterflies can maintain desired flapping-of-wings flights stably against environmental uncertainties and variations, e.g. gust, weight gain and loss, etc., because of their adaptation-capability, even though they cannot repeat the same movement with sufficient accuracy like robots. There are some features in the flight of the butterfly, e.g. small flapping frequency (Azuma 1992) and small Reynolds number. Moreover, the flow to the wings is unsteady because of the flapping motion. Those features may have some effect on the flight.

The objective of this study is to clarify the principle that enables the flapping-of-wings flight. The flapping-of-wings flight of a butterfly might be realized by intricately-intertwining with many factors, i.e. air flow over wings, motion of the butterfly's wings and body, and so on. There are many unknown points for the stable flight, e.g. "how the butterfly moves," "how the control is realized," etc.

To clarify the unknowns, this research proceeds by the following plan (Senda et al. 2003, 2004):

1. derivation of dynamics model and simulator construction,
2. experimental measurement of butterfly motions and aerodynamic forces, and
3. discussions for the stable flapping-of-wings flight through simulations.

This paper reports on the results obtained from the plan. Item 1 derives some models with aerodynamics formulation using a lumped-vortex method and a panel method. In item 2, experimental data of the flapping motions and the aerodynamic forces are measured and validity of the mathematical models is examined by the good agreement with the measured data. In item 3, a periodic trajectory of flapping-of-wings flight is searched in order to fly each butterfly model. The flapping-of-wings flights obtained from the models are instable. Instability is decreased in the panel method by considering free-vortices in wakes.

The rest of this paper is organized as follows. A butterfly is considered as a rigid multi-body system, and a dynamics model is developed by Lagrange's method in the second section. In the third section, aerodynamic forces are formulated by a lumped-vortex method and a panel method, and the characteristic flapping-of-wings aerodynamic effects are considered. In the fourth section, an experimental system with a low-speed wind tunnel is constructed for fundamental data acquisition of flapping-of-wings motion, where the system measures aerodynamic force and motion simultaneously by a measure and an optical measurement system. In the fifth section, measured data are compared with numerical results and validity of the mathematical model is examined. In the sixth section, periodic orbits of flapping-of-wings flights are searched in order to fly the butterfly model, and stabilities of the flights are discussed. The final section gives concluding remarks.

2 Mechanism and Mechanics of Butterfly

2.1 Outline of Flapping Mechanism

The flapping mechanism can be considered as an almost one degree-of-freedom vibrating mechanism. A butterfly vibrates the mechanism for flapping of the wings. According to experimental observation, each butterfly has an almost constant flapping eigenfrequency. It may be because to vibrate sympathetically at eigenfrequency of the mechanism realizes efficient flapping motion (Azuma 1992). Other direct muscles can add slight motions, which are in the extent of modification to the basic vibration.

2.2 Mechanics as Rigid Multi-body System

A butterfly is modeled by a rigid multi-body system as well as typical robot modeling, which is simple dynamics modeling. For more precise modeling, it will be desirable to consider a flexible multi-body system because experimental observations have found that butterfly wings are elastically deformed by aerodynamic forces.

The butterfly model is a multi-body system with 4 links as shown in Fig. 1, which is composed of the thorax *Bt* considered as the main body, the abdomen *Ba*, the left wings *WL*, and the right wings *WR*. Both the model and its motions are supposed to be symmetrical. A pair of fore and hind wings on each side is modeled by a plate as shown in Fig. 1. The joint between the thorax and wing has 3 rotational degree-of-freedom (DOF) and the joint between the thorax and the abdomen has 1 DOF.

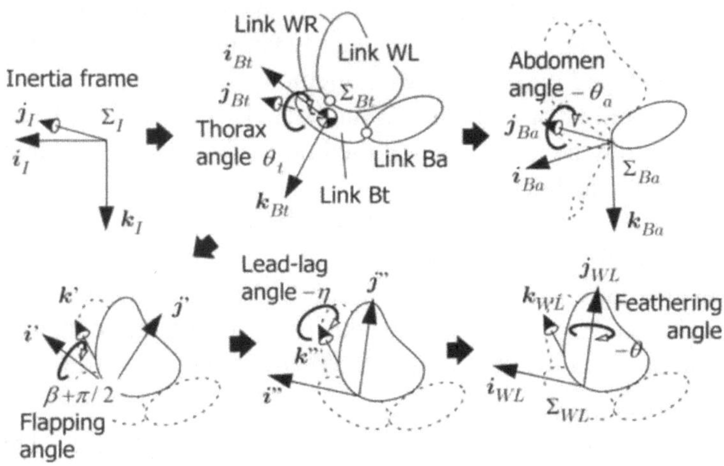

Fig. 1. Frames and coordinates for modeling of a butterfly

In the figure, the frame Σ has a set of orthonormal bases $[\boldsymbol{i}\ \boldsymbol{j}\ \boldsymbol{k}]$, coordinates of which are referred as x, y, z. The Σ_I is the inertial reference frame, Σ_{Bt} thorax-fixed frame, Σ_{Ba} abdomen-fixed frame, Σ_{WL} left-wing-fixed frame, and Σ_{WR} right-wing-fixed frame. The $\boldsymbol{i}_I$ corresponds to the butterfly's traveling direction, and $\boldsymbol{i}_I$-$\boldsymbol{k}_I$ is the sagittal (symmetry) plane. Rotating Σ_I about $\theta_t \boldsymbol{j}_I$ yields Σ_{Bt}, rotating Σ_{Bt} about $-\theta_a \boldsymbol{j}_{Bt}$ outputs Σ_{Ba}, a sequence to rotate Σ_{Bt} via 1-3-2 Euler angles with $\beta + \pi/2$, $-\eta$, and $-\theta$ results in Σ_{WL}.

The following Lagrangian equations of motion are obtained:

$$\boldsymbol{M}\ddot{\boldsymbol{\theta}} + \dot{\boldsymbol{M}}\dot{\boldsymbol{\theta}} - \frac{1}{2}\frac{\partial}{\partial \boldsymbol{\theta}}\left(\dot{\boldsymbol{\theta}}^{\mathrm{T}}\boldsymbol{M}\dot{\boldsymbol{\theta}}\right) + \frac{\partial V}{\partial \boldsymbol{\theta}} = \boldsymbol{\tau} \tag{1}$$

where the generalized coordinates are $\boldsymbol{\theta} = [x\ \ z\ \theta_t\ \ \theta_a\ \beta\ \eta\ \theta]^{\mathrm{T}}$. As illustrated in Fig. 1, x, z, and θ_t are the x, z-positions and the attitude angle of the thorax, θ_a the abdomen angle, β the flapping angle in up-down direction, η the lead-lag angle, and θ the feathering angle representing torsion angle. The $\boldsymbol{M}$ and V are the inertia matrix and the gravitation potential energy, respectively. The vector $\boldsymbol{\tau} = \boldsymbol{\tau}_d + \boldsymbol{\tau}_{control}$ is the generalized forces corresponding to $\boldsymbol{\theta}$, where $\boldsymbol{\tau}_d$ and $\boldsymbol{\tau}_{control}$ are the aerodynamic torque and the joint control torque, respectively, developed in the following sections.

3 Aerodynamics of Butterfly

3.1 Fundamental Computation Methods

The aerodynamic force to a wing is the sum of the inertia force of added mass and aerodynamic force $\boldsymbol{\tau}_d$ that is a function of the flow velocity. The mass of fluid moving with the wing is called added mass. The added mass has been considered in the left-hand side of Eq. 1, and $\boldsymbol{\tau}_d$ will be derived in following subsections. For fundamental aerodynamic forces applied to wings, we have developed two computation methods, i.e. a simple method (Sunada et al. 1993) and a vortex method (Senda et al. 2003, 2004). In addition to them, this study develops a panel method model. This paper does not mention the simple method.

Viscosity should be considered in the model because of the low Reynolds number $Re \approx 10^3$ for the butterfly. However, both computational and experimental evaluations show that the viscous drag is approximately 0.5% of the maximum drag. Since the viscous effect is negligible, the potential flow theorem might result in good agreement with the real flow.

3.1.1 Lumped-vortex method

A lumped-vortex method (Katz and Plotkin 2001) is introduced. This method divides the wing into thin strip segments along the wingspan.

Then τ_d is computed by the sum of the aerodynamic forces applied to all the thin strip segments. The vortex method allocates a bound vortex for each strip segment, determining its circulation so as to satisfy the Kutta condition at trailing edge, and obtains aerodynamic forces using the two-dimensional potential flow theorem.

3.1.2 Panel method

Wing-tip vortices greatly effect the aerodynamic characteristics because a butterfly has flat wings of a small aspect ratio. In addition, the flapping motion generates very strong wing-tip vortices. The wing-tip vortices may combine to the free-vortices leaving from the trailing-edge, make complex structures, and greatly influence the aerodynamic characteristics. Therefore, this study models the aerodynamics of a flapping-of-wings butterfly using a panel method (Katz and Plotkin 2001).

3.2 Effects of Flapping-of-Wings

There are the characteristic aerodynamics of beating and peeling phases of wings (Wootton 1990). Experimental results will show the strong wing tip vortex wakes that might have an effect on aerodynamics.

Therefore the following four effects in Fig. 2 are modeled and considered. Those are (a) downwash induced by wing tip vortices, (b) interaction of right-and-left wings, (c) peel mechanism, and (d) damming effect. Those effects are ever-changing because the relative position and orientation of the wings vary. The lumped-vortex method additionally considers all those effects in simulations later. The panel method automatically considers those effects other than the peel mechanism. Only the panel method can consider wakes and free-vortices. See (Senda, et al. 2006) for the detail of the modeling of the characteristic effects. Dickinson et al. (Dickinson et al 1999) analyzed the characteristic flapping-of-wings aerodynamics of flies through experiments and numerical simulations.

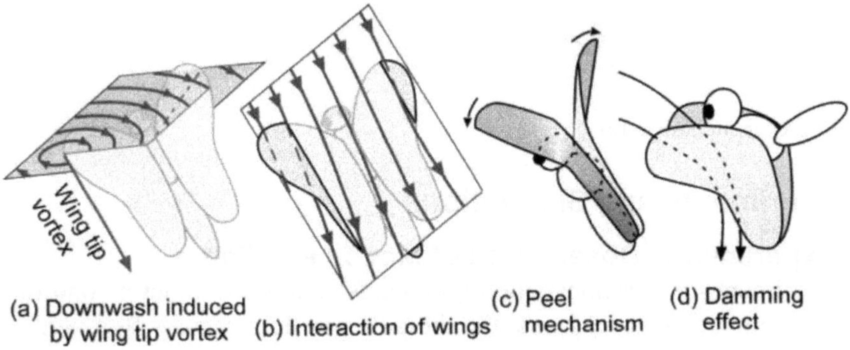

Fig. 2. Aerodynamics of flapping-of-wings effects

4 Experimental System

Verification of the models and parameter setting require fundamental data of flapping-of-wings motions and aerodynamic forces. For this purpose, an experimental system with a low-speed wind tunnel is constructed and a wind tunnel experiment is conducted using actual butterflies, Parantica sita niphonica, which is similar to Danaus in the US. It is 0.24 g weight and its half wing span is 50 mm long.

The simultaneous measurement system is constructed as illustrated in Fig. 3, which measures butterfly's motion and the forces applied to the butterfly using an optical measurement system and a force measurement system. The butterfly is gummed up on the back of the thorax to the tip of the measure, i.e. the force/torque sensor. The measure with the butterfly is put into the wind tunnel, the flapping-of-wings motion in the flow is captured as a video image, and the state vector $\boldsymbol{\theta}$ is measured using the video images. Simultaneously, the forces applied to the butterfly, i.e. lift L, drag D and pitching moment M, are sensed by the measure. There is a smoke wire for airflow visualization.

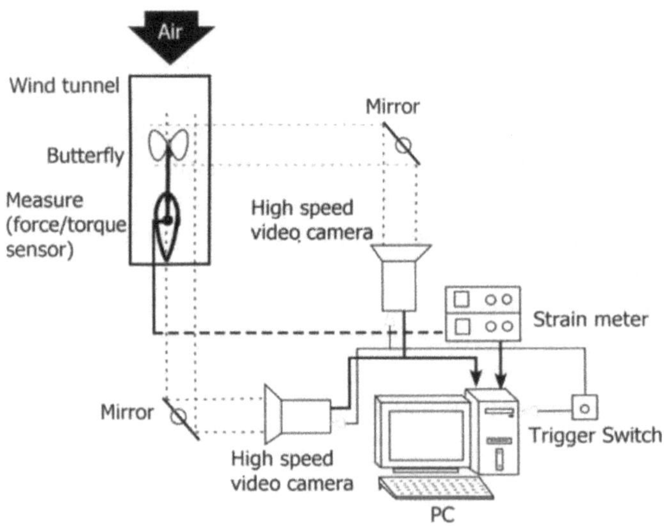

Fig. 3. Experimental setup

5 Experiments and Simulations

5.1 Experimental Observation of Motion and Force

An experiment for Parantica sita niphonica is conducted under a condition that the mainstream is 1.64 m/s, the thoracic positions $x = z = 0$ m, and its angle $\theta_t = -1°$. Fig. 4 (a) shows the generalized coordinates $\boldsymbol{\theta}$ of an experimental result measured by the above optical measurement system.

Figs. 4 (b)-(d) illustrate the measured forces, i.e. the lift L, the drag D, and the pitching moment M. The butterfly repeats this cyclic motion in a period of 0.136 seconds when it continues the periodic flapping motion. Measured aerodynamic forces are also cyclic during the motion.

The flapping angle β is a cosine-like curve whose maximum and minimum are $81°$ and $-45°$. The left and right wings overlaps perfectly at the maximum position and almost overlaps at the minimum. The abdomen angle θ_a is also a sinusoidal curve, but is out of phase to the flapping angle β. When the butterfly varies amplitude, it changes the minimum of β and does not change the maximum. The fixed butterfly swings the abdomen unnaturally in a larger amplitude than butterflies flying freely. Even if β or θ_a changes the amplitude depending on the flying condition and/or the individual, they keep similar smooth cosine curves. Lead-lag angle η has a vibration mode whose frequency is twice as large as β. Feathering angle θ seems to be varied passively in small by the aerodynamic forces. Fry et al. (Fry et al., 2003) similarly observes fly's flapping-of-wings motion.

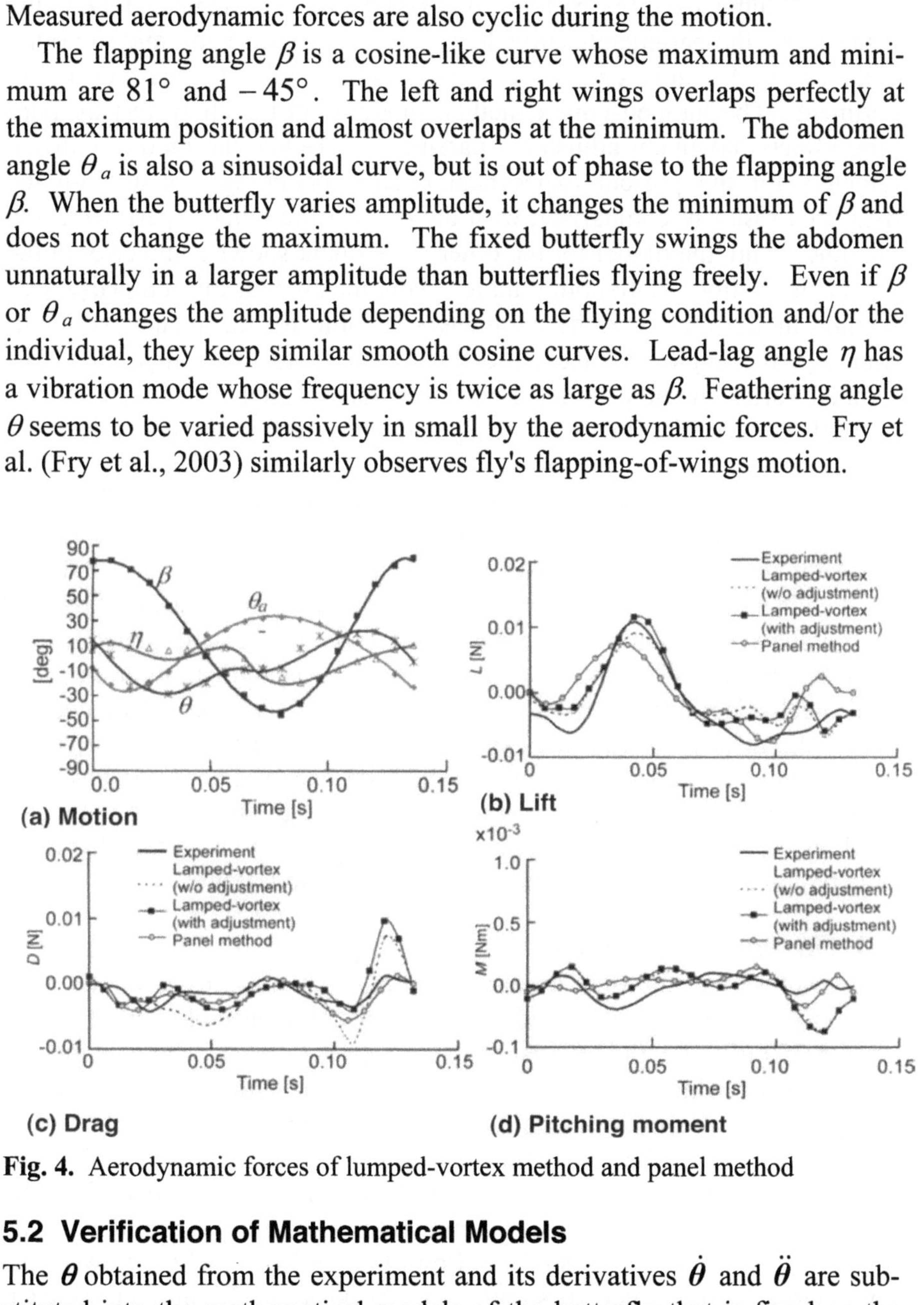

Fig. 4. Aerodynamic forces of lumped-vortex method and panel method

5.2 Verification of Mathematical Models

The $\boldsymbol{\theta}$ obtained from the experiment and its derivatives $\dot{\boldsymbol{\theta}}$ and $\ddot{\boldsymbol{\theta}}$ are substituted into the mathematical models of the butterfly that is fixed on the back of thorax to the tip of the measure. Its aerodynamic forces are illustrated in Figs. 4 (b)-(d). The aerodynamic models are the lumped-

illustrated in Figs. 4 (b)-(d). The aerodynamic models are the lumped-vortex method and the panel method. Validity of the model can be evaluated by comparing the simulation results with the experimental data.

For the lumped-vortex method, two results are represented. One is the result using the parameters suitable for the past experiment, and the other is the result after adjusting the parameter to be suitable for the present experiment. The lumped-vortex method model with the characteristic flapping effects has many adjustable parameters. Hence the model with parameters for the present experiment obtains better agreement with the measured result in comparison with the one for the other experiment. On the other hand, the model for the other experiment shows some error at the top or the bottom of flapping angle. This means that the lumped-vortex method model must use the parameters suitable for instantaneous condition, e.g. the angle of attack, the flow velocity to the wings, etc. The lumped-vortex method model is not necessarily suitable for a free-flying butterfly because the flow is unsteady and the condition always varies.

On the other hand, the panel method can calculate the aerodynamic forces with the same degree of accuracy for any experiment. The panel method model always agrees well with experimental results by considering unsteady flow induced by wakes. The calculation accuracy of the developed code has been evaluated by using an oval wing in the steady flow that has an analytical solution. The developed code respectively generates the drag and the lift 17% and 5% smaller than the analytical solution. The peel mechanism has not been considered in the panel method model yet. Hence the difference from the experimental result is growing at the beginning of downstroke where the peel mechanism becomes considerable.

In terms of drag, the panel method is in better agreement with the measured result. In terms of lift, the lumped-vortex method with the characteristic flapping effects after the parameter adjustment is better.

5.3 Observation of Flow

Fig. 5 shows that the flow around the wings visualized by a smoke wire. There are four images of the side view. They are images of every 4 frames shot in 250 FPS, i.e. the interval of the images is 0.016 s. The peel mechanism and the wing tip vortex wakes are observed during the flapping-of-wings motion. It shows qualitative validity of the proposed model.

Fig. 6 illustrates the visualized flow of the experiment and the instantaneous stream lines of the numerical simulation, which are at scene 3 in Fig. 5 and on cross sections at $x = -100$ mm and $y = -30$ mm. A strong wing tip vortex is observed in the visualized flow at (1) of Fig. 6, where flow velocity is large in the numerical simulation. At position (2), another vortex is observed that was left from the wing tip at upstroke of a period before the vortex (1). The simulation also captures the vortex at the position as same as the visualized flow. The area (3) is a vortex that was left from the wing tip at downstroke of a period before the vortex (1). Accordingly, the numerical simulation analyses the flowfield corresponding to the experiment.

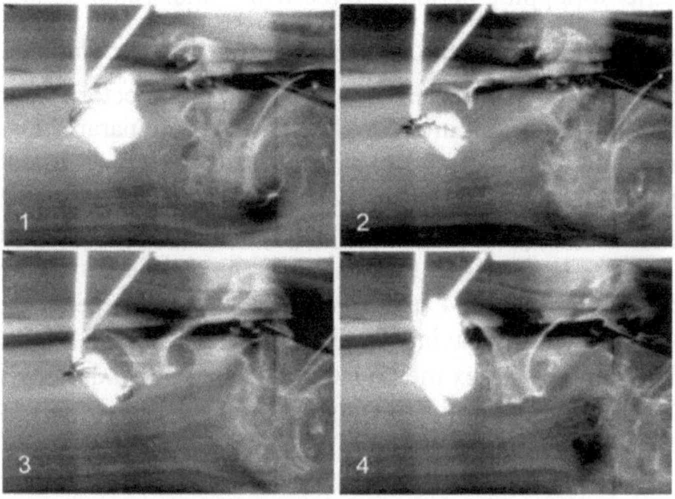

Fig. 5. Visualized flow by experiment (V = 1.0 m/s, θ_t = 10 deg, every 0.016 s)

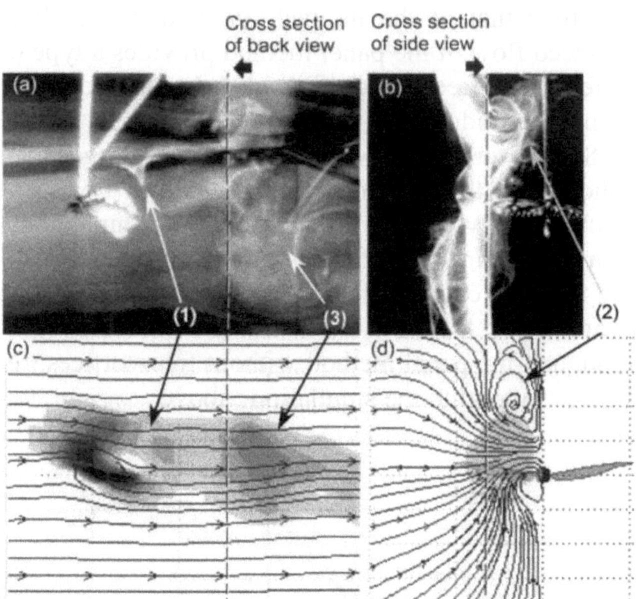

Fig. 6. Comparison of flow fields of experimental visualization and numerical simulation (x = -100 mm, y = -30 mm)

6 Stabilizing Effect of Wakes to Flapping Flight

6.1 Trajectory Search

The periodic flapping-of-wings flight is defined as "all other than x in state vector $\boldsymbol{\theta}_f$ and $\dot{\boldsymbol{\theta}}_f$ after a flapping period T agree with initial value $\boldsymbol{\theta}_s$ and $\dot{\boldsymbol{\theta}}_s$." The trajectory of motion is searched to realize a periodic flapping flight of each model. Fourier parameters describing joint trajectories for cyclic motions and initial conditions are considered as learning parameters.

Both models can obtain almost periodic trajectories, though they do not perfectly correspond to the definition of the periodic flight. The obtained flights by the lumped-vortex method and the panel method are illustrated in Figs. 7 (a) and 8 (a). The obtained flapping motions are similar to the trajectory of the experiment in Fig. 4 (a). Because they are not perfectly periodic, the flapping flights are leaving gradually from the initial flights and destabilized. The lumped-vortex method and the panel method models go down after two and eleven flapping-of-wings motions, respectively.

6.2 Stability Analysis of Models

Figs. 7 (b) and 8 (b) are the flights started from the initial states of Figs. 7 (a) and 8 (a) with same small perturbations. Degree of instability of the panel method is smaller than that of the lumped-vortex method. It is thought that the wake-induced flow of the panel method provides a type of feedback stabilization effect. The free-vortices in wakes are then removed from the panel method model, and Fig. 9 shows the simulations on the same conditions as Fig. 8. The unstable level of Fig. 9 (b) is larger than Fig. 8 (b). This result shows that the free-vortices in the wakes are essential for the stabilization of butterfly flights. In addition, the sampled-data systems are derived from the original continuous periodic systems by observing their states every flapping cycle. The highest unstable poles of the sampled-data systems are 61.8, 8.6, and 17.0 for the lumped-vortex method, the panel method, and the panel method without free-vortices, respectively. This result gives support to the simulations above.

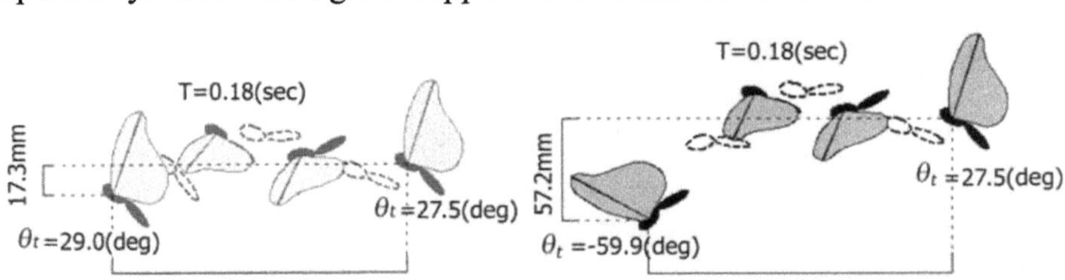

(a) Almost periodic flight (b) Flight with initial perturbation

Fig. 7. Flapping flights of lumped-vortex method model

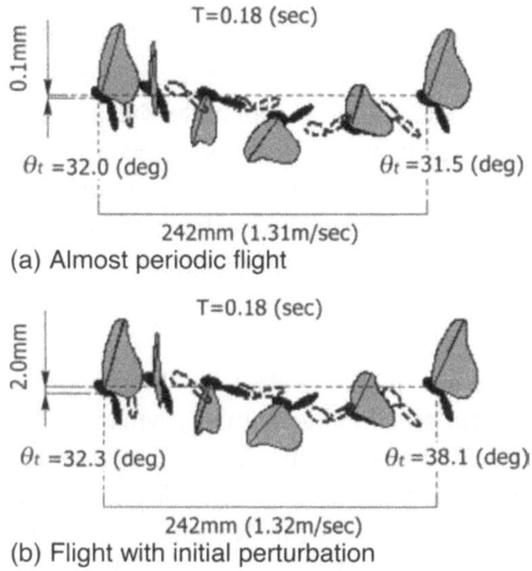

Fig. 8. Flapping flights of panel method model

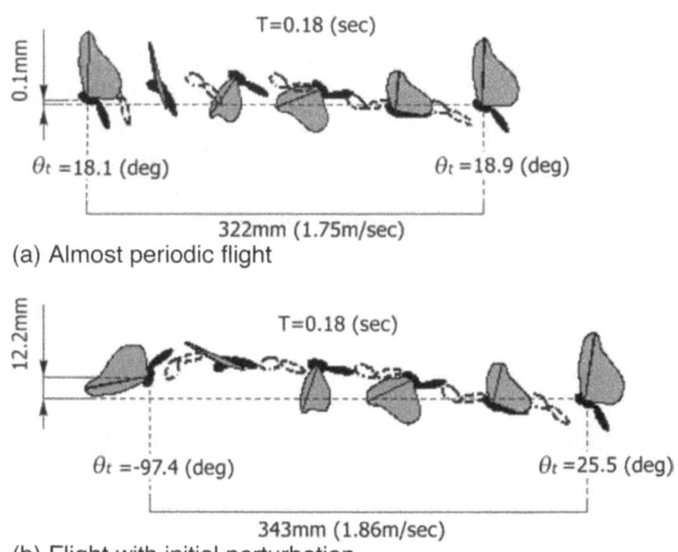

Fig. 9. Flapping flights of panel method model without wakes

7 Concluding Remarks

In order to clarify the stabilization phenomenon in flapping-of-wings flights, this study has performed the modeling of the butterfly, the experiment for the quantitative data acquisition and qualitative observation, the simulator formulation, and the comparison between actual and modeled butterflies. The lumped-vortex method and the panel method have been applied to make the simulator, where the methods calculate the aerodynamic force. All models in the flapping-of-wings flight have been unstable, where the lumped-vortex method is highly unstable. The unstable level of the panel method model has become small. It has been shown that the wake-induced flow has a type of feedback stabilization effect.

Acknowledgement

The authors would like to express sincere thanks to Prof. M. Ishii and Dr. N. Hirai of the Osaka Prefecture Univ. and Ishikawa Insect Museum for their help making use of butterflies. The authors thank Mr. T. Shibahara and Mr. N. Matsumura for their help of experimental work. A part of this work was financially supported by a Grant-in-Aid for Advanced Scientific Research from Kanazawa Univ. and a Grant-in-Aid for Scientific Research from Ministry of Education, Science, Culture, and Sports of Japan.

Reference

Azuma A (1992) The Biokinetics of Flying and Swimming. Springer-Verlarg

Senda K, Tanaka T, Sawamoto M (2003) Measurement and Numerical Simulation of a Flapping Butterfly. In: Proc. of 2nd International Symposium on Adaptive Motion of Animals and Machines. March 4-8, Kyoto, WeP-II-1, pp 1-7

Senda K, Sawamoto M, Shibahara T, Tanaka T (2004) Study on Flapping-of-Wings Flight of Butterfly with Experimental Measurement. Providence, AIAA-2004-5368, pp 1-14

Brodsky A. K (1994) The Evolution of Insect Flight. Oxford University Press

Sunada K, et al. (1993) Performance of a Butterfly in Take-off Flight. J. of Experimental Biology. vol. 183, pp 249-277

Katz J, Plotkin A (2001) Low-Speed Aerodynamics. Cambridge University Press. 2nd ed.

Wootton R J (1990) The Mechanical Design of Insect Wings. Scientific American. vol. 263, no. 5, pp. 114-120

Senda K, Sawamoto M, Kitamura M, Tanaka T (2006) Study on Flapping-of-Wings Flight of Butterfly Considering Wakes. In: CD Proc. of ISABMEC, S11, pp 1-8

Dickinson M H, Lehmann O F, Sane P S (1999) Wing Rotation and the Aerodynamic Basis of Insect Flight.Science. vol. 284, pp 1954-1960

Fry N S, Sayaman R, Dickinson H M (2003) The Aerodynamics of Free-Flight Maneuvers in Drosophila. Science. vol. 300, pp 495-498

Chapter 17

3-D Unsteady Computations of Flapping Flight in Insects, Fish, and Unmanned Vehicles

William C. Sandberg and Ravi Ramamurti[1]

[1] Laboratory for Computational Physics and Fluid Dynamics, Naval Research Laboratory, Washington, DC

Summary. 3-D unsteady computations have been carried out for a swimming tuna with an oscillating caudal fin, the flapping flight of the fruit fly and a pectoral fin swimmer, the bird wrasse, and a variety of unmanned air vehicles. Such computations for creatures or vehicles with moving and deforming surfaces can provide information on the dynamics of force production that is quite useful for vehicle design. Novel biomimetic vehicles have been designed and built and their performance is described. As vehicle size decreases there has also been a need for incorporation of novel materials, sensors, and control systems. Computational challenges for coupling novel sensor designs, vehicle time-varying force and moment computations, and self-consistent vehicle trajectory computations are discussed.

Key words. Biomimetic pectoral fin, UUV, MAV, 3-D Unsteady CFD, fish swimming, insect maneuvering, nanosensor, artificial muscle

1 Introduction

If we stand quietly and observe nature in action around us it is humbling to realize how little we understand of the dynamics that are taking place. It is easy to lose track of the time and become mesmerized by the apparent effortlessness with which living creatures perform the most incredible maneuvers or the apparent ease with which they counter large external, rapidly varying forces. They appear to instantaneously react and harmoniously accommodate to their changing environment. This is particularly striking when creatures, in large numbers, move as one, such as schools of reef fish which appear to move as a thin iridescent sheet, executing a complex folding maneuver without ever colliding with each other or striking the coral around them. If we seek to attain, in our designed vehicles, that level of performance, then it is obvious we still have much to learn about sensory mechanisms, dynamic force production, novel biomimetic materials with passive control response and the integration of all of these in a coherent manner in active control systems. We describe below the results of computational investigations we have carried out to gain insights into one aspect of this challenge, namely the unsteady aerodynamic performance of flying insects and swimming fish, and to incorporate that understanding into the design of unmanned underwater and unmanned air vehicles.

2 Natural Swimmers and Fliers; 3-D Unsteady Computational Investigations

2.1 Underwater Flight

We obviously must have in mind a performance objective for an underwater vehicle in order to begin the design process, particularly if the design process includes some aspects of biomimetics. In our case we decided long ago that operational interest in the development of a flexible vehicle was not as likely as it was for a rigid vehicle and therefore chose to look at living swimmers that propelled themselves primarily with fins. Hence we excluded undulatory swimmers from further consideration, even though we had been looking at eel swimming dynamics (Ramamurti et al. 1996). We explicitly selected high force production control surfaces as the ultimate design objective, as opposed to the vehicle hull itself. We chose to investigate first, the tuna caudal fin as a candidate for superior thrust production. We assumed that body deformation was absent forward of the peduncle but was continuously varying beyond it. The first 3-D unsteady computations of the flow past a continuously deforming geometry with an oscillating control surface were then carried out (Ramamurti et al. 1996). The geometry used for these computations and an instantaneous pressure distribution is shown in Fig 1a.

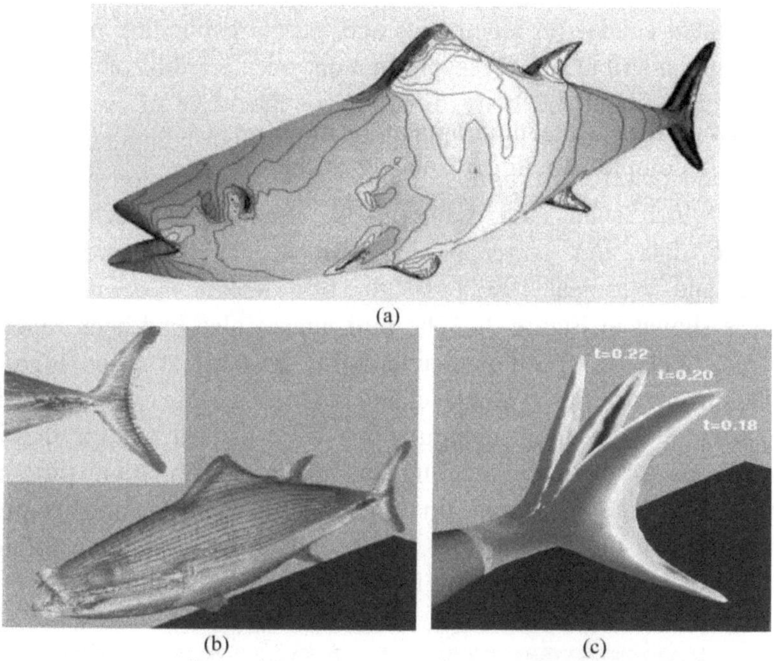

Fig. 1. (a) Instantaneous pressure distribution on a swimming tuna. (b) Instantaneous velocities on surface of swimming tuna. (c) Instantaneous pressure distributions on oscillating caudal fin at selected times during oscillation cycle (From Ramamurti et al. 1996, with permission)

Instantaneous velocities are shown in Fig. 1b. It is clear that the flow past the closed mouth has a significant effect all the way to the leading edge of the caudal fin. We did not have, for this computation, the geometry of the gill openings to permit investigation of flow into the mouth and out of the gills. It would be interesting to carry this out since it may be that this is a flow control means of reducing overall body drag which we did not properly account for. The pressure distribution variation on both sides of the caudal fin throughout the caudal fin stroke is given in Fig. 1c for a few selected time instants during the computation. These computations were carried out for tuna swimming speeds of one and ten body lengths per second, as possible low cruise speed and high burst speed values. The pressure time-histories were integrated to give body and caudal fin force time-histories throughout the fin oscillation stroke period. One of the significant findings from this investigation was that inertial forces were critically important. 3-D quasi-unsteady computations, even with the inclusion of the correct surface normal velocity imposed, drastically underpredicted the time-varying force production magnitudes throughout the stroke cycle.

At the time these computations were carried out in 1994 there did not exist detailed caudal fin kinematics and the corresponding measured data on pressure distribution, force production, acceleration, or velocity field time histories about the body, fins, and in the wake of swimming tuna, hence there could not be any validation made of our computational results. We decided therefore to choose for our next computation a flapping fin fish for which such data did exist. A candidate for which both fin kinematic data and corresponding fish accelerations were available (Walker and Westneat 1997) was the bird wrasse (*Gomphosus varius*) which is shown in Figure 2a. The bird wrasse also had been reported to have excellent low-speed maneuverability as well as good high speed propulsion capabilities, vehicle characteristics we are very interested in possessing. 3-D unsteady computations were carried out for the wrasse with its' pectoral fins executing the shape deformation time-history obtained from the measured fin kinematics. An instantaneous pressure distribution and the time history of lift and thrust production throughout several stroke cycles is given in Figure 2b.

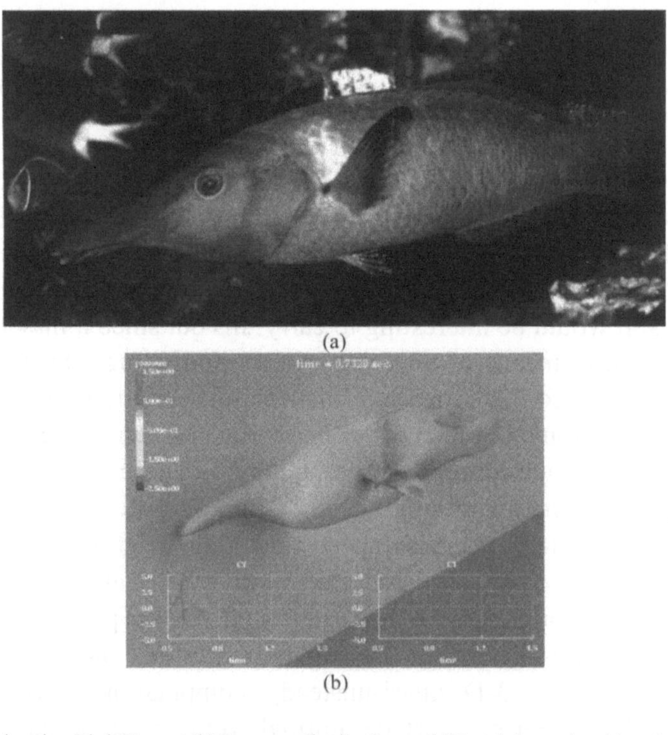

Fig. 2. (a) Swimming Bird Wrasse. (b) Pressure distribution and lift and thrust time histories. (From Ramamurti and Sandberg 2002, with permission)

An important objective, as we studied the flow field evolution about the flapping fin and attempted to gain insights into the force production mechanisms underlying the impressive dynamic performance of living creatures such as the bird wrasse, was understanding how variation of major design parameters changed the force production dynamics. For example, how important is it that the precise deformation time history, as obtained from kinematics of live swimming wrasses, be maintained? We needed to understand what performance would be sacrificed if we were to significantly alter the ability of the fin to deform. This question was answered computationally. We investigated the effect on the thrust and lift production time history for the flapping fins as we went from a completely rigid fin to the actual realistic deformation time history. It was clear from Ramamurti and Sandberg (2004) that the penalty incurred by substantially restricting fin deformation is high. Whether it is unacceptably high is a design decision that is a function of the vehicle performance goals. We have designed an unmanned underwater vehicle for the purpose of demonstrating the propulsive performance of the NRL actively controlled curvature fins. The details of the controlled curvature fin design, the fin construction and testing, and the vehicle design are presented in this volume (Palmissano et al 2007). The test vehicle with it two fins is shown in Figure 3.

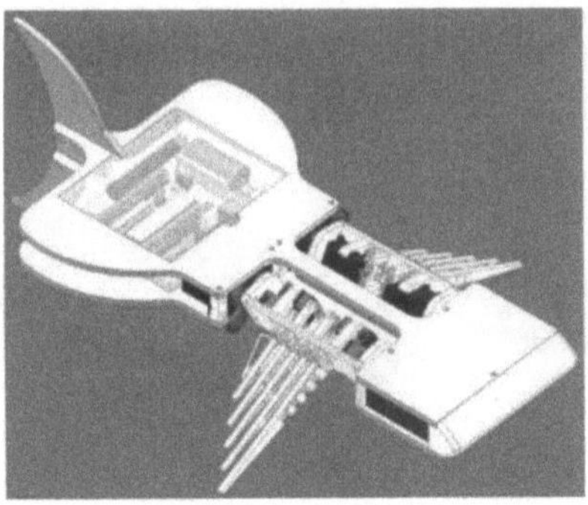

Fig. 3. Test vehicle for carrying a set of two NRL actively controlled curvature fins

2.2 Air Fliers

In addition to underwater vehicles we are also interested in the development of small unmanned air vehicles, usually referred to as MicroAirVehicles (MAV's) or more recently, NanoAirVehicles (NAV's).

As for the underwater vehicles we begin with an operational performance objective to focus our efforts. We have been interested in developing small vehicles with good low-speed maneuverability, obstacle avoidance, and possibly a hovering capability. The extensive investigations of the hawkmoth and the fruit fly are of great value in this regard. We chose to pursue the fruit fly for computational investigations since both detailed wing kinematics and corresponding force data were available. As for the bird wrasse, 3-D unsteady computations were carried out for the hovering Drosophila. The lift, thrust, and drag time histories and wing flow field evolution were computed first about a single wing executing the specified kinematics. The results of these computations were compared with experimental data and also with computations which included the body of the Drosophila since we often observe, as in the wrasse, extremely complicated body-wing root interactions which need to be studied, in addition to wing alone results. The pressure distribution on the body and both wings at particular instants is shown in Figure 4.

Kinematics and force measurements also exist for the Drosophila saccade, a 90 degree turning maneuver. The saccade maneuver was therefore computed for the complete 2-wing and body, with each wing kinematics being separately specified. Computational investigations can be useful here in attempting to understand the nature of the force and moment variations and the asymmetries in force and moment production needed by the insect, or an MAV, to enable a turning maneuver.

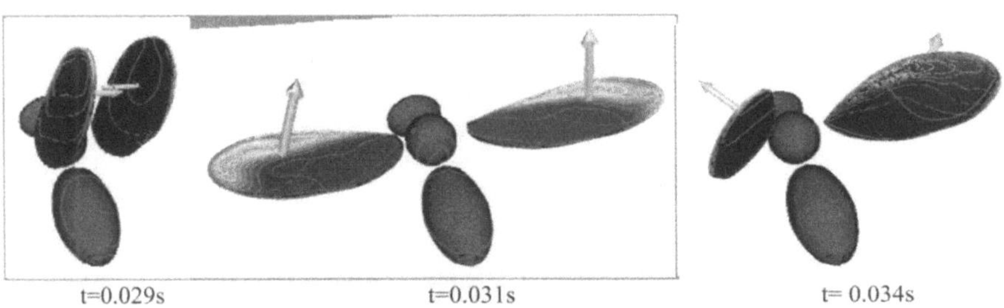

Fig. 4. Instantaneous pressure distributions through the middle of the saccade maneuver (From Ramamurti and Sandberg 2007, with permission)

The pressure distribution (Fig. 4) is integrated over the wing surfaces at each time step, from the beginning of the maneuver, which is the hovering phase, through the complete 90 degree turn, to obtain the lift and thrust time-histories. The computed lift and thrust forces are compared with the measured values in Fig. 5.

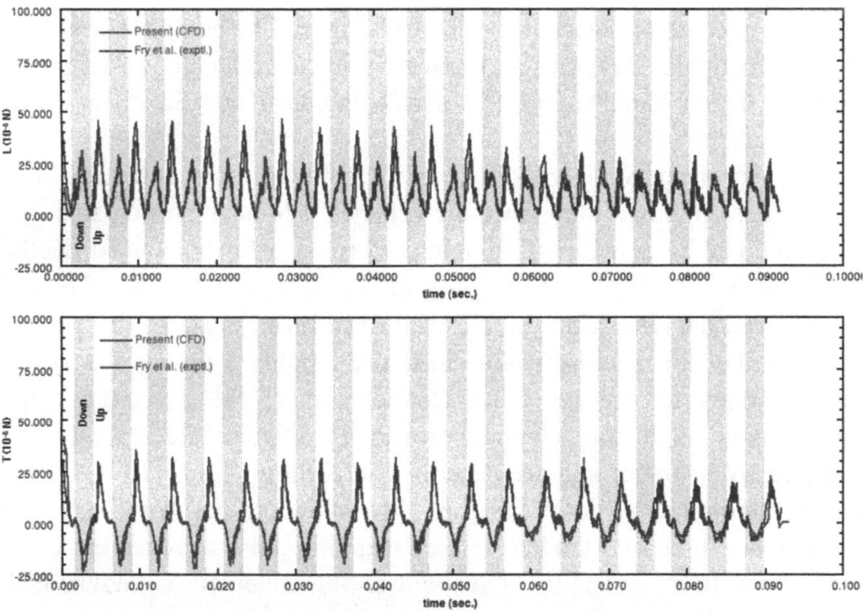

Fig. 5. Comparison of experimental and computed hovering and maneuvering force time histories (From Ramamurti and Sandberg 2007, with permission)

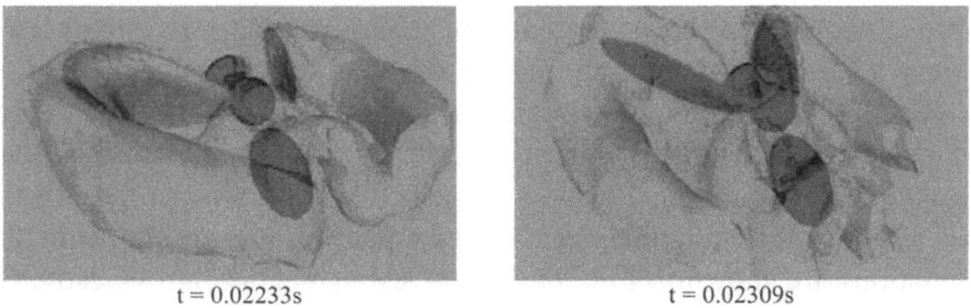

Fig. 6. Isovorticity surfaces as seen from the back of the fruitfly (From Ramamurti and Sandberg 2007, with permission)

The computed lift is very close to the measured lift, on both the upstroke and the downstroke, as the saccade maneuver progresses. The thrust force peaks to a value of nearly 25 μN during the upstroke during the hover and early stages of the saccade and decreases during the later part of the maneuver. The computed thrust and drag force time history is nearly identical to the measured thrust throughout the entire maneuver. The thrust force peaks to a value of nearly 25 μN during the upstroke during the hover and early stages of the saccade and decreases during the later part of the maneuver. An example of the vorticity about the Drosophila as it begins to execute the saccade is shown in Fig. 6. The complete description of the Drosophila hovering and maneuvering computations are given in Ramamurti and Sandberg (2007) and Ramamurti and Sandberg (2001), respectively.

There has been an increased level of activity in recent years in the development of small unmanned air vehicles. We have participated in the computational assessment of vehicle performance for both fixed (Kellogg et al. 2001) and flapping (Kellogg et al. 2003) wing concepts. The initial hybrid fixed/flapping wing vehicles had fixed lifting surfaces that provided most of the lift necessary for sustained flight. Two of the original configurations to use reversing camber wings, the reversing camber Delphinopter, and the clapping reversing camber Pectenopter were designed by Cylinder and are described in Kellogg et al. (2003). A more recent configuration, also designed by Cylinder, has no fixed lifting surfaces and uses two tandem sets of clapping pairs of reversing camber wings for both thrust and lift. The 4 wings of this vehicle, called the BITE-WING, are shown in Fig. 7a. Each flapping foil uses a reversing camber mechanism. Early flight tests achieved stable, slow, and controllable flights. Subsequent design modifications have led to further performance improvements. We have carried out computational investigations to achieve a better understanding of the action of the reversing camber foils, the effects of the clapping interaction, and the interaction between the front and rear pairs of wings. In particular we carried out numerous unsteady simulations investigating the interactions between the wings of each pair as well as the interactions between the two pairs. Some results from these simulations are shown in Fig. 7b. The MAV vehicle design investigations are described in Ramamurti et al (2005) and Shyy et al. (1997).

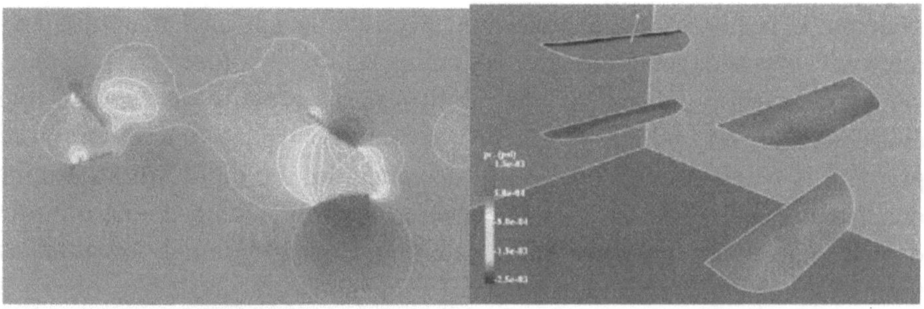

(a) (b)

Fig. 7. (a) Instantaneous pressure distribution on four flapping wings of BITE vehicle. (b) Instantaneous vorticity field for 4 flapping wings of BITE vehicle

3 Novel Biomaterials, Biosensor Dynamics, and Computational Challenges for Integration into Unmanned Vehicles

As the need for smaller vehicles intensifies, we are driven to novel lightweight materials, functionally tailored materials, an increasing number of MEMS components, and now even sensors at the nanometer scale. This is new design territory with new computational challenges. Our initial prototype demonstration vehicle incorporates the NRL actively controlled curvature fins which themselves are an integration of individually designed compliant structural fin rays, individual linear actuation servos, and a polymeric membrane skin to produce the desired surface curvature time-histories resulting in the production of good propulsive force time-histories. The linear actuation servos may subsequently be replaced with the NRL Liquid Crystal Elastomer (LCE) artificial muscles, which have been shown to replicate human muscle quite well (Naciri et al. 2003), are very lightweight, and require small power. Coupled fluid-structure interaction computations have been carried out for some materials (Shyy et al. 1997, and Shyy and Smith 1997), but such computations will become increasingly difficult, if not intractable, as new functional materials are created whose dynamic behavior is not completely understood over the vehicle or device operating range. In addition to the advances in biomimetic materials, the development in MEMS actuators is enabling new strategies for local flow control (Ho 2003). Computing microfluidic device performance and its effect on the vehicle aerodynamics due to distributed local flow modification is another challenging area which will become more important in the development of specialized vehicle control systems. It is interesting to note that, even while artificial materials and microsensors and actuators are beginning to be incorporated into new vehicles, possibly ahead of the ability to thoroughly understand the consequences on vehicle performance and controllability, even more pressure is being exerted on vehicle designers by the rapid breakthroughs occurring at the nanoscale. Nanosensors, possibly within a microfluidic device environment, present daunting computational challenges themselves. New atomic and molecular methods are needed to obtain a clearer picture of the multi-scale dynamics of complex systems, such as the core part of biosensors that combine solid surfaces, fluid and free/tethered biomolecules in solvents. DNA-based biosensors that use short tethered ssDNA chains of 20 ~30 bases to capture target ssDNA molecules are becoming feasible.

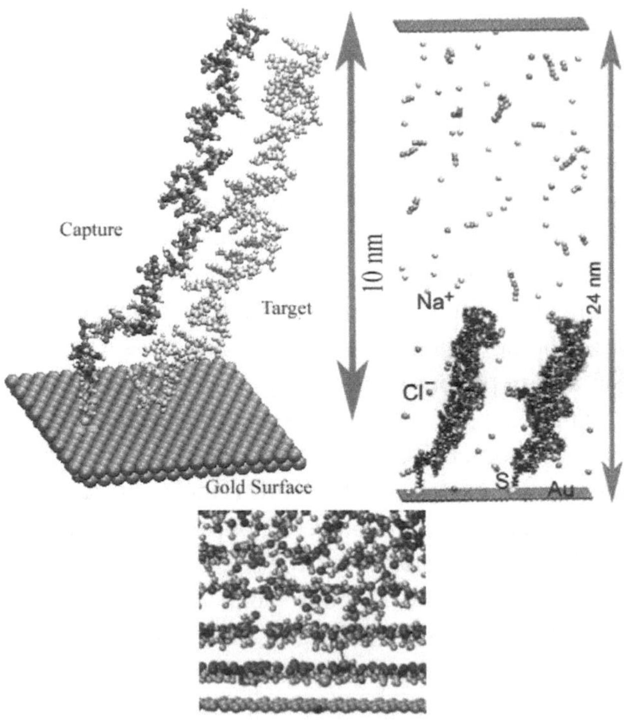

Fig. 8. (a) One free single strand (ss) target DNA and one capture ssDNA that is tethered onto a gold surface. (b) MD simulation box 24nm high with 3D periodic boundary conditions. Two dsDNAs are tethered to a gold surface. Water molecules (not drawn) fill the rest of the box, including 160 Na^+ and 80 Cl^- ions. (c) Shear-induced layering in water near the tethered DNA molecule. (From Wang and Sandberg 2007, with permission)

Accordingly, as length and time scales have been brought to the nano/atomic scale and to ns/ps, respectively, it becomes imperative to develop new computational methods to deal with the details directly at the atomic level. One must compute first at the quantum scale, then the atomic scale, then the molecular scale and couple these to a Stokes flow computation for the microfluidic device. Progress has been made in such multiscale computations at the device scale (Wang and Sandberg 2006, Wang et al. 2007), but it is only the beginning of what needs to be done. For example, a typical element of a current biosensor is shown in Fig. 8. It includes single or double-stranded DNA molecule of about 10 nm in length attached to a gold nanosurface substrate. Short chain DNA molecules and functionalized quantum dots are of great utility to the experimental nanosensor community. The element shown here is part of a microflow chamber through which biomolecules are flowing in an ionic water solvent. Understanding the dynamics of the DNA molecules in the shear flow experienced in the device is crucial for the design of responsive biosensors. Embedding multiple such sensors, microfluidic local flow controllers, artificial muscle actuators, and novel functional vehicle structural materials into a vehicle itself represents a major design challenge. Coupling the computations of sensors, actuators, and material dynamics with unsteady aerodynamics and hydrodynamic for complete vehicle performance is another challenge that will need to be accomplished first, before the goal of a unified unsteady autonomous system flow prediction capability, integrated with self-consistent vehicle trajectory prediction and active control is reached. So, while substantial progress has been made on several fronts, there is still much more to do, apparently, before our vehicles will be capable of harmoniously swimming with the fishes and flying with the winged creatures.

Acknowledgments

The support of and collaborations with J. Kellogg of NRL and D. Cylinder of Princeton University on the MAV design investigations have been critical to our unsteady aerodynamic computations. The insights and experimental data provided to us by Profs. M. Dickinson of the California Institute of Technology and M. Westneat of the Field Museum of Natural History in Chicago enabled us to participate in the very exciting field of insect and fish swimming dynamics. We also gratefully acknowledge the continuous collaboration with Prof. R. Löhner of George Mason University who has worked closely with us to develop new computational technology. We also acknowledge the collaborative experimental efforts of Drs. L. Whitman of the NRL Chemistry Division and Guan Wang of the NRL LCPFD on the Nanosensor/Microfluidic multi-scale computational technology developments.

References

Ho S (2003) Unsteady aerodynamics and adaptive flow control of micro air vehicles. PhD dissertation, University of California, LA

Kellogg J, Bovais C, Cylinder D, Dahlburg J, Foch R, Platzer M, Ramamurti R, Sandberg WC (2001) Non-conventional aerodynamics for Micro-UAV's. 16th Intnl UAV Systems Conference, Bristol, UK

Kellogg J, Bovais C, Foch R, Cylinder D, Ramamurti R, Sandberg WC, Gardner J, Srull D, Piper G, Vaiana P (2003) Development and testing of unconventional micro air vehicle configurations. Aerospace Sciences Meeting, AIAA 2003-6356, Reno, NV

Naciri J, Srinivisan A, Sandberg WC, Ramamurti R, Ratna B (2003) Nematic liquid crystal elastomers as artificial muscle materials. Proc 13th International Symposium on Unmanned Untethered Submersibles Technology, Durham, NC

Ramamurti R, Sandberg WC (2001) Computation of the aerodynamic characteristics of a micro air vehicle. In: Mueller TJ (Ed) Fixed and Flapping Wing Aerodynamics for Micro Air Vehicle Applications. Progress in Astronautics and Aeronautics 195:537-554

Ramamurti R, Sandberg WC (2002) Fluid dynamics of flapping aquatic flight in the bird wrasse: three-dimensional unsteady computations with fin deformation. J Exp Biol 205:2997-3008

Ramamurti R, Sandberg WC (2004) The influence of fin rigidity and gust on force production in fishes and insects: a computational study. Aerospace Sciences Meeting, AIAA 2004-404, Reno, NV

Ramamurti R, Sandberg WC (2007) A computational investigation of the 3-D unsteady aerodynamics of Drosophila maneuvering and hovering. J Exp Biol 210:865-880

Ramamurti R, Löhner R, Sandberg WC (1996) Computation of unsteady flow past a tuna with caudal fin oscillation. In: Rahman M, Brebbia CA (Eds) Advances in Fluid Mechanics Vol. 9. Computational Mechanics Publications Southampton, UK, pp 169-178

Ramamurti R, Sandberg WC, Vaiana P, Kellogg J, Cylinder D (2005) Computational fluid dynamics study of unconventional air vehicle configurations. The Aero J of the Roy Aero Soc, 337-347

Shyy W, Smith RW (1997) A study of flexible airfoil aerodynamics with application to micro aerial vehicles, AIAA-97-1933

Shyy W, Jenkins DA, Smith RW (1997) Study of adaptive shape airfoils at low Reynolds numbers in oscillatory flows. AIAA Journal 35:1545-1548

Wang GM, Sandberg WC (2007) Non-equilibrium all-atom molecular dynamics simulations of free and tethered DNA molecules in nano-channel shear flows. Nanotechnology accepted

Wang GM, Sandberg WC, Kenny SD (2006) Density functional study of a typical thiol tethered on a gold surface: rupture under normal or parallel stretch. Nanotechnology 17:4819-4824

Walker JA, Westneat MA (1997) Labriform propulsion in fishes: kinematics of flapping aquatic flight in the bird wrasse, *Gomphosus varius*. J Exp Biol 200:1549-1569

Part III

Biomimetic Swimming or Flying Robots

Chapter 18

Design and Simulations of a Virtual Fishlike Robot Actuated by a Muscle Model

Chunzhao Guo[1] and Zengfu Wang[1]

[1] Department of Automation, University of Science and Technology of China, P.O. Box 4, Hefei 230027, China

Summary. Motivated by trying a new top-down research mode of the study of fish-like robot related problems, we use a Chinese sturgeon (*Acipenser sinensis*) as the base model and rebuild the structures and functions of its swimming apparatus in a virtual fish-like robot, especially for its shape, backbone, fins, and muscular system. The mechanical model of fish's superficial red muscle is established, which is used as the actuator of the virtual fish-like robot. Besides, the virtual fish-like robot also has the same physical and kinematic parameters as the specimen of Chinese sturgeon used in the experiment, such as body length, weight, mass distribution, and undulation parameters etc. Thus, the virtual fish-like robot can mimic the real fish structurally and functionally as realistically as possible, which could be used to perform various virtual experiments and serve as a platform to strengthen the intercrossing between theory and engineering studies of fish-like robot related problems. Furthermore, kinematics simulations of the virtual fish-like robot actuated by fish's superficial red muscle model are conducted, and the experimental results proved the veracity, rationality, and availability of the design of the virtual fish-like robot.

Key words. Fish-like Robot, Biomimetics, Muscle Model, Kinematics Simulation, Virtual Experiment

1 Introduction

Fish have astonishing swimming abilities after millions of years of evolution, which are very close to optimal with regard to their habitats of lives. They can propel themselves by the coordinate motion of the body, fins, and tail, achieving tremendous noiseless, effective, and maneuverable propulsion that has the advantages over the conventional rotary propeller used in ship or underwater vehicles.

Inspired by biomimetics, the fish-like robot is one of the hotspots in the field of bionics in recent years. Many researchers and engineers focus on the design and realization of the mechanical structure of a fish-like robot with the guidance of fish-like robot related theory studies contributed by biologists and fluid mechanics experts, and achieved valuable results. In 1994, MIT developed an eight-link fish-like robot named RoboTuna and subsequent RoboPike (Streitlien et al, 1996). The Northwestern University (2004) applied Shape Memory Alloy (SMA) on the robotic lamprey. In Japan, Kato's lab realized a few kinds of fish-like robots actuated by various mechanical pectoral fins (Kato, 2000, 2006). Tokyo Institute of Technology studied the three-dimensional movements of a dolphin robot (Nakashima et al, 2004, 2006). But the current fish-like robot systems cannot give a good solution due to the restrictions on materials and techniques. On the one hand, the current materials as well as the components used for making fish-like robots cannot satisfy the needs of the theory studies. On the other hand, the swimming mechanism of a fish, especially the kinematics and dynamics characteristics of it, is not well studied. What more important is that the feedback for the engineers on their demands and ideas etc. cannot be effectively returned to the biologists and fluid mechanics experts because of the huge differences in structures between the real fish and their fish-like robots. These problems limited the progress of the fish-like robot.

The motivation of this paper is to try a new top-down research mode of the study of the fish-like robot related problems. As shown in Fig. 1.

We investigate systematically the structures and functions of the swimming apparatus of fish, and a few conclusions are drawn to be the principles of the design of the virtual fish-like robot, especially for the muscular system of fish. The internal reasons of high efficiency of fish swimming are discussed and the mechanical model of fish's superficial red muscle is established, which is used to be the actuator of the virtual fish-like robot. Besides, the physical and kinematic parameters of the specimen of Chinese sturgeon are examined to help us design the virtual fish-like robot. Synthesize the achievements obtained above; we can build a virtual fish-like robot in ADAMS by imitating the specimen as realistically as possible structurally and functionally. In the experiment, kinematics simulations of the virtual fish-like robot actuated by fish's superficial red muscle model are conducted. The experimental results proved the veracity, rationality, and availability of the design of the virtual fish-like robot.

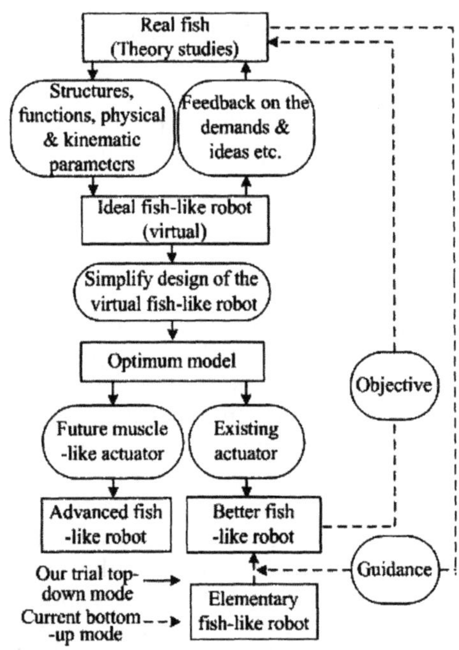

Fig. 1. Research mode

2 The Base Model of the Virtual Fish-like Robot

We use a Chinese sturgeon

(*Acipenser sinensis*) as the base model to rebuild the structures and functions of its swimming apparatus in a virtual fish-like robot, especially for its shape, backbone, fins, and muscular system. Besides, the virtual fish-like robot also has the same physical and kinematic parameters as the specimen of Chinese sturgeon, such as body length, weight, mass distribution, and undulation parameters etc., which are the necessary preconditions to mimic a real fish as realistically as possible.

2.1 Structures and Functions of the Swimming Apparatus

Efficient swimming requires that (Videler, 1993):
1. most of the energy expended by the swimming muscles be turned into appropriate motions of the propulsive surfaces;
2. the generation of propulsive forces from the fish to the water leaves as little kinetic energy in the water as possible;
3. the transmission of forces from the water to the fish body be achieved with a minimum of energy loss through elastic deformations;
4. the resistance of the moving body be as small as possible.

These four requirements are directly related to the structures and functions of the swimming apparatus of fish, which mainly include the shape, fins, backbone, and muscular system.

2.1.1 Shape

The shape of the swimming body should ideally be streamlined, offering the lowest drag for the largest volume. Nevertheless, the shapes of fish bodies vary considerably. Functional anatomists assume that organisms are adequately designed to perform the functions needed to live. So the various shapes of fish bodies are determined by the needs to survival in various habitats of different fish. Spindled-shaped, so called streamlined bodies usually belong to the fast swimmers, and the lateral compression of many diamond-shaped fishes betrays extreme maneuverability.

2.1.2 Fins

Fins of fish are used for propulsion, breaking and steering. Their structure should allow both a subtle interaction with the water and the exchange of large forces. Fins have two major swimming related properties which are the stiffness and shape. For slow-swimming fish in spatially complex environment, steering capacity and maneuverability are more important properties than the ability to generate large propulsive forces. The fins of these fish are flexible, highly adjustable, and with a small aspect ratio (AR); while fast swimmers have stiff, stout fins with high ARs because there are no obstacles in the open ocean where agility probably underlies high speed as a survival factor.

2.1.3 Backbone

The backbone of fish must have two swimming related mechanical properties (Videler, 1993):

1) It should be flexible in lateral directions to allow undulatory swimming motions.
2) It must be incompressible in the longitudinal direction to receive the forward-directed propulsive force from the tail without large losses through elastic deformation.

2.1.4 Muscular System

Increasing demand for higher efficiency to allow longer missions in autonomous underwater vehicles (AUVs) cannot be satisfied by the current actuator materials. An effective way of developing innovative technological solutions is to search for alternative propulsive mechanisms in biological systems, since evolutionary processes have already performed the design iterations and the cost-benefit analysis to optimize particular design for specific functions (Shinjo & Swain, 2004). What's the reason that fish can swim with so high efficiency? Triantafyllou et al. (2000) proposed an explanation, which is that fish are able to utilize the energy that exists in the eddies of an oncoming flow and the oscillating motion of fish swimming can induce flow relaminization which serves to reduce the body drag. As a matter of fact, this is more like one part of the whole explanation from the point of view of external effects or hydrodynamic perspective. The internal effects, i.e. fish's muscular system and the way it drives the swimming motion, should be the other part of the whole explanation.

A. Multiple actuators

Swimming motion is generated by activating alternate sides of segmented muscles in sequential manner. The myomeres and vertebras of the backbone are of staggered arrangement. Thus, one myomere's shortening will affect two vertebras simultaneously with myoseptum as the connective tissue that separates two adjacent myomeres and transmits the force to the vertebra.

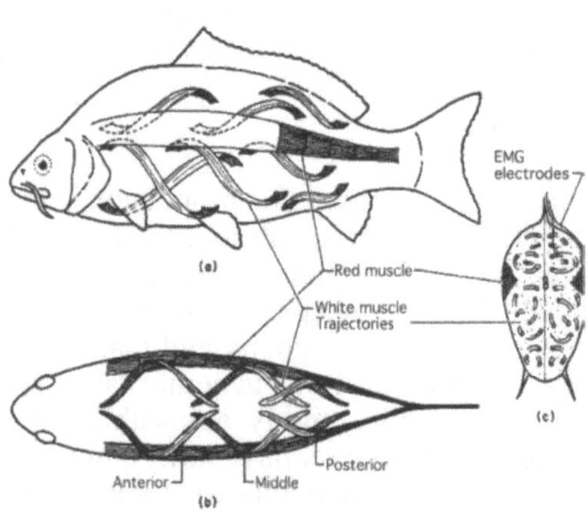

Fig. 2. Different muscle fiber types of fish

A myomere is composed of two types of muscles: oxidative red muscles and glycolitic white muscles, as shown in Fig. 2. Red muscles located along the mid lateral line just under the skin in a wedge formation (superficial red muscle) are primarily for low-intensity and long-duration activities such as steady swimming; thus, it is the primary focus of this study. The remaining muscle volume is occupied by white muscles, which are used for short bursts of high-intensity activities, such as fast starts and escape maneuvers. In other words, a fish has multiple actuators to produce different motions. This fact provides another reason why fish have so high efficiency. They, like AUVs, have to be able to perform their motor activities for a low energetic cost (food is the limitation for fish, battery life for AUVs). Increase muscle speed results in increased energetic cost, a basic tenet in the design of muscular system is that muscle speed is set to be sufficiently fast to perform a given activity, but not faster than necessary, as this would entail a large waste of energy. Thus, although the white muscle could easily generate the mechanical power needed for slow steady swimming, it would require far more energy to do so than the red muscle, and hence the red muscle is used for slow swimming (Rome, 2005).

B. Elastic energy storage

Several studies have suggested that fish can conserve the energy dissipated during deceleration of locomotors by elastic materials mainly consisted of muscle and tendon, and recycle it in subsequent motion, as supplemental energy, to reduce energy input (Shinjo & Swain, 2004). In fact, muscles often perform cyclical activities so that the kinetics of muscle activation and deactivation (relaxation) are equally important. When shortening occurs during relaxation, the elastic elements in muscle can release its stored energy to recoil and reduce energy input from the source. If there are no elastic elements in muscle, all this energy will be converted to heat and dissipated in muscle and water, and cause energy waste.

C. Mechanical model of fish's superficial red muscle

The classic Hill model of muscle (Hill, 1938) is used to analyze the mechanical properties of fish's superficial red muscle. As shown in Fig. 3, *CE* is the force generator, which composed of the actin-myosin cross-bridges, *SEE* is the elastic element in series which is composed of 2 fractions including active fraction (localized in actin-myosin cross-bridges) and passive fraction (localized in tendons). *PEE* is the passive elastic element, which is derived from muscle membranes.

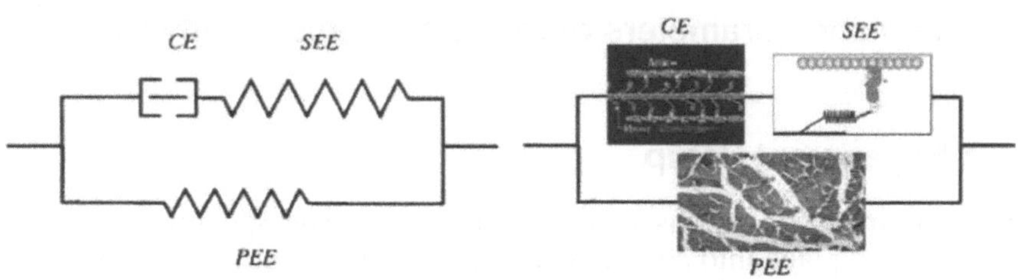

Fig. 3. Mechanical model of fish's superficial red muscle

The mechanical properties of the elements can be described by the following equation:

$$F_{jCE} = a_j (b_j + v_{CE})^{-1} + c_j \quad j = \begin{cases} ec & v_{CE} > 0 \\ co & v_{CE} < 0 \end{cases} \tag{1}$$

$$F_{SEE} = k_{SEE}(l_{SEE} - l_{0SEE}) \tag{2}$$

$$F_{PEE} = k_{PEE}(l_{PEE} - l_{0PEE}) + \eta v_{PEE} \tag{3}$$

where ec, co represent eccentric and concentric shortening, respectively, l_{SEE}, l_{PEE} and l_{0SEE}, l_{0PEE} are the length and rest length of SEE and PEE, respectively, v_{CE}, v_{PEE} are the velocities of CE and PEE, respectively, r_0, r_1, r_2, a_j, $b_j, c_j, k_{SEE}, k_{PEE}, \eta$ are fitting constants, which can be obtained by fitting the force-length, force-velocity curves in Rome (2005).

2.2 Physical Parameters of the Specimen Used in Experiment

A BCF (Body and/or Caudal Fin) swimmer—Chinese sturgeon is chosen as the base model of the virtual fish-like robot because of its body rigidity and cruising ability. The specimen of Chinese sturgeon is shown in Fig. 4(a), which is 0.48m long, 0.59kg weight. After the experiment, the specimen was deeply frozen and then cut into thin slices with a fine diamond saw to obtain the mass distribution between the head and tail, as shown in Fig. 4(b).

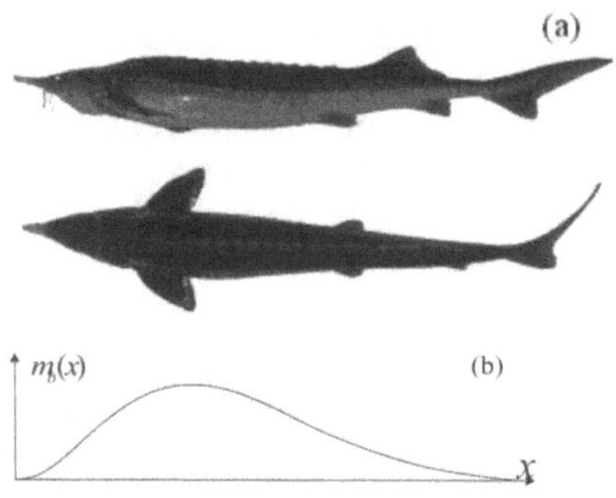

Fig. 4. (a) Lateral and top views of the specimen of the Chinese sturgeon **(b)** Mass distribution

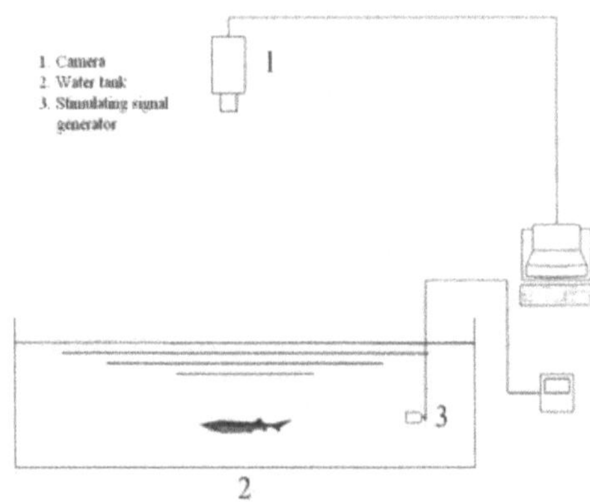

Fig. 5. Experiment apparatus

2.3 Kinematic Parameters of the Specimen's Steady Swimming

2.3.1 Experiment Setup

The experiment apparatus is shown in Fig. 5. The specimen of Chinese sturgeon was place into a water tank with the measurement section of 2.1m long, 1.75m wide and 0.6m deep. Video tapes of the steady swimming of the specimen were obtained by using Nikon 7900 digital camera operating at a rate of 30 frames/s. The stimulating signal generator was used to make the specimen start swimming.

2.3.2 Kinematics Model of the Chinese Sturgeon's Steady Swimming

The steady swimming of the Chinese sturgeon involves the undulation of the entire body, whose large amplitude undulation is mainly confined to the posterior half of the body, and thrust is produced by a rather stiff caudal fin. The amplitude of this undulation is close to zero in the anterior portion of the fish, increasing drastically in the immediate vicinity of the trailing edge. Based on this information, in the body-fixed coordinate shown in Fig. 6, the undulation of fish body can be described by the equation as below, which is originally suggested by Lighthill (1960),

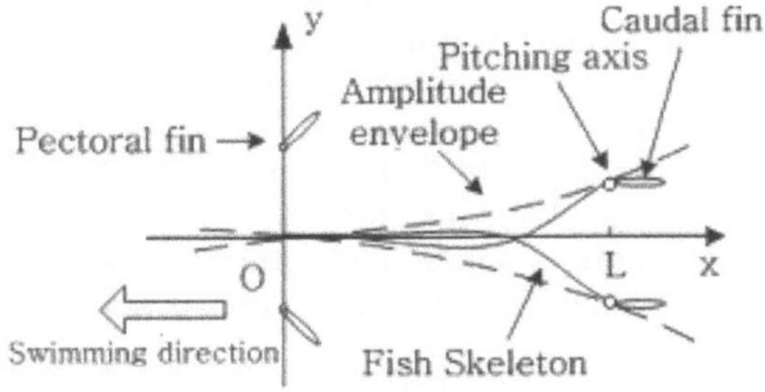

Fig. 6. Physical model of fish swimming

$$y(x,t) = (c_1 x^2 + c_2 x)\sin[2\pi(x/\lambda - t/T)] \quad L/2 \le x \le L \tag{4}$$

where $y(x,t)$ denotes the lateral displacement of the fish body, x denotes the displacement along x-axis which ranges in the posterior half of the body, c_1 is the quadratic wave amplitude envelope, c_2 is the linear wave amplitude envelope, λ and T are body wave length and period, respectively.

The caudal fin plays a very important role in fish swimming. The motion of the caudal fin is a combination of heave and pitch. And the pitching movement leads the heaving motion by ψ. The heaving motion equation of the pitching axis is:

$$y_{PA}(L,t) = (c_1 L^2 + c_2 L)\sin[2\pi(L/\lambda - t/T)] \tag{5}$$

And the pitching movement can be described by the equation:

$$\theta(t) = \theta_{\max}\sin[2\pi(x/\lambda - t/T) + \psi] \tag{6}$$

where $\theta_{\max}$ is the pitch angle amplitude.

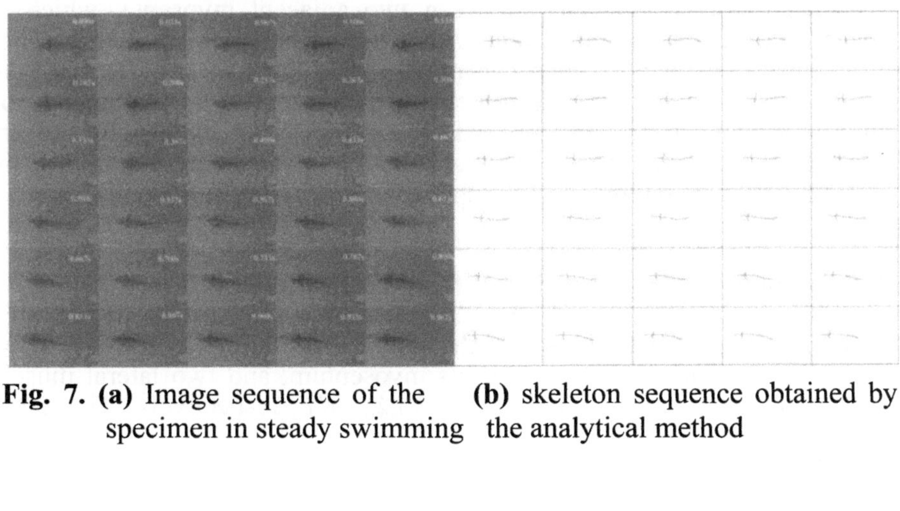

Fig. 7. (a) Image sequence of the specimen in steady swimming **(b)** skeleton sequence obtained by the analytical method

Given the mathematical model of the steady swimming, the dorsal view image sequence of the swimming specimen of Chinese sturgeon at a speed of 17.8cm/s (shown in Fig. 7(a)) was processed to obtain the kinematics parameters automatically by the analytical method based on image sequences (Guo & Wang, 2005). First the silhouette of the bass body was extracted, and then an energy function was employed to obtain the skeleton of the fish body (shown in Fig. 7(b)). In this way, we could get the precise deformation data and the precise displacement data of the bass body in swimming at each sampling moment.

By using a spline based curve fitting technique to fit the data above, we could obtain $c_1 = 0.002103$, $c_2 = 0.02379$, $\lambda = 43.43cm$, $T = 1s$, $L = 36.2cm$, $\psi = 72°$, $\theta_{\max} = 30.5°$.

3 Design and Simulations of the Virtual Fish-like Robot

3.1 Structure of the Virtual Fish-like Robot

Fig. 8. Structure of the virtual fish-like robot

The virtual fish-like robot is built in ADAMS, as shown in Fig. 8. Its backbone is made up of 40 links and joined together by Revolute Joints, just imitating the vertebral column of real fish. Polylines are formed to the shape of myoseptum and attached to each of the links. The muscle model as actuator of the virtual fish-like robot is connected between two adjacent myosepta, which is made up of a Translational Joint with a Spring in series and two parallel viscoelastic elements. It also can move inside the fish body to serve as white muscle to perform high-intensity motions after the parameters of muscle model are properly revised. The geometrical dimensions and arrangement of all the virtual fish-like robot components are determined by examining the counterparts of the specimen of Chinese sturgeon.

3.2 Propulsive Model of the Virtual Fish-like Robot

Considered that the backbone of the virtual fish-like robot consists of N links, let the length of each link, its myoseptum, and two lateral muscle models be l_i, a_i, $l_{il}(t)$, and $l_{ir}(t)$ $(i = 1,2,...,N)$, respectively. The angle between link l_i and x axis is $\theta_i(t)$, and angle between two adjacent links is $\phi_i(t)$. To mimic a fish swimming defined by $y(t)=f(x(t),t)$, the virtual fish-like robot need to move all its joints to coincide to the body curve of the real fish at every sampling moment, especially for the propulsive oscillatory part of the robot. Therefore, the following equation set must be satisfied.

$$\begin{cases} (x_{i-1}(t)-x_i(t))^2+(y_{i-1}(t)-y_i(t))^2=l_i^2 \\ y_i(t)=f(x_i(t),t) \\ tg\,\theta_i(t)=(y_{i-1}(t)-y_i(t))/(x_{i-1}(t)-x_i(t)) \\ \phi_i=\theta_{i+1}-\theta_i \\ c_{i-1}=\sqrt{(\frac{a_{i-1}}{2})^2+(\frac{l_{i-1}}{2})^2} \quad c_i=\sqrt{(\frac{a_i}{2})^2+(\frac{l_i}{2})^2} \\ l_{il}(t)=\sqrt{c_{i-1}^2+c_i^2-2c_{i-1}c_i\cos[\pi+\phi_i(t)-arctg\,(\frac{a_{i-1}}{l_{i-1}})-arctg\,(\frac{a_i}{l_i})]} \\ l_{ir}(t)=\sqrt{c_{i-1}^2+c_i^2-2c_{i-1}c_i\cos[\pi-\phi_i(t)-arctg\,(\frac{a_{i-1}}{l_{i-1}})-arctg\,(\frac{a_i}{l_i})]} \end{cases} \tag{7}$$

3.3 The Superficial Red Muscle Model Actuator and Its Control Strategy

The structure of the superficial red muscle model is shown in Fig. 9.

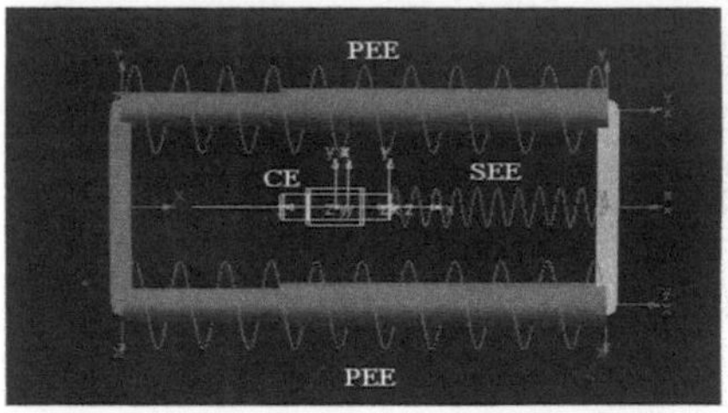

Fig. 9. Structure of the superficial red muscle model

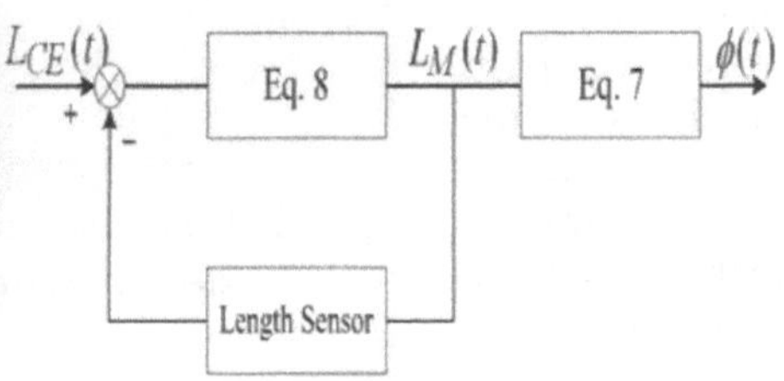

Fig. 10. The closed-loop control circuit of the muscle model actuator

The length and force relationship between each element can be described as follows,

$$\begin{cases} L_M(t)=L_{PEE}(t)=L_{CE}(t)+L_{SEE}(t) \\ F_M(t)=F_{PEE}(t)+F_{CE}(t) \\ F_{CE}(t)=F_{SEE}(t) \end{cases} \tag{8}$$

where $L_M(t), L_{PEE}(t), L_{CE}(t), L_{SEE}(t), F_M(t), F_{PEE}(t), F_{CE}(t), F_{SEE}(t)$ are the length and force of the corresponding element, respectively. The length of the muscle model is chosen as the control variable. And the closed-loop control (shown in Fig. 10) is adopted to decrease the steady-state response time caused by the *SEE* and increase the robustness of the system.

3.4 Kinematics Simulations of the Virtual Fish-like Robot Actuated by Fish's Superficial Red Muscle Model

Kinematics simulations of the virtual fish-like robot actuated by fish's superficial red muscle model are conducted in ADAMS. Fig. 11 gives the image sequence of the virtual fish-like robot in steady swimming, which coincides exactly with the image sequence shown in Fig. 7. Fig. 12 is the superimposed body bending sequence of the virtual fish-like robot, shown every 0.1s. Fig. 13 gives the traces of the joints in the propulsive part of the fish-like robot. Fig. 14 shows the lengths of the muscle model actuators in the propulsive part. As we can see, the more posterior it is, the more resistance it will endure, the smaller change the muscle model is, the bigger stress it will generate. Fig. 15 gives the angles of the joints in the propulsive part, which shows the body should be more flexible near the caudal peduncle.

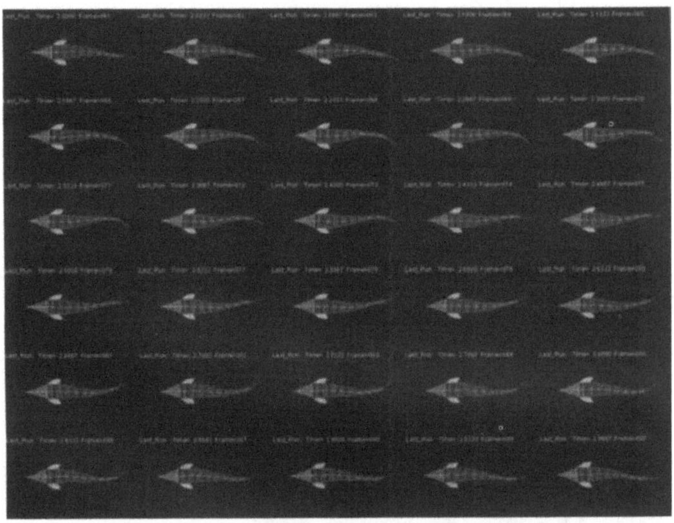

Fig. 11. Image sequence of the virtual fish-like robot in steady swimming

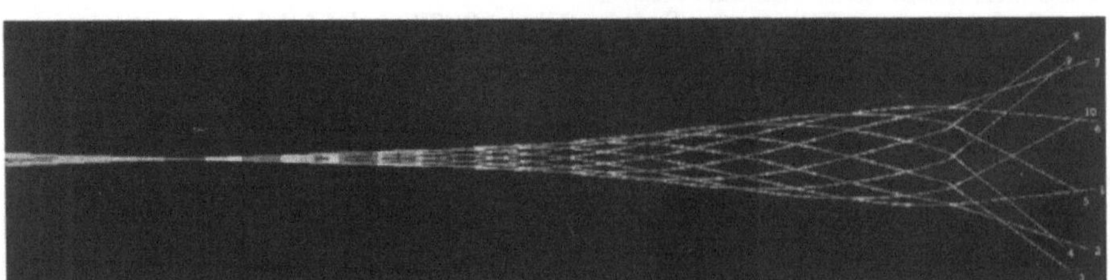

Fig. 12. Superimposed body bending sequence of the virtual fish-like robot

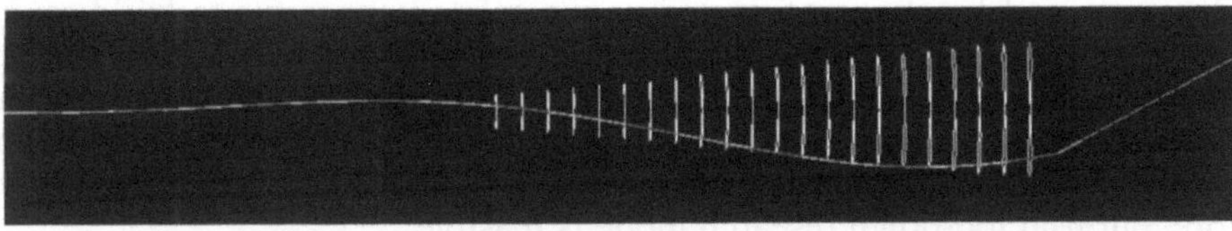

Fig. 13. Traces of the joints in the propulsive part of the fish-like robot

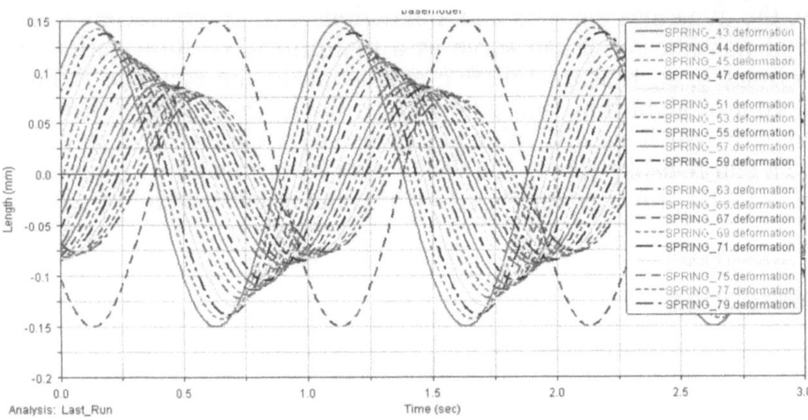

Fig. 14. Lengths of the muscle model actuators in the propulsive part

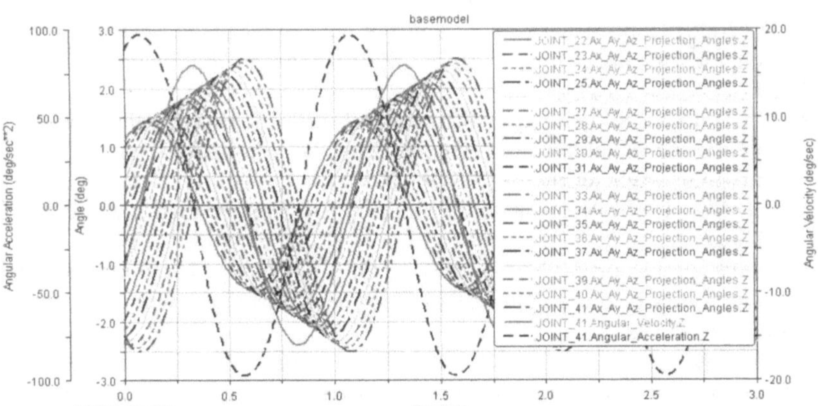

Fig. 15. Angles of the joints in the propulsive part, and the angular velocity, angular acceleration of the last joint

4 Conclusions

In this paper, the design and kinematics simulations of a virtual fish-like robot actuated by fish's superficial red muscle model is presented. In order to develop a fish-like robot that can mimic a real fish as realistically as possible, we use a Chinese sturgeon as the base model and examine the structures and functions of its swimming related apparatus to build a virtual fish-like robot with the fish's superficial red muscle model as the actuator in ADAMS, which also has the same physical and kinematic parameters as the specimen of Chinese sturgeon. The kinematics simulation results proved the virtual fish-like robot can swim like a real fish. Therefore, our trial top-down research mode is feasible.

This paper is the first step of the study of virtual fish-like robot. In future, the dynamics analysis and simulations will be conducted so that the virtual fish-like robot can serve as a platform to perform various virtual experiments, such as quantitative evaluation of the structure design, motion control algorithm, and autonomous navigation method of a fish-like robot etc., which are difficult and expensive in real world, even though we have a real fish-like robot. The virtual fish-like robot built in this paper can deal with the problems easily, efficiently and reliably with low cost.

References

Guo CZ, Wang ZF (2005) Analysis and modeling of fish swimming locomotion mechanism based on image sequences, Proceedings of The Fifteen(2005) International Offshore and Polar Engineering Conference, 4: 565-570

Hill AV (1938) The heat of shortening and the dynamic constants of muscle, Proc. R. Soc. Lond. B. Biol. Sci. 126: 136-195

Kato N (2000) Control performance in the horizontal plane of a fish robot with mechanical pectoral fins, IEEE Journal of Oceanic Engineering, 25(1): 121-129

Kato N, Ando Y, Shigetomi T, Katayama T (2006), Biology-inspired precision maneuvering of underwater vehicles (part 4), International journal of offshoreand polar engineering, 16(3): 195-201

Lighthill, MJ (1960) Note on the Swimming of Slender Fish, J.Fluid Mech., 9: 305-317.

Nakashima M, Takahashi Y, Tsubaki T, Ono K (2004), Three-dimensional maneuverability of the dolphin robot, Bio-mechanisms of swimming and flying, Springer, 79-92

Nakashima M, Tsubaki T, Ono K (2006), Three-dimensional movement in water of the dolphin robot – control between two positions by roll and pitch combination-, Journal of robotics and mechatronics, 18(3): 347-355

Rome LC (2005) Principles of actuation in the muscular system of fish, IEEE Journal of oceanic engineering, 30(3): 630-646

Shinjo N, Swain GW (2004) Use of a shape memory alloy for the design of an oscillatory propulsion system, IEEE Journal of Oceanic Engineering, 29(3): 750-755

Streitlien K, Triantafyllou GS, Triantafyllou MS (1996) Efficient foil propulsion through vortex control, AIAA Journal, 34: 2315-2319

Triantafyllou MS, Triantafyllou GS, and Yue DKP (2000) Hydrodynamics of fish-like swimming, Annual Review of Fluid Mechanics, 32(1): 33-53.

Videler JJ (1993) Fish swimming, London, U.K. Chapman & Hall

Chapter 19

Development of Fish Robots Powered by Fuel Cells : Improvement of Swimming Ability by a Genetic Algorithm and Flow Analysis by Computational Fluid Dynamics

Yogo Takada[1] , Toshiaki Tamachi[1], Satoshi Taninaka[1], Toshinaga Ishii[1], Kazuaki Ebita[1] and Tomoyuki Wakisaka[1]

[1] Department of Mechanical and Physical Engineering, Graduate School of Engineering, Osaka City University, 3-3-138 Sugimoto, Sumiyoshi-ku, Osaka 558-8585, Japan

Summary. A fish robot with image sensors is useful to seek disaster victims in flood areas. In order to perform rescue operation for long time, the method of supplying electric power is important. The purpose of this study is to develop fish robots powered by polymer electrolyte fuel cells (PEFCs).

For investigating the propulsion mechanism of a fish, the water flow around a fish was analyzed with an original computational fluid dynamics (CFD) code "GTT" developed by the authors. The flow velocity distributions around a real swimming fish were measured by particle image velocimetry (PIV). From both results, it has been confirmed that the tail fin generates the reverse Karman vortex street in the downstream region.

The authors created two fish robots powered by PEFCs. Type I fish robot wriggled its body variously with six actuators and the operation sequence of the actuators was determined for faster or more efficient swimming by means of a genetic algorithm (GA). Type II fish robot with a tail fin oscillating simply was powered by an original small-sized passive-type PEFC called "Power Tube". The water flow around this robot was numerically analyzed, and it has been confirmed that the tail fin of the fish robot generates the reverse Karman vortex street to get propulsive force.

Key words. Fish Robot, Fuel Cell, GA, CFD, PIV

1 Introduction

It is useful to use a small fish robot with image sensors in that it can access to any small area where a rescue team finds it difficult to access when a flood disaster occurs due to downpour, earthquake, tidal wave, etc. A fish robot as small as a killifish is well suited for such purpose because it would otherwise take too much time to reach the disaster area if it is too small. Furthermore, since it is difficult to supply energy to the fish robot in a remote disaster area by wireless means, an energy source has to be provided in the fish robot. In this study, it is attempted to use a polymer electrolyte fuel cell (PEFC) as the long-lasting power source for a small fish robot.

In general, a fish swims in various ways depending on different circumstances. An example is Type I which refers to the way that a horse mackerel swims by wriggling all over the body, and another example is Type II which refers to the way that a globefish swims by oscillating its tail fin alone. This study aims to investigate the propulsion mechanism of fishes by using the two types of fish robots. First of all, in order to conduct experiments in various ways of swimming achievable by means of a fish robot of Type I, a fish robot (260 mm in length) with six actuators which can wriggle its body variously is created for a trial purpose. Then, using a genetic algorithm (GA)(Ishida et al. 1997), it is attempted to optimize the operation sequence of the actuators by experiments for swimming with high acceleration or for swimming with high propulsion efficiency, and the difference in swimming between the two optimized cases is investigated.

Furthermore, by using a small fish (a young carp: about 120 mm in length), the water flow around the fish is investigated by flow velocity measurement based on particle image velocimetry (PIV) and also by flow analysis based on computational fluid dynamics (CFD) with the GTT (Generalized Tank and Tube) code (Wakisaka et al. 1995), which had been developed originally by the authors on the basis of the finite volume method and the SIMPLEC algorithm.

In addition, a prototype small fish robot (110 mm in length) of Type II is created. The fish robot moves only its tail fin from left to right by using an ultra-small motor and a small-sized and light-weighted PEFC (called "Power Tube"(Ishii et al. 2005)) developed originally by the authors. In order to investigate the propulsion mechanism by the tail fin, CFD-based analysis of the water flow around the fish robot is conducted.

2 Attainment of High-speed or High-efficiency Swimming for the Type I Fish Robot by Means of GA

Figure 1 shows the structure of the Type I fish robot created in this study. This fish robot is composed of a 260 mm long fish-shaped rubber sheet (including a fish head, trunk and tail fin) to which six DC motors are fixed, and is moved by six motor-string units. Once a string is wound on a motor axle, the end of a stainless steel pipe is pulled toward the motor to bend the rubber sheet in this direction. This fish robot can wriggle its body in various ways by changing the operation sequence of the motors. However, it was difficult to artificially determine the optimum operation sequence of the six motor-string units mounted on the fish robot. Therefore, its operation sequence was determined using GA.

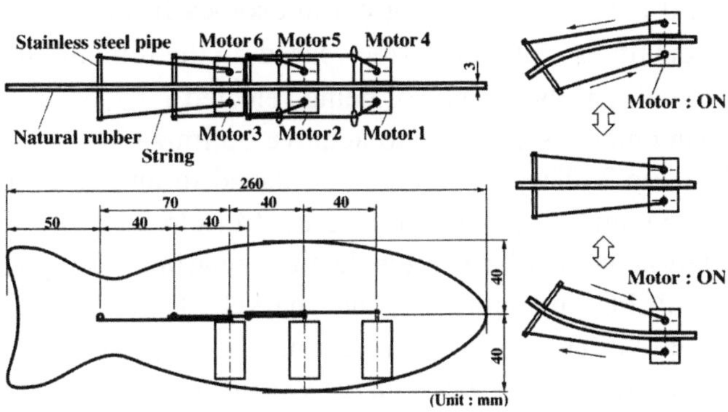

Fig.1. Structure of a fish robot

A genetic algorithm (GA) is an optimization method that imitates the evolution and natural selection of organisms. In the GA, a quasi-optimal solution is sought by repeating the hereditary operation in the evolutional process of an organism.

One cycle of the fish robot motion was divided into 6 phases, and then 30 pieces of chromosomes in 18-bit length (3 phases × 6 motors) were generated per one generation. Calculation based on the GA was conducted for up to several tens of generations. Chromosomes were prepared only for 1st, 2nd and 3rd phases, because the fish robot motion during the first half period (1st–3rd phases) and that during the latter half period (4th-6th phases) were symmetrical each other.

A fitness function f_{iv} was defined as shown in Eq. (1) in consideration of "speed" alone.

$$f_{iv} = V / V_{\max} \qquad (1),$$

where V referred to the fish robot speed, and $V_{\max}$ referred to the maximum speed ($V_{\max}$ = 63 cm/3.6 sec=17.5 cm/sec) at which the fish robot reached the aquarium wall in 5 cycles (3.6 sec).

In addition, another fitness function f_{ip} was defined as shown in Eq. (2) in consideration of both "speed" and "electric power consumption".

$$f_{ip} = \frac{V / V_{\max}}{P / P_{\max}} \qquad (2),$$

where P referred to an average output of the power source, and P_{max} referred to the maximum output of the power source when all the motors were turned on. The cross-over probability in chromosomes was set at 0.3, and the mutation probability in chromosomes was set at 0.05 in the process of selecting 30 pieces of surviving chromosomes from the population according to their fitness function values based on the roulette rule (the better the fitness, the bigger the chance of being selected).

The experimental system used to achieve optimization by means of GA is outlined in Fig.2. The fish robot was placed in an aquarium sized 900 mm × 600 mm, and a power source, a control circuit and a computer for GA computation were placed outside the aquarium. A hydrogen-fueled passive-type PEFC available on the market (Daido Metal Co., Ltd., PFC1212, rated voltage: 12 V, rated output: 12 W (8 W in continuous operation)) was used as the power source. Various operation schedules of the six motors were generated by GA. Therefore, it was necessary to repeat several hundred times the procedure that the fish robot started swimming and stopped, and then returned to the initial position. Two motor-string units with weights shown in the right part of the aquarium in Fig.2 were used for taking the fish robot back to the initial position. The traveling distance of the fish robot was measured with an optical sensor installed on the float of the fish robot for measuring the distance between the float and the reflection plate installed slantingly. The signal concerning the fish robot location was sent to the computer via an A/D board, and the average swimming speed of the fish robot was obtained by measuring the traveling distance in 3.6 seconds after start of swimming. The operation schedule of the motors was generated by GA on the personal computer, and the driving signal was sent to each motor driver via a digital I/O board.

When the actuator sequence of the fish robot was optimized for "speed" alone by GA, the fish robot swam as shown in Fig.3, where the fish robot photographs at every 0.13 sec were shown.

It had been considered that it would be best to use motors 1 and 2 in phase 1, motors 2 and 3 in phase 2, and motors 3 and 4 in phase 3 until GA was applied. The experimental result with this prescribed operation schedule and the experimental results with the above-mentioned two schedules obtained for the two optimization targets by means of GA were compared. Evaluation results are shown in Fig.4, where 'A' refers to the prescribed schedule, 'B' refers to the GA-generated schedule focusing on 'speed', and 'C' refers to the GA-generated schedule focusing on 'propulsion efficiency'. The results show that 'B' is far better in terms of speed, and 'C' is even much better in terms of propulsion efficiency. In the case of the schedule 'B', namely the speediest case, all motors are used actively in turn. This means that the motion of a Type I fish which swims by deforming all over the body like a snake is advantageous when speed or acceleration is important. In the case of the schedule 'C', namely the most efficient case, few motors are used for saving electric power consumption.

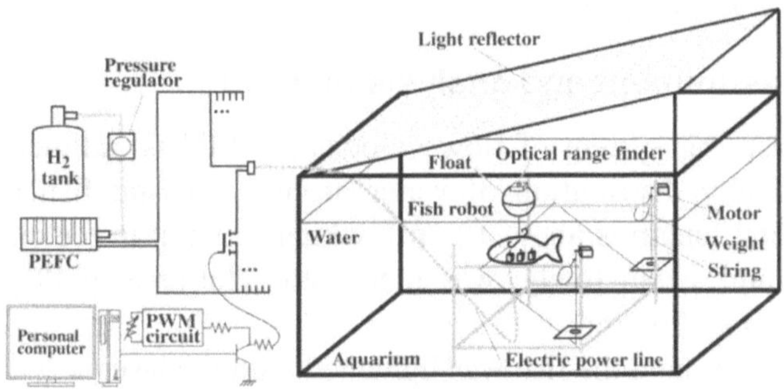

Fig.2. Experimental system for improving the performance of the fish robot by means of GA

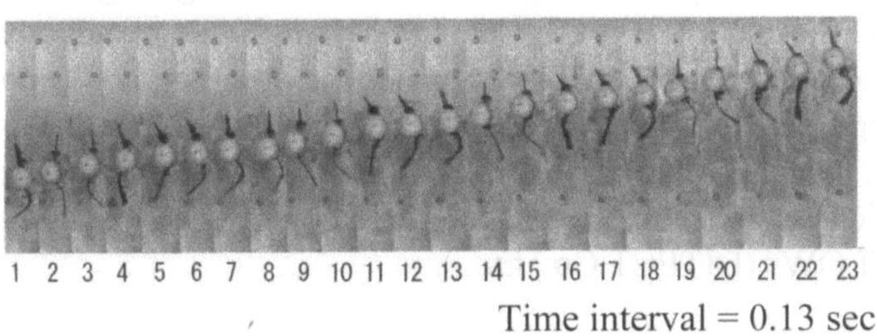

Fig.3. Swimming of the fish robot optimized for "speed" alone by GA

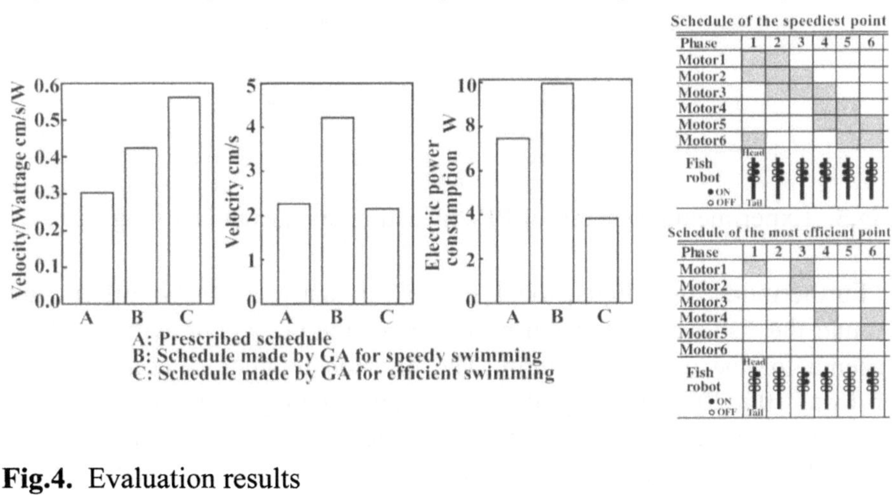

Fig.4. Evaluation results

3 PIV Measurement and CFD Analysis

The water flow around a real swimming fish was measured by PIV and analyzed two-dimensionally by CFD in order to make its swimming mechanism clear and to create a fish robot with better propulsive performance for a trial purpose.

3.1 Measurement and analysis method

The measuring system of flow velocity by PIV is outlined in Fig.5. The flow velocity distribution of water around a living fish (a young carp, whose total length is about 12 cm), which was put in the circulating water channel and swam freely, was measured. The area of PIV measurement was 110mm×76.2mm.

The area around the fish was lightened with the laser seat from the side of the measurement chamber. An LD-excited YAG laser (KATOUKOUKEN Co., Ltd., Green Laser Sheet 50m/G, beam power: 50 mW) and a semiconductor laser (beam power: 16 mW,) were used as lighting sources. A high-speed CCD camera (DITECT Co., Ltd., HAS-200R, photographing speed: 200 frame per second) was placed under the observation chamber. Successive images of tracer particles (Daicel-Degussa Ltd., Daiamid (Nylon 12 Powder), diameter: 40 microns) were taken by the camera, and flow velocity distributions were obtained through correlation calculations using a software "DIPP-FLOW" (DITECT Co., Ltd.,).

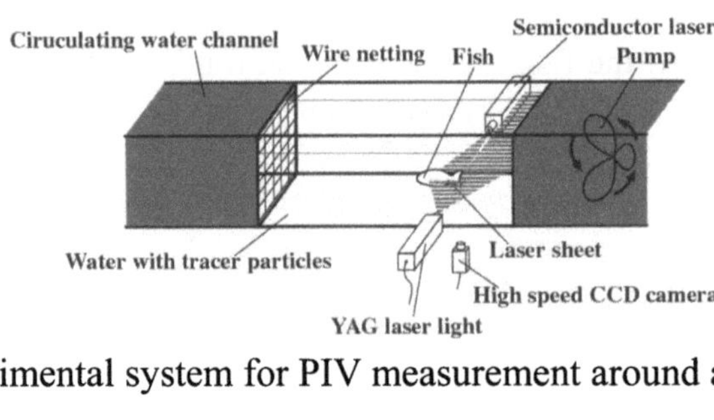

Fig.5. Experimental system for PIV measurement around a living fish

Furthermore, two-dimensional numerical analysis of the water flow around the fish was conducted by means of the GTT code with the TVD scheme for the convection term. The division number of computational grid was 105×203, and the grid spacing in the vicinity of the fish was made fine. To generate moving grids in accordance with the tail fin motion, plural grids were prepared based on the moving tail fin images taken by the high-speed camera. Coordinates of grid points around the tail fin at each computational time step were determined by interpolating the coordinates of the two grids before and after the time step. As the upstream boundary condition in computation, the velocity of water flow at the upstream end was uniformly set at 320 mm/s (representative velocity U_m).

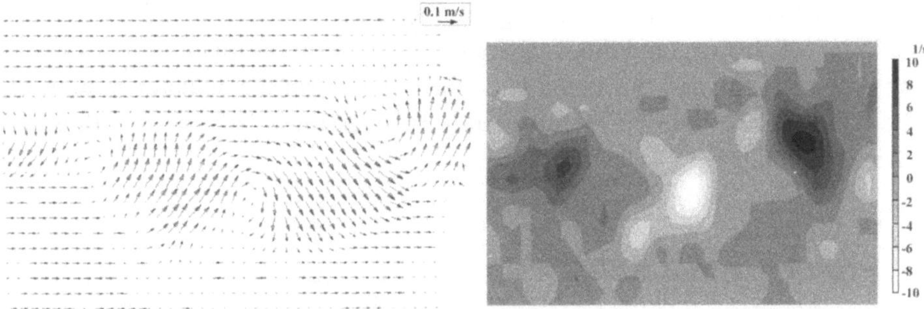

(a) PIV measurement (left: velocities, right: vorticities)

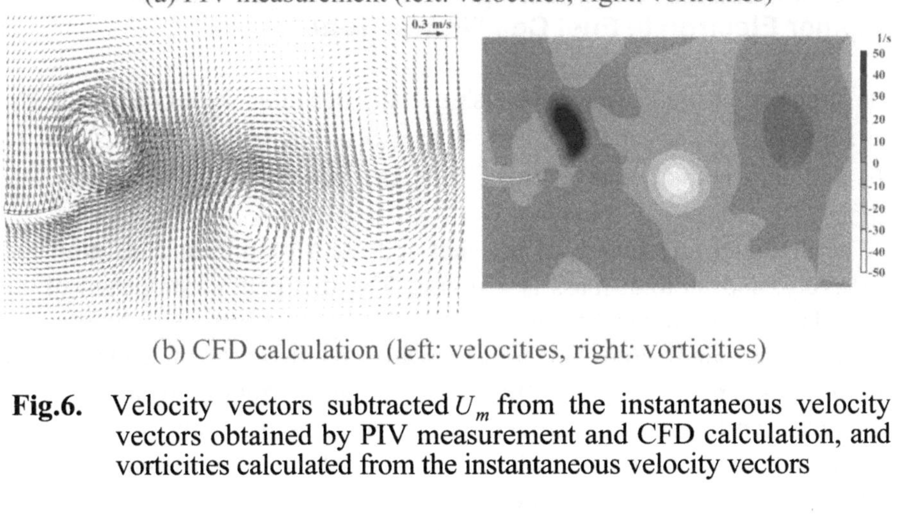

(b) CFD calculation (left: velocities, right: vorticities)

Fig.6. Velocity vectors subtracted U_m from the instantaneous velocity vectors obtained by PIV measurement and CFD calculation, and vorticities calculated from the instantaneous velocity vectors

3.2 Measurement and calculation results

The relative velocity vectors obtained by subtracting the representative velocity U_m from the instantaneous flow velocity vectors around the living fish and the vorticities calculated from the instantaneous flow velocity vectors are shown in Figs. 6(a) and (b), which were obtained by the measurement based on PIV and the numerical analysis based on CFD, respectively.

The results of PIV measurement and CFD calculation show that the reverse Karman vortex street is ejected in the main flow direction in the downstream region. In addition, the measurement and calculation results have shown that the positions of the vortices behind the tail fin almost ac-cord with each other. But the rotation speeds of the vortices differ from each other. The reason for this may be that the CFD calculation was conducted in two-dimension.

The operation of a tail fin which can generate the clear reverse Karman vortex street has been explored on the basis of CFD analysis. As a result, it has been found that the swimming number which is determined by the velocity of a fish robot and the frequency of the tail fin oscillation, the flexibility of the tail fin, and the signal waveform to control the actuator motion of the fish robot have to be set adequately for creating a fish robot with good swimming ability.

4 Prototype of a Small Fish Robot

4.1 Small-sized and Light-weighted Passive-type Cylindrical Polymer Electrolyte Fuel Cell "Power Tube"

The structure of the Power Tube (Ishii et al. 2005) developed originally by the authors is shown in Fig.7, and its specifications are shown in Table 1. The authors tried to use this Power Tube as a power source of a small fish robot.

The diameter and length of the fuel cell are 13.5 mm and 30 mm, respectively, and its total mass is 2 grams. The fuel cell is composed of an ABS plastic tube (outer diameter: 8 mm, thickness: 1 mm, length: 30 mm, having many small holes on its side wall) and two sets of a membrane electrode assembly (MEA), diffusion layers and current collectors, which are connected in series. Hydrogen was supplied as fuel into the tube, and air was fed from the outside of the tube by natural convection. This fuel cell is a passive type, namely it does not necessitate any auxiliary devices.

The MEAs were prepared using the authors' original "Cast Method" (Koike et al. 2003). For uniting two catalytic layers (they were formed in carbon cloths) with an electrolyte membrane (ASAHI KASEI Co., Ltd., Aciplex-S S-1002, thickness: 56 microns) on its each surface, a hot press machine (TESTER SANGYO, Co., Ltd., SA-401-M) was used. The hot press condition was as follows; time: 10 minutes, temperature: 140 deg.C, pressure: 5.1 MPa. As catalyst for anode, Pt-Ru/C (Electro Chem Co. Ltd., EC-20/10-Pt/Ru; Pt: 20 wt%, Ru: 10 wt%, Vulcan XC-72R Carbon) was used, and that for cathode, Pt/C (Electro Chem Co. Ltd., EC-20-Pt; Pt: 20 wt%, Vulcan XC-72R Carbon) was used. As electrolyte solution, Aciplex-S solution (ASAHI KASEI Co., Ltd., Aciplex-SS, SS-950) was used. The platinum density in the catalytic layer was 1.0 mg/cm^2. As diffusion layers, Teflon-treated carbon cloths were used.

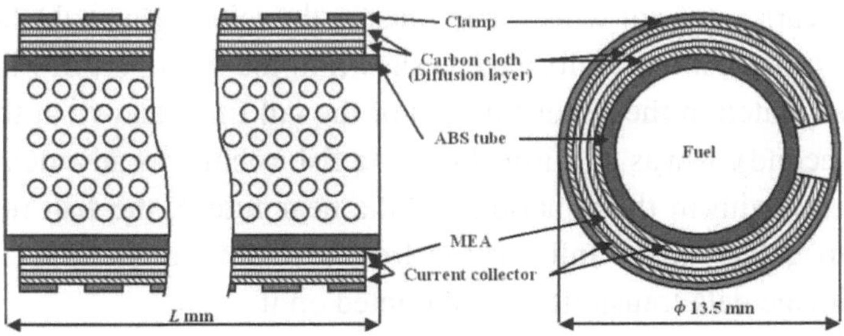

Fig.7. Configuration of the Power Tube

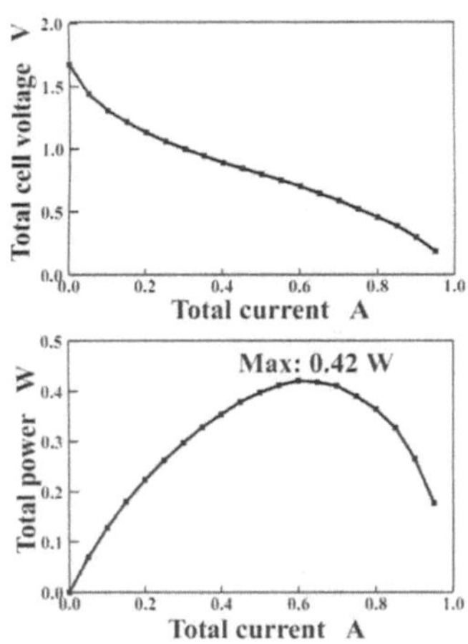

Table 1. Specifications of the Power Tube

	H_2-type
Length	L = 30 mm
MEA (in series)	8 mm × 26 mm, 2 pcs.
Current collector	Ti plated with Au
Clamp	O-ring (6 pcs.)
Total mass	2.0 g
Fuel	Hydrogen

Fig.8. Performance of the H_2-type Power Tube

The performance of the Power Tube is shown in Fig.8. It has been confirmed that the Power Tube created here has sufficient output power for a small fish robot, since it can generate electric power up to 0.42W.

4.2 Creation of a small fish robot for a trial purpose

A small fish robot of Type II (total length: 110 mm) was created for a trial purpose. Its photograph and structure are shown in Figs. 9 and 10, respectively. As shown in these figures, this fish robot has an ultra-small motor and its controller, a wireless signal receiver, and a power source which is the H_2–fueled Power Tube mentioned above. Since this fuel cell was designed to feed air from outside, it was mounted on the upper part (above the water surface) of the fish robot. Hydrogen was filled into a bag and supplied to the fuel cell using a tube. Motor rotation was transformed into oscillating motion of the tail fin (its period was 0.18 seconds) through reduction gears, a pair of wheels and links in the robot. Since the tail fin was made of rubber, it was deformed as shown in the right part of Fig.10_when it was oscillated in the water (motion of the tail fin is shown at the interval of 0.01second). It was confirmed that the fish robot was advancing when it was set to swim in the aquarium. In the near future, the fish robot is expected to gather image information from a remote area if a CCD camera and an image data transmitter are mounted on it.

Fig.9. Photograph of a small fish robot

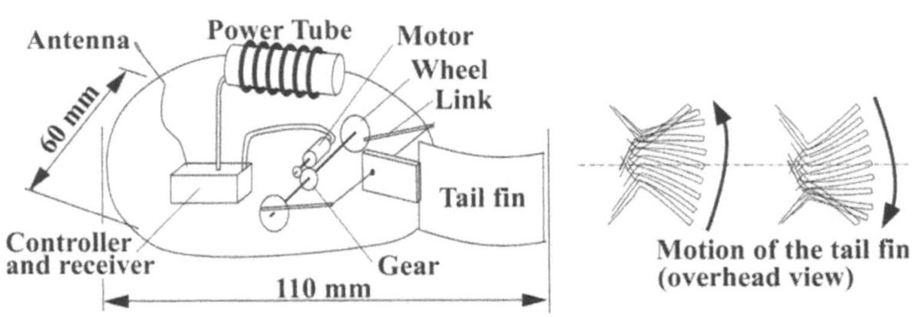

Fig.10. Configuration of the small fish robot and the motion of a tail fin

4.3 CFD analysis of the propulsion mechanism of the small fish robot

With the aim to create a far more superior fish robot, the propulsion mechanism of the prototype small fish robot shown in Fig.10 was investigated. Two-dimensional CFD analysis was conducted by means of the GTT code to examine the water flow around the fish robot (fixed in the stream), especially in the rear part. The division number of computational grid was 78×130, and the grid spacing in the vicinity of the fish robot was made fine. The images of moving tail fin in the water (shown in the right part of Fig.10) were taken by the high-speed CCD camera (200 frames per second), and the computational grid around the tail fin at each computational time step was generated on the basis of the image data in the same way as described in the section 3.1. The velocity of water flow at the upstream boundary was uniformly set at 70 mm/s (representative velocity U_m), which was nearly the same velocity at which the small fish robot swam. Numerical analysis of water flow was carried out for the two types of fish robots (Type A and Type B). Type A refers to the prototype small fish robot, and Type B refers to a fish robot whose width is half of the width of Type A, but its tail fin motion is the same as that of Type A.

Figure 11 shows the relative velocity vectors obtained by subtracting the representative velocity U_m from the calculated instantaneous flow velocity vectors for each case. From these figures, it has been found that both types of fish robots generate similar reverse Karman vortex streets (the spacing of the vortices is nearly the same for both types) in the downstream regions. In the case of Type B, however, the flow velocity in the downstream boundary is larger than that for Type A. From such calculated flow velocity distributions, the net momentum of water pushed out in the downstream region was calculated to estimate the net propulsive force of each fish robot per unit thickness (1 cm).

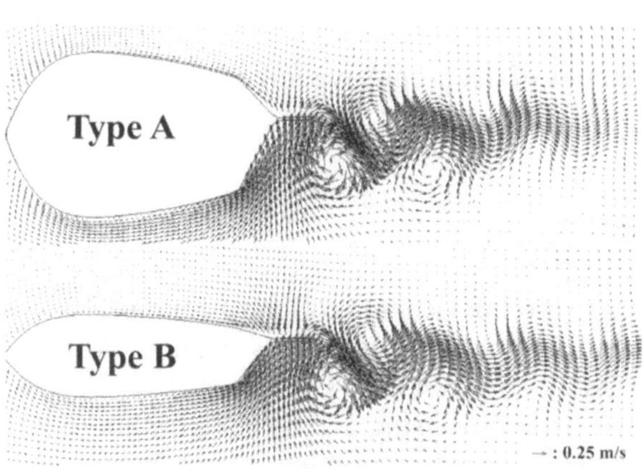

Fig.11. Velocity vectors subtracted U_m from the instantaneous velocity vectors obtained by CFD calculation

Figure 12 shows the calculated results for the period from 0.8 through 1.8 seconds after the tail fin begins to move. From the comparison of net propulsive force between Type A and Type B, it has been confirmed that the slender fish robot of Type B attains higher propulsion performance because of its lower flow resistance.

The drag forces of Type A and Type B fish robots were estimated on the assumption that their underwater shapes were elliptic cylinders (height = 1 cm) with aspect ratios of 1:1.5 (Type A) and 1:3 (Type B). The estimated drag forces were 1.143 mN for Type A and 0.323 mN for Type B. The difference of these drag forces is nearly the same as that of the net propulsive forces of Type A and Type B in Fig.12. This implies that the propulsive forces by the tail fins of Type A and Type B fish robots are nearly the same, and the difference of the drag forces depending on their shapes appears as the difference of the net propulsive forces.

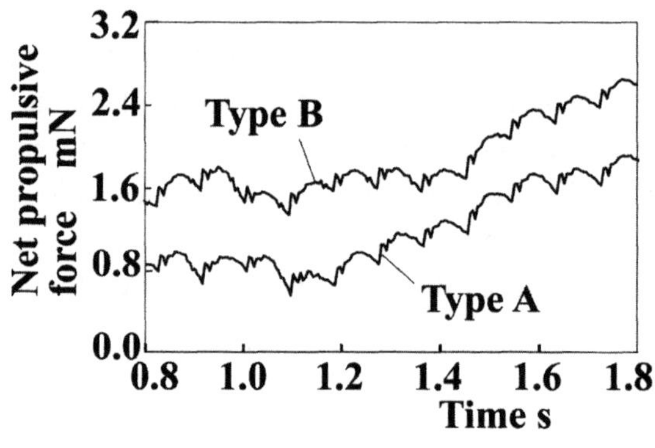

Fig.12. Variations of net propulsive force calculated by CFD

5 Conclusions

(1) A flexible fish robot (length: 260 mm) equipped with six motor-string actuators to mimic muscle motion was created. As its power source, a polymer electrolyte fuel cell (PEFC) was used. The operation schedule of six motors was optimized using a genetic algorithm (GA) toward two different targets. High-speed swimming was achieved by using the GA-generated schedule focusing on speed, in which the fish robot deformed the body like a snake. Furthermore, efficient swimming with less consumption of electricity was achieved by using the GA-generated schedule focusing on efficiency.

(2) As to a living fish, which swims freely in the water, the water flow around the fish was investigated by means of particle image velocimetry (PIV) and computational fluid dynamics (CFD). The results of the PIV measurement and the two-dimensional numerical analysis by means of the authors' GTT code have shown that the reverse Karman vortex street is generated in the downstream region of the tail fin. Therefore, in order to create a fish robot with good swimming ability, it is important to control the tail fin motion so that it can generate the clear reverse Karman vortex street.

(3) A passive-type cylindrical polymer electrolyte fuel cell (called 'Power Tube') having very simple structure has been developed originally. It has been confirmed that the Power Tube is suitable as a power source for a small fish robot, since this fuel cell is small-sized (13.5 mm in diameter and 30 mm in length) and light-weighted (2 grams), and generates high electric power (maximum output power: 0.42 W) when hydrogen is used as fuel.

(4) A prototype small fish robot of Type II with the hydrogen-fueled Power Tube was created. It has been confirmed that this fish robot was advancing on the surface of water. Through the two-dimensional CFD analysis of water flow around the fish robot, it has been found that the reverse Karman vortex street is generated in the downstream region of the tail fin and the fish robot acquires propulsive force by this vortex street which pushes the water toward the downstream region.

References

Ishida, R., Murase, H. and Koyama, S., (1997), Fundamental Theories and Application Programs for Genetic Algorithms by Personal Computer, Morikita Shuppan Co., Ltd.

Wakisaka, T., Isshiki, Y., Shimamoto, Y. and Magarida, N., (1995), Numerical Analysis of the Diffusion Process of Intake Mixture in Dual-Intake Valve Engines, JSME International Journal, Series B, Vol.38, No.2, pp.213-221

Ishii, T., Taninaka, S., Takada, Y. and Wakisaka, T., (2005), Improvement of the Performance of a Small-size and Light-weight Passive-type Polymer Electrolyte Fuel Cell, Proc. JSME Symp., Vol. 3, No.05-1, pp.7-8

Lau, T., Kelso, R. and Hassan, E., (2004), Flow Visualisation of a Pitching and Heaving Hydrofoil, Proc. 15th Australasian Fluid Mechanics Conference, 00203

Koike, G., Nishioka, K., Nagahara, K., Wakisaka, T., Takada, Y. and Takiyama, T., (2003), Investigation on the Structure of a Membrane Electrode Assembly for a Polymer Electrolyte Fuel Cell, Proc. JSME Symp., Vol.3, No.03-1, pp.295-296

Chapter20

Design and Control of Biomimetic Robot Fish FAC-I

Chao Zhou[1], Long Wang[1], Zhiqiang Cao[1], Shuo Wang[1], and Min Tan[1]

[1] Laboratory of Complex Systems and Intelligence Science, Institute of Automation, Chinese Academy of Sciences, Beijing 100080, China

Summary. A novel prototype of biomimetic robot fish named FAC-I with 3-D locomotion ability is developed in this paper. The locomotion of the caudal fin conforms to the law of dolphin motion to drive the robot fish's movement, and pectoral fins are applied to provide auxiliary locomotion ability and each can flap separately and rotate freely, which compose the original three-fin structure. The dynamic analysis of the caudal fin is discussed based on the 2-D waving plate theory. In order to complete descending and ascending motions, a distinctive piston-based artificial swim bladder is designed, which simulates the air bladder of fish. Several biomimetic motions are designed, such as accelerating, decelerating, turning, diving. An information relay system on water is designed to connect the robot fish under water and upper console. The information processing framework of the robot fish and the relative information processing methods are discussed for the purpose of automatic control and environmental monitoring. Some experiments show the performance of FAC-I.

Key words. Biomimetic robot fish; 3-D locomotion; Three-fin structure; Artificial swim bladder; Information relay

1 Introduction

There are more and more productive underwater activity and research work. Underwater tasks are changing, and many challenges are put forward. It is paid more attentions on the Autonomous Underwater Vehicle (AUV). In many cases, autonomous underwater vehicles are expected to be small-size, efficient, maneuverable, and low noise. However, most of AUVs are mainly driven by screw propeller, which was accompanied with some limitations to efficiency or maneuverability of the systems. So roboticists explore other approaches to solve these problems.

Fishes is widely distributed aquatic biology and can adjust itself to underwater environment perfectly. In the locomotion, fishes and cetaceans make effective use of the characteristics of the water, and exhibit the advantages of the high efficiency, high maneuverability and low noise. Fishes with left-right oscillating tails and cetaceans with up-down oscillating tails, propel themselves by the combination of their body, fins and tails, which has the advantages over conventional marine vehicles powered by rotary propellers with the same power consumption.

Generally speaking, biomimetic robot fish is defined as a fish-like aquatic vehicle based on the swimming skills and anatomic structure of a fish: the undulatory/oscillatory body motions, the highly controllable fins and the large aspect ratio lunate tail. Because of the similarity of fishes and cetaceans, in this paper, they are no longer discriminated. Investigation on biomimetic robot fish systems has provided significant insights into both theory and application of robotics in recent years, whose appealing nature involves higher efficiency, more remarkable maneuverability and quieter actuation[1]–[4]. These advantages are of great benefit to applications in marine and military fields.

After the Gray's paradox [5], which says that a fish does not seem to have enough power to propel itself at the speed it does, many theories are proposed to explore secrets of fish swimming mechanisms and summarize driving modes of fish motions[7]-[12]. Based on the theories, many prototypes of biomimetic robot fish have been developed.[1][4][13]-[19]. From the view point of AUV, the 3-D locomotion is very important for mission execution. There are several descending/ascending methods for biomimetic robot fish [6] [20]: Water tank method, Pectoral fins method, Changing body shape method, Changing the barycenter, and they are adopted on many prototypes [6][17][20]-[25]. [26] deals with the guidance and control of an underwater vehicle equipped with 2 pairs of oscillating fins in waves and in water currents, which endows the robot with the ability of ascending and descending. [27][28] give a prototype with three pectoral fins and a caudle fin, which realize three dimensional maneuverability.

In this paper, we develop a robot fish named FAC-I (Fish-like Aquatic Cruiser-I). An original three-fin structure (TFS) that is composed of a simulating-dolphin tail, two pectoral fins with 3-DOF (Degrees of Freedom) is proposed. This structure is more similar to fishes and provides more adaptability to motion control. An artificial swim bladder is designed for descending and ascending motions, which is a piston mechanical structure to simulate fish air bladder function, by which, the robot fish moves up and down vertically without the forward velocity. Compared with other descending/ascending methods, it is the only one to directly imitate the fish's 3-D motion, which is formed in nature selection. Several kinds of sensors are mounted and the related information processing methods are given to endow FAC-I with certain autonomy. In order to deal with the communication problem, the concept of information relay is introduced into the robot fish system. The robot fish provides a new means for underwater environment observation and it may recognize and localize goals actively.

2 The Design of FAC-I

Influenced by the characteristics of the fishes, we design a robot fish FAC-I, which is a combination of the pectoral fins, caudal fin and the artificial swim bladder. These three parts of FAC-I endow the robot fish more flexible locomotion ability.

2.1 Design of the Caudal Fin

The caudal fin is the main part for driving robot fish forward. By observing dolphin, the motion of caudal fin is swinging up and down. In order to endow caudal fin of robot fish with ability of pitching and heaving, we design the caudal mechanism structure as shown in Fig. 1(a). The tail swings up and down to propel the robot fish body forward by a DC motor and linkage mechanism. The supporting frames fixed on the tail link also swing up and down with the caudal fin. Although it distorts when swinging, the rubber tube keeps the streamline shape because of the supporting frames. The crescent-shaped caudal fin is shown in Fig. 1(b).

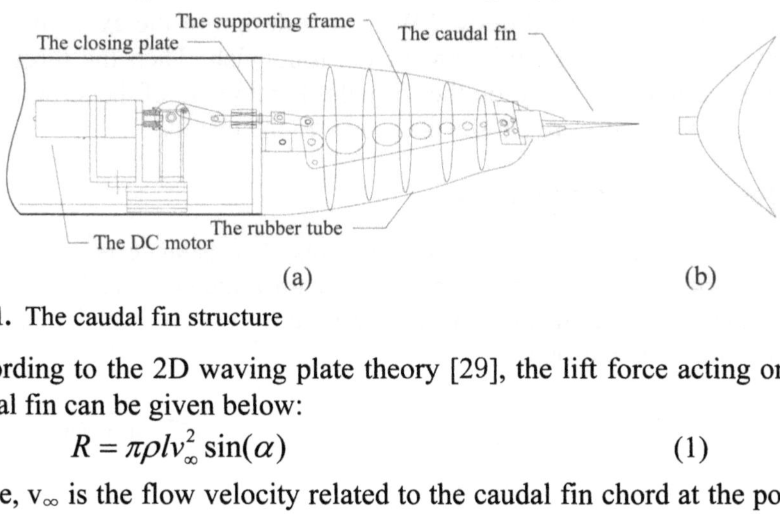

Fig. 1. The caudal fin structure

According to the 2D waving plate theory [29], the lift force acting on the caudal fin can be given below:

$$R = \pi \rho l v_{\infty}^{2} \sin(\alpha) \tag{1}$$

where, v_∞ is the flow velocity related to the caudal fin chord at the point c which is quarter of chord. α is the attack angle of caudal fin.

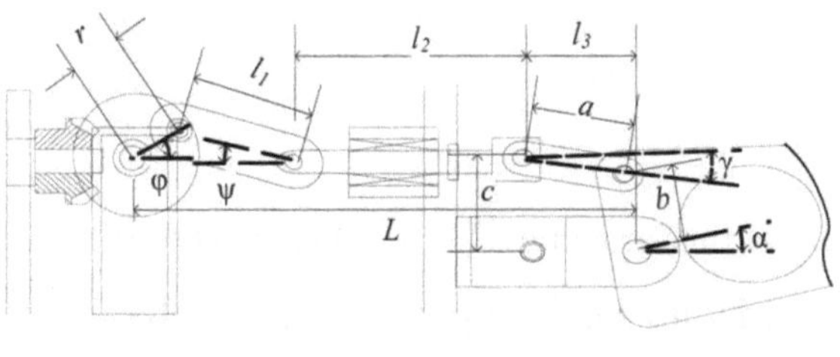

Fig. 2. The gearing part of the caudal structure

Fig. 2 is the enlarged local of caudal structure, and the relationship between the posture of caudal fin and rotating angle of motors is listed in follows:

$$\begin{cases} l_3 = a\cos\gamma + b\sin\alpha \\ c = a\sin\gamma + b\cos\alpha \\ L = r\cos\varphi + l_1\cos\psi + l_2 + l_3 \\ r\sin\varphi = l_1\sin\psi \end{cases} \tag{2}$$

where l_1, l_2, l_3, a, b, c are parameters of structure. Then (3) is gotten:

$$L - l_2 = r\cos\varphi + \sqrt{l_1^2 - r^2\sin^2\varphi} + \sqrt{a^2 - (c - b\cos\alpha)^2} + b\sin\alpha \tag{3}$$

And the required output power of the motor is:

$$P(t) = T(t)\dot{\varphi}(t) \tag{4}$$

where T denote the output torque of the motor, P is the power of the motor. According to (1) (3) (4), we can calculate the relationship between the forward velocity, the lift force and the motor's angle velocity, torque and power by numerical method. The result is useful for the selection of the motor and the mechanical design. We give a design of the executing components according to the mission for the robot fish. Then the torque and power are calculated based on the formula. Parameters of the structure are modified, and motors are chosen according to the results.

2.2 Design of Pectoral Fins

Fishes swim and keep balance by using fins. The balance is mainly controlled by pectoral fins and they may also control the posture. The locomotion of the fish pectoral fins can be simplified to two kinds of fundamental motions: rotation and flap.

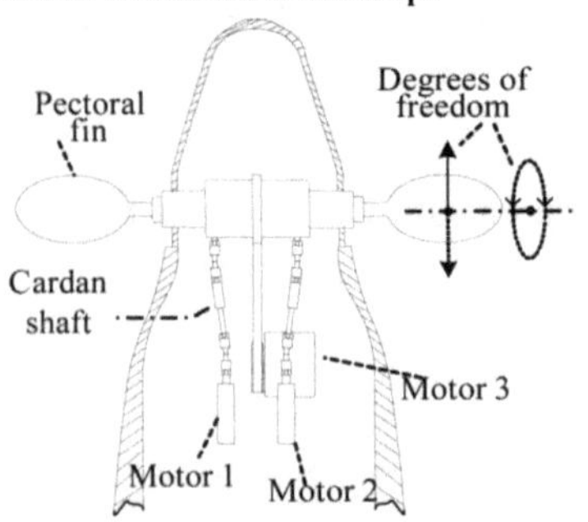

Fig. 3. The pectoral fins structure

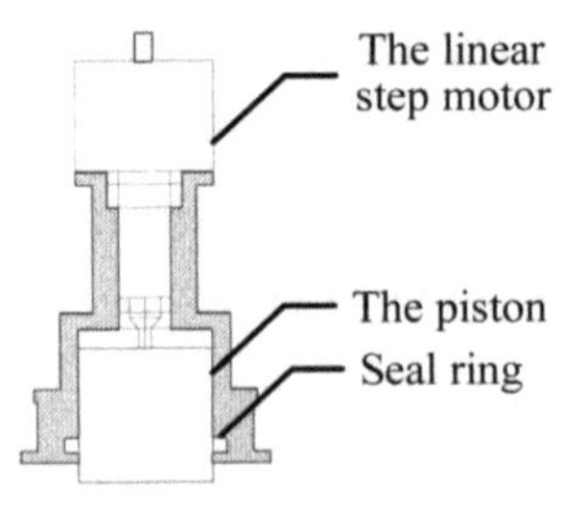

Fig. 4. The artificial swim bladder

FAC-I is equipped with a pair of pectoral fins, each of which can flap separately and rotate freely, as shown in Fig. 3. In order to simplify the mechanical structure and control, the rotation motions of two pectoral fins may be uniform, thus there are three degrees of freedom in total and each is corresponding to a step motor. The motor 1 and motor 3 rotate forward and backward in a certain angle range to make the left/right pectoral fin flap continuously. When the motor 2 rotates, both pectoral fins roll synchronously to change the attack angle. This structure satisfies the maneuverability of the locomotion.

2.3. Design of the Artificial Swim Bladder

As shown in Fig. 4, a piston driven by a linear step motor simulates fish air bladder function, which is installed at the center of buoyancy. Motions of the piston adjust the volume of the robot fish to change the buoyancy of the robot fish, and the robot fish may descend and ascend. The closing plate, as shown in Fig. 1, hermetically separates the soft caudal tube from the abdominal cavity where the piston is installed and thus the volume change of robot fish caused by moving the piston is regardless of the caudal tube.

Compared with *Water tank method*, the motion of piston system changes the volume instead of the gravitation, which is much simpler in structure. The control of the piston position is much easier than the water level control in the tank by a pump, and it avoids the affection of water undulation in the tank when the robot fish moves. Compared with other methods except *Water tank method*, this method may make the robot fish descend and ascend vertically, and even suspend. Energy is saved because it do not depend on the velocity.

2.4. Foundational Bio-motions

According to behaviors of fishes and the function of the robot fish, a library of foundational motions of the robot fish is designed based on the mechanical structure and the control system. The robot fish adjusts its poses in water by controlling the oscillatory frequency, the attack angle of the fins and the position of the piston. The motions are implemented and integrated in a library.

Go ahead: Including the acceleration, deceleration and uniform motions, which is realized by sending different signals to the motor of the robot fish tail.

Crash stop: pectoral fins steady vertically while the tail stops oscillating. The velocity of the robot fish reduces quickly because of the resistance of water.

Turn: The robot fish turns right/left by flapping the left/right pectoral fin. The flapping frequency is decided by the turning radius required: the higher the frequency is, the smaller the turning radius is.
Dive: There are two approaches to dive for FAC-I, the pectoral fins and the artificial swim bladder.
Dive through pectoral fins: It is achieved by adjusting the rotation angle of pectoral fins. When pectoral fins rotate clockwise to a specific angle, the robot fish descends because of the water pressure. Similarly, the fish ascends when pectoral fins rotate counterclockwise to the ascend-angle.
Dive through the artificial swim bladder: This method makes use of the difference between the buoyancy and gravity. The piston adjusts the volume of the robot fish to perform descending and ascending. In practice, a suspending position has to be determined firstly and it is a little different in experiments. When the piston stretches out the suspending position, both the robot fish volume and its buoyancy increase, which make the robot fish ascend. When the robot fish wants to descend, it shrinks the piston to reduce the volume until the buoyancy is smaller than the gravity.
These two dive methods may perform descending and ascending independently, however, each has its advantages and disadvantages. They may be cooperated to perform the dive and the posture control much better. By adjusting the attack angle of the pectoral fins and the oscillating frequency of the tail, the robot fish may change the pitching, and descend or ascend in movement. Changing the position of the piston can make the robot fish move up and down vertically and the diving depth of the robot fish may be controlled accurately without forward movement.

3. The Control and Information Process System of FAC-I

The control system of FAC-I includes MCU circuit, wireless module, the power supply, the control and drive modules of motors. The MCU circuit sends a PWM signal to the tail motor to control its motions, and sends four clock signals to control pectoral fins step motors and the piston linear step motor to realize the motions of pectoral fins and the piston. The MCU circuit communicates with outside by an information relay system. In order to gather environmental information, several kinds of sensors (infrared, video, and pressure sensors) are equipped with the robot fish for the purpose of autonomic control and environmental monitoring. Infrared sensors are used to detect obstacles with the rapid response, by which the biomimetic robot fish may swim safely. The water depth information can be gained by the pressure sensor mounted on the side of the body and thus the robot fish can control its diving depth. The camera is mounted on the fish head to monitor the front region. The video information is sent to the upper console by an information relay system on water for displaying and processing.

3.1 The Framework for Information Relay and Process

It is necessary to develop a communication system between the robot fish and the outside for practical applications. There are several approaches on the information transmission of the robot fish. One is to utilize radio equipments. Because of the severe attenuation of the electromagnetic wave in water, the equipments can only work on water surface, which forces the robot fish swim on water or ascend to water surface to exchange the data. To a great extent, this method destroys the continuity of underwater work and the realtime of communication, which limit the application. Another method is to connect the robot fish and an upper console by a cable or optical fiber cable directly, which can assure the transfer speed and accuracy. However, this method will limit the range of locomotion to the neighborhood of the upper console. In order to resolve these problems, the concept of information relay, which combines the wired and wireless communication, is introduced into FAC-I. The information relay system of the robot fish adopts a buoyage on water to connect to the robot fish by the cable and optic cable, and communicates with the upper console on shore by wireless signal, which may satisfy the requirement of extended locomotion. The buoyage is small and light, which is suitable to be dragged by the robot fish. As shown in Fig. 5, the video signal and digital signal are transmitted separately by the optical fiber cable and the cable.

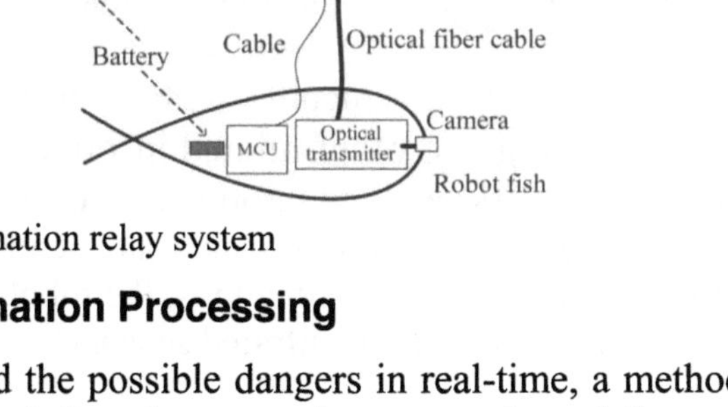

Fig. 5. The information relay system

3.2 The Information Processing

In order to avoid the possible dangers in real-time, a method to avoid obstacles based on infrared sensors is presented as follows. As described above, there are three infrared sensors mounted on right, middle, left sides of the fish head, and the outputs are R_{IR}, F_{IR} and L_{IR}, respectively. The robot fish may make decisions based on the inputs of sensors. The logic-rule set is given in (5), which depicts the relationships between motions of robot fish and states of sensors.

$$\begin{bmatrix} \mathrm{UNIFORM} \\ \mathrm{DECE} \\ \mathrm{LEFT}_{\mathrm{TURN}} \\ \mathrm{RIGHT}_{\mathrm{TURN}} \\ \mathrm{LINEAR} \end{bmatrix} = \begin{bmatrix} \overline{F_{IR} L_{IR} R_{IR}} \\ F_{IR} + \overline{F_{IR}} L_{IR} + \overline{F_{IR} L_{IR}} R_{IR} \\ \overline{F_{IR} L_{IR}} R_{IR} + F_{IR} \overline{L_{IR} R_{IR}} + F_{IR} \overline{L_{IR}} R_{IR} \\ \overline{F_{IR}} L_{IR} \overline{R_{IR}} + F_{IR} L_{IR} \overline{R_{IR}} + F_{IR} L_{IR} R_{IR} \\ \overline{F_{IR}} L_{IR} R_{IR} + \overline{F_{IR} L_{IR} R_{IR}} \end{bmatrix} \tag{5}$$

where UNIFORM, DECE, $LEFT_{TURN}$, $RIGH_{TURN}$, LINEAR are the Boolean variables of keeping changeless, decelerating, turning left, turning right, swimming forward , respectively.

After the robot fish determines the proper motion based on the logic rules in (5), it chooses the combination of the foundational motions described in section III to drive its mechanical structure.

The vision system of the robot fish FAC-I is composed of a CCD camera, signal conversion, the transmission subsystem, a video capture card and the upper console. The information on the environment and goal are extracted for the high-level decision. An image segmentation algorithm based on edge detection is presented to extract the under water goal information. In this paper, the color space HLS is selected during segmentation. Considering that the changing environmental conditions affect the appearance of an image, and ripples in surface of water disturbed by free-swimming fish reflect light, an adaptive threshold has to be adopted. Firstly, a region of the image is selected as the initial recognition characteristic. Then the change of the environment lighting is calculated and it is added to the threshold. Thereby a closed-loop threshold is created in each segmentation cycle. After basic segmentation and filtering, a basis morphological algorithm, closing operation is used on the binary image to smooth the boundary and fill small hole. The result of vision processing and the target identification reflect the information of environment and goals.

The robot fish may execute the instructions from the operator based on the results of image processing and other related information. Also, the operator may only give the task, and the information fusion is conducted by the upper console, which is the decision-maker and sends control instructions to robot fish. In some emergent situations, the robot fish makes reactive responses to outside based on the infrared sensors.

4. Experiments

Fig. 6 gives a functional prototype of the biomimetic robot fish FAC-I with a tail for propulsion, 3-DOF pectoral fins, the artificial swim bladder mechanical structure and the information relay system. Some parameters of the prototype are shown in Table 1.

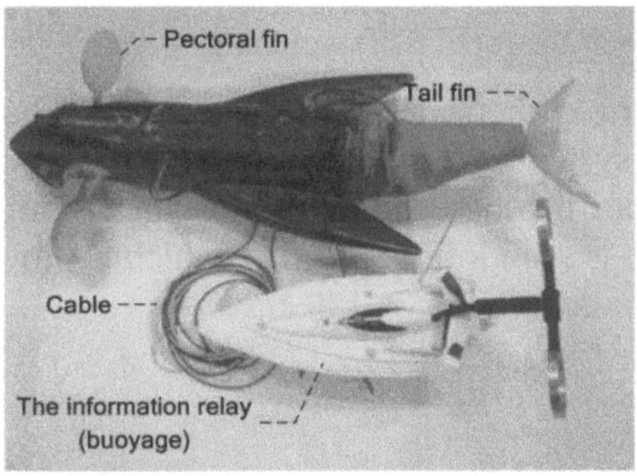

Fig. 6. The functional prototype of FAC-I

Table 1. Parameters of the prototype

Items	Parameters
Size	87cm*29cm*25cm
Weight	9.4kg
Pectoral fins flap frequency	2.4Hz
Caudal fin flap frequency(underwater)	2.8Hz
Duration of swimming	1.5~2h
Velocity (max)	0.25m/s
Angular velocity of turn	~0.175rad/s
Acceleration of descending	~0.004m/s^2

Some experiments are conducted to testify the locomotion ability of FAC-I and we choose several representative experiments in this paper. The first one is done in an experiment pool of 1.8m*1.0m*0.6m safely and the results are shown in Fig. 7. Fig. 7 (a)-(c) describe the process of turn by the right pectoral fin and Fig. 7 (d)-(f) give the sequence of the robot fish swimming forward by the tail oscillating. The average angular velocity of turn is about 0.1rad/s, which is lower than the maximum because it starts at velocity of zero.

(a) t=0s (b) t=5.5s (c) t=7.7s

(d) t=9.2s (e) t=12.4s (f) t=14.4s

Fig. 7 The sequence of the robot fish swimming

Fig. 8(a)-(c) describes the swimming sequence of robot fish FAC-I dragging the buoyage system in a swimming pool. The velocity is 0.25m/s. The process of the robot fish descending in the experiment pool by the adjustment of the piston is seen in Fig. 9(a)-(c), which is proved to be valid. The average velocity of descending is 1.86cm/s.

(a) t=0s (b) t=2.9s (c) t=8.2s

Fig. 8. The sequence of swimming with the relay

(a) t=0s (b) t=10.4s (c) t=11.1s

Fig. 9. The descending image sequence of the robot fish

Another experiment is related to goal recognition based on the video signal sent to the upper console. Fig. 10 gives several selected images from the camera of the robot fish, where the robot fish swims in the blue experiment pool with the red goal A, and C describes the given color for recognition. The corresponding recognition results are seen in B. Fig.10 (a), (b) and (c) is a sequence of the video pictures when robot fish swims forward.

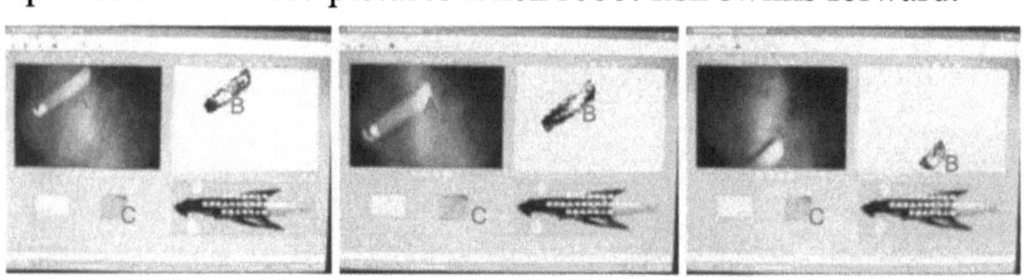

(a) (b) (c)

Fig. 10. The selected images and the corresponding recognition results

5 Conclusions

In this paper, a novel biomimetic robot fish with a tail for propulsion, 3-DOF pectoral fins, and a piston system as an artificial swim bladder is constructed based on the propulsion and maneuvering mechanisms for dolphin swimming. An information relay system on water is designed for information transmitting. Several foundational motions are given based on the mechanical structure and the control system to realize the underwater motion. The sensors information processing of the fish are then studied. Experiments have demonstrated the good performance of the novel mechanical structure and control system of the biomimetic robot fish.

Acknowledgement

This work was supported in part by the National Natural Science Foundation of China (No. 60635010, No.60605026, No.50475179), 863 Project (No.2006AA11Z225, No.2006AA04Z258), and CASIA Innovation Fund for Young Scientists.

References

[1] M. S. Triantafyllou and G. S. Triantafyllou, "An efficient swimming machine," Sci. Amer., vol. 272, no. 3, pp. 64–70, Mar. 1995.
[2] M. Sfakiotakis, D. M. Lane, and J. B. C. Davies, "Review of fish swimming modes for aquatic locomotion," IEEE J. Oceanic Eng., vol. 24, no. 2, pp. 237–252, Apr. 1999.
[3] Y. Terada and I. Yamamoto, "An animatronic system including lifelike robotic fish," in Proc. of the IEEE, vol. 92, no. 11, Nov. 2004, pp. 1814–1820.
[4] Barrett D S, Triantafyllou M S, Yue D K P, et al., "Drag reduction in fish-like locomotion," J. Fluid Mech. 1999, vol.392, 183-212
[5] J.Gray, "Studies in Animal Locomotion, VI. The Propulsive Powers of the Dolphin," J. Exp. Bi ol., No. 13, 192-199 (1936)
[6] http://www.nmri.go.jp/eng/khirata/fish/index_e.html
[7] Lighthill M J., "Note on the swimming of slender fish," J. Fluid Mech. 1960, vol.9: 305-317
[8] Lighthill M J., "Aquatic animal propulsion of high hydromechanical efficiency," J. Fluid Mech. 1970, vol.44, 265-301
[9] Lighthill M J., "Large-amplitude elongated-body theory of fish locomotion," Proc. R. Soc. Lond, Ser. B. 1971, vol.179, 125-138
[10] Wu T Y., "Swimming of a waving plate," J. Fluid Mech. 1961, vol.10, 321-344
[11] Triantafyllou M S, Barrett D S, Yue D K P., "A new paradigm of propulsion and maneuvering for marine vehicles," Trans. Soc. Naval Architects Marine Eng. 1996, vol.104, 81-100
[12] B. Tong, "Propulsive mechanism of fish's undulatory motion," Mechanics In Engineering. 22: 69-74 . 2000.

[13] Biomimetic Underwater Robot Program. Available at: http://www.neurotechnology.neu.edu/
[14] N. Kato, M. Furushima, "Pectoral fin model for maneuver of underwater vehicles," Proc. 1996 IEEE AUV Symp. New York: IEEE Press, 1996, 49-56
[15] N. Kato, "Control Performance in the Horizontal Plane of a Fish Robot with Mechanical Pectoral Fins," IEEE J. Ocean. Eng., vol. 25, pp.121-129, Jan. 2000
[16] T. Fukuda, A. Kawamoto, F. Arai, H. Matsuura, "Mechnism and swimming experiment of micro mobile robot in water," Proceedings, IEEE Workshop on Micro Electro Mechanical Systems. January 1994, pp273-278
[17] J. Liu, H. Hu, "Building a 3D simulator for autonomous navigation of robotic fishes," IROS 2004. 613 – 618
[18] J. Liang, T. Wang, H. Wei, W. Tao, "Researchful development of under water robofish II- development of a small experimental robofish," Robot, 2002, 24(3): 234-238
[19] J. Yu, S. Wang, M. Tan, "A Simplified Propulsive Model of Biomimetic Robot Fish and Its Realization," Robotica 2005, volume 23, 101-107
[20] C. Zhou, Z. Cao, S. Wang and M. Tan, "The Posture Control and 3-D Locomotion Implementation of Biomimetic Robot Fish," IEEE IROS, 5406-5411, 2006
[21] http://birg.epfl.ch/page57462.html
[22] Yamamoto I. and Terada Y., "Research on flexible oscillating fin propulsion system and robotic fish," Proc. IFAC CAKS, U.K. (2001)
[23] Shen Zhizhong, Wang Shuo, Cao Zhiqiang, Tan Min, Wang Long, "An Image Segmentation Algorithm Based on Edge Detection and Its Application to Robotfish," Robot, 2006
[24] M. Sfakiotakis, D. M. Lane, and J. B. C. Davies, "Review of fish swimming modes for aquatic locomotion," IEEE J. Oceanic Eng., vol. 24, pp. 237–252, Apr. 1999.
[25] Z. Zhang, S. Wang, M Tan, "3-D Locomotion control for a biomimetic robot fish," Journal of Control Theory and Applications, 2004, 2(2): 169-174.
[26] Kato,N., Liu, H. and Morikawa, H., Biology-inspired Precision Maneuvering of Underwater Vehicles—Part 3, Int. J. of Offshore and Polar Engineering (ISOPE), 2005, Vol. 15, No. 2, pp. 81–87
[27] M. Nakashima, Y. Takahashi, T. Tsubaki and K. Ono, "Three-Dimensional Maneuverability of the Dolphin Robot", in "Bio-mechanisms of Swimming and Flying" Springer, 2004, pp79-92.
[28] M. Nakashima, T. Tsubaki, and K. Ono, "Three-dimensional Movement in Water of the Dolphin Robot – Control Between Two Positions by Roll and Pitch Combination ", Journal of Robotics and Mechatronics, Vol.18, No.3, 2006, pp.347-355.
[29] J. Yu, "The kinetic analysis of biomimetic robot fish,"Doctor Paper of University of Science and technology of China, 2004.
[30] H. Sang, S. Wang, M. Tan, Z. Zhang, "Research on Patrol Algorithm of Multiple Behavior-Based Robot Fishes," IJOPE, 2005, 15(1):1-6
[31] C. Zhou, Z. Cao, L. Wang, S. Wang and M. Tan, "Study on design and control of a 3-D locomotion biomimetic robot fish," ISABMEC, Japan, 2006.

Chapter 21

Thrust Force Characteristics of Propulsion Mechanism in Fluid Using Fin with Dynamic Variable-Effective-Length Spring

Shunichi Kobayashi, Masataka Nakabayashi, Reiji Kobayashi, Ji Jie, and Hirohisa Morikawa

Department of Functional Machinery and Mechanics, Shinshu University, 3-15-1 Tokida, Ueda 386-8567, Japan

Summary. We aimed to develop a propulsion mechanism using a variable-bending-stiffness fin of which stiffness can be changed dynamically. As the one such variable-stiffness fin, we have developed a fin with a variable-effective-length spring. The apparent bending stiffness of the fin can be changed dynamically. We have measured the thrust force of the propulsion mechanism using a fin with a variable-effective-length spring and visualized flow around the fin. In this paper, we describe the structure of the propulsion mechanism in fluid using a fin with a variable-effective-length spring, and the thrust force characteristics with changing effective length of the spring dynamically. Furthermore, we discuss the relationship between thrust force and flow around the fin.

Key words. Propulsion Mechanism, Fin, Variable-Effective-Length Spring, Flow Visualization, PIV

1 Introduction

The conventional screw propeller is the general propulsion system for ships or underwater vehicles. As an alternative propulsion mechanism for higher propulsion efficiency and better safety than the screw propeller, propulsion by oscillating an elastic fin resembling a caudal fin or pectoral fin of fish has been proposed, and a basic study on the bioinspired oscillating fin and its development as a propulsion method for ships or underwater vehicles/robots have been carried out (Morikawa et al.1980, Nakashima et al. 2000, Watanabe et al. 2002). The optimum elasticity of a fin is not constant and changes according to the movement task and environment, such as swimming speed and oscillating frequency (Nakashima et al. 2000, Watanabe et al. 2002). However, it is very difficult to exchange fins of different bending stiffnesses while moving. On the other hand, fish are able to move their fins dynamically and can change their resistance to external force. Fish may be able to change the apparent stiffness of their fins by muscle contraction, though further investigations of fin stiffness for fish are needed.

We aimed to develop a propulsion mechanism using a variable-bending-stiffness fin of which stiffness can be changed dynamically (Kobayashi et al. 2003). As the one such variable-stiffness fin, we have developed a fin with a variable-effective-length spring. The apparent bending stiffness of the fin can be changed dynamically. We have measured the thrust force of the propulsion mechanism using a fin with a variable-effective-length spring and visualized flow around the fin. In this paper, we describe the structure of the propulsion mechanism in fluid using a fin with a variable-effective-length spring, and the thrust force characteristics with changing effective length of the spring dynamically. Furthermore, we discuss the relationship between thrust force and flow around the fin.

2 Structure of Propulsion Mechanism in Fluid

Figure 1 shows the principle of the variable-effective-length spring. Effective length of the spring H is changed by varying the length of rigid plates supporting the plate spring made of polyethylene terephthalate, (PET). The relationship between the displacement of the plate spring and effective length H (constant lateral load: 1.96 N) is shown in Fig. 2. Apparent stiffness changes with H. Figure 3 shows the fin with a variable-effective-length spring. The fin system consists of an aluminum box (length: 130 mm, height: 60 mm, width: 25 mm), a fin made of a chloride plate (length: 120 mm, height: 60 mm, thickness: 3 mm), and a variable-effective-length spring (length L: 20 mm, height: 43 mm, thickness: 0.5 mm). A DC motor, waterproof bearing, slipping screw, resin nut and rigid plate are housed in the aluminum box. A transparent chloride plate cover is attached on the open side of the aluminum box to seal it. Figure 4 shows the experimental system. The fin is moved in a stationary water tank (length: 2020 mm, width: 730 mm, depth: 250 mm). The fin base is positioned at the middle of the length and width of the water tank, and the upper surface of the fin is 90mm below the water level. The propulsion mechanism is supported by a linear guide and is able to move. For the measurement of x-directional thrust force while the mechanism is stationary on the linear guide, the mechanism is connected to a load cell. The pitching movement of the fin is driven by the geared DC servomotor. We can change the maximum pitching angle θ_{max} and movement cycle T. The signal from the load cell is amplified by a strain amplifier and recorded on a personal computer in synchrony with the control of the DC servomotor. Figure 5 shows the devices for flow visualization. We used two projectors for the light sheet and a high speed video camera to take images of flow. Particles of diaion (250 μm) were used for the tracer. For the analysis of flow field, we used the particle image velocimetry (PIV) system.

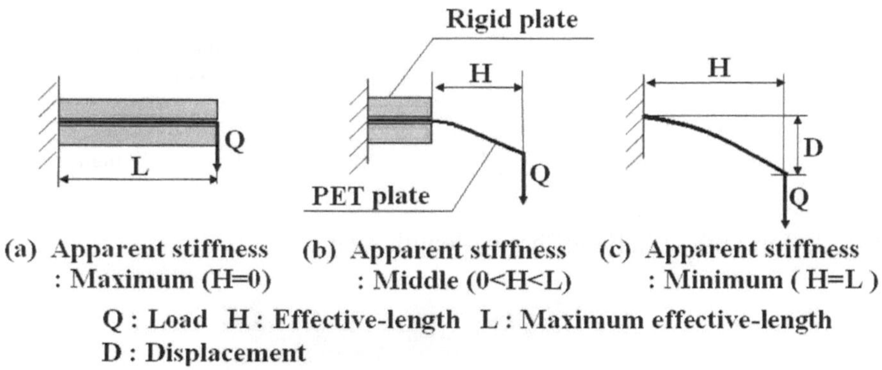

Fig. 1. Principle of effective length spring

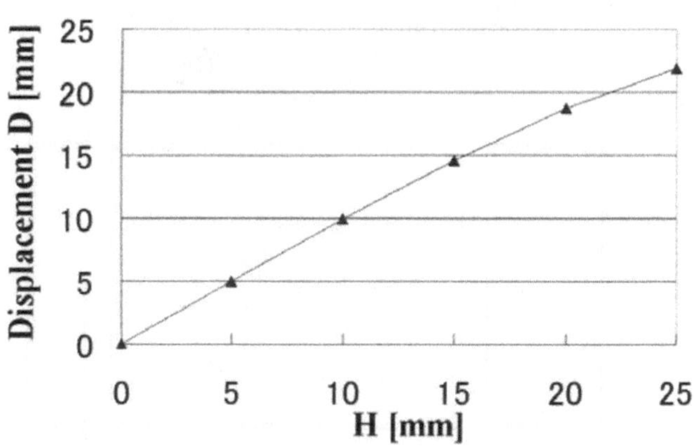

Fig. 2. Relationship between displacement of plate spring D and effective length of spring H. Total length of spring $L = 25$ mm. Thickness of spring $t = 0.5$ mm. Constant lateral load $Q = 1.96$ N.

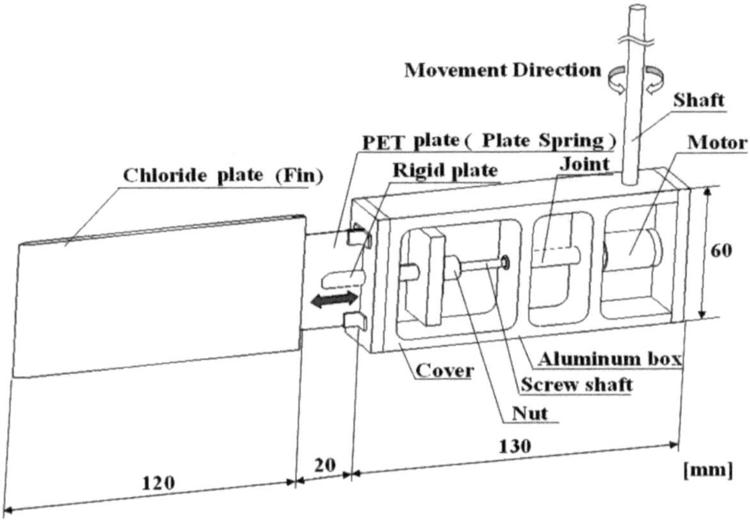

Fig. 3. Fin with variable-effective-length spring

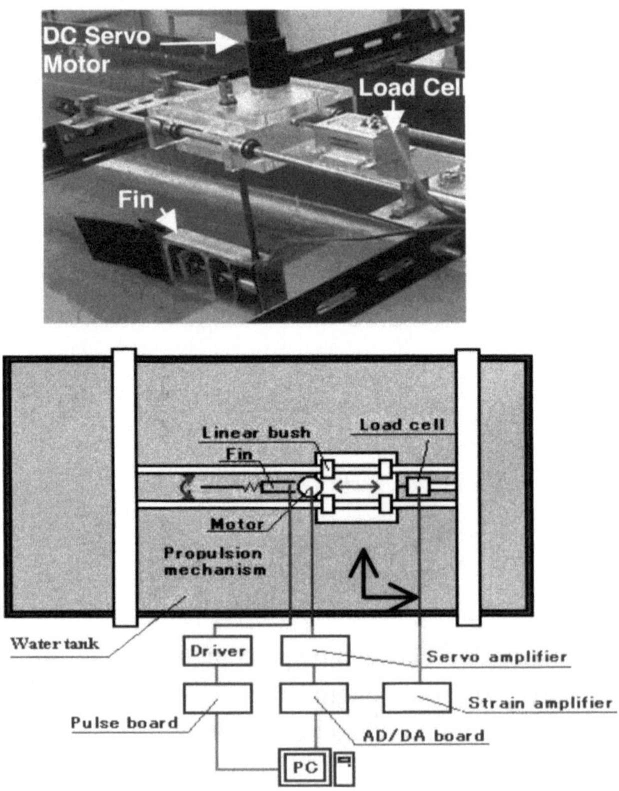

Fig. 4. Experimental system

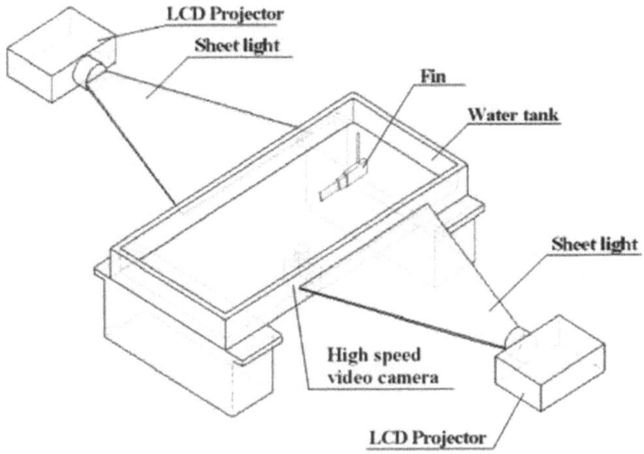

Fig. 5. Devices for flow visualization

3 Results and Discussion

3.1 Static Condition of Effective Spring Length

3.1.1 Relationship between Effective Spring Length and Thrust Force

Figure 6 shows the variation of x-directional thrust force in one movement cycle. The effective length of spring H was fixed during motion (static condition of H). With $H = 0$ as the maximum apparent stiffness condition, the difference between maximum thrust force F_{xmax} and minimum thrust force F_{xmin}, and the magnitude of the negative thrust force are greater than those when $H = L$. Average thrust forces when $H = L$ (F_{xavg} =0.334 N) is greater than that when $H = 0$ (F_{xavg} = 0.332 N). These results indicate that an elastic fin is favorable realizing the effective thrust force. We have changed H closely and found the optimum effective length to generate the greater average thrust force H_{opt}. In this condition ($T = 3$ s, $\theta_{max} = 30$ deg), H_{opt} was $9L/20$ (F_{xavg} = 0.416 N). The phases of F_{xmax} and F_{xmin} when $H = L$ and $H_{opt} = 9L/20$ are delayed compared with those when $H = 0$. This is because the phase of the fin shape to generate a great magnitude of thrust force is delayed by the elastic bending.

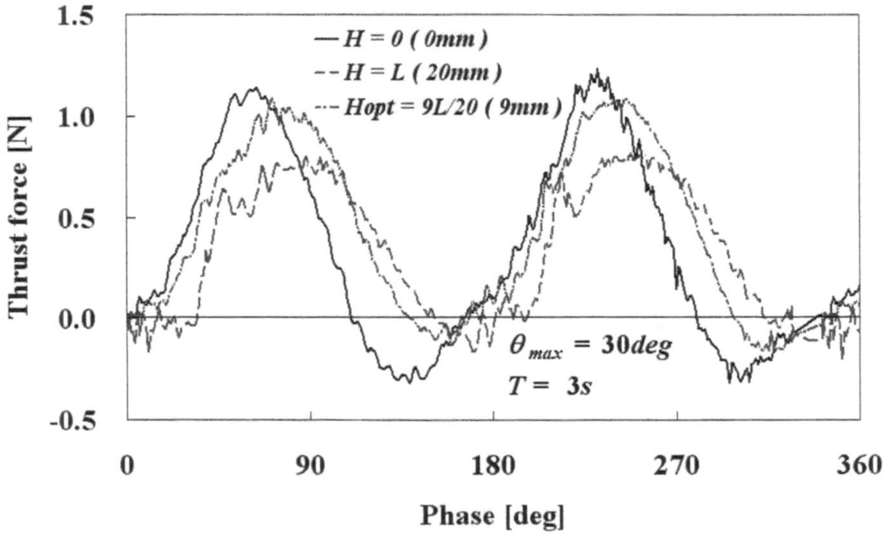

Fig. 6. Variation of x-directional thrust force in one movement cycle ($T = 3$ s, $\theta_{max} = 30$ deg).

3.1.2 Fin Shape and Flow Around the Fin

Figures 7 and 8 show the change in fin shape and flow around the fin on the flat plane at the center of height of the fin in one movement cycle. The effective spring length H is fixed during motion (Static condition of H). Bending of fin increases with increasing H, that is, with a decrease in apparent stiffness.

***H* = 0 (no effective length of spring, maximum apparent stiffness)**
When $H = 0$ (no effective length of spring), bending of fin is small but not 0. This small bending is due to small gap between the top of the rigid plate and the chloride plate (fin) when $H = 0$. Phase = 60 deg (phase for the maximum thrust force): Area of high flow velocity on the left side of fin is greater than that on the right side. Thus the pressure on left side of fin would be smaller than that on the right side of fin, consequently, this pressure difference would produce the positive x-directional (downward direction in Fig.4) thrust force. Phase = 90 deg: Flow velocity on the left side of fin is greater, but since the fin position is almost parallel to x-axis, the positive thrust force component would be reduced.

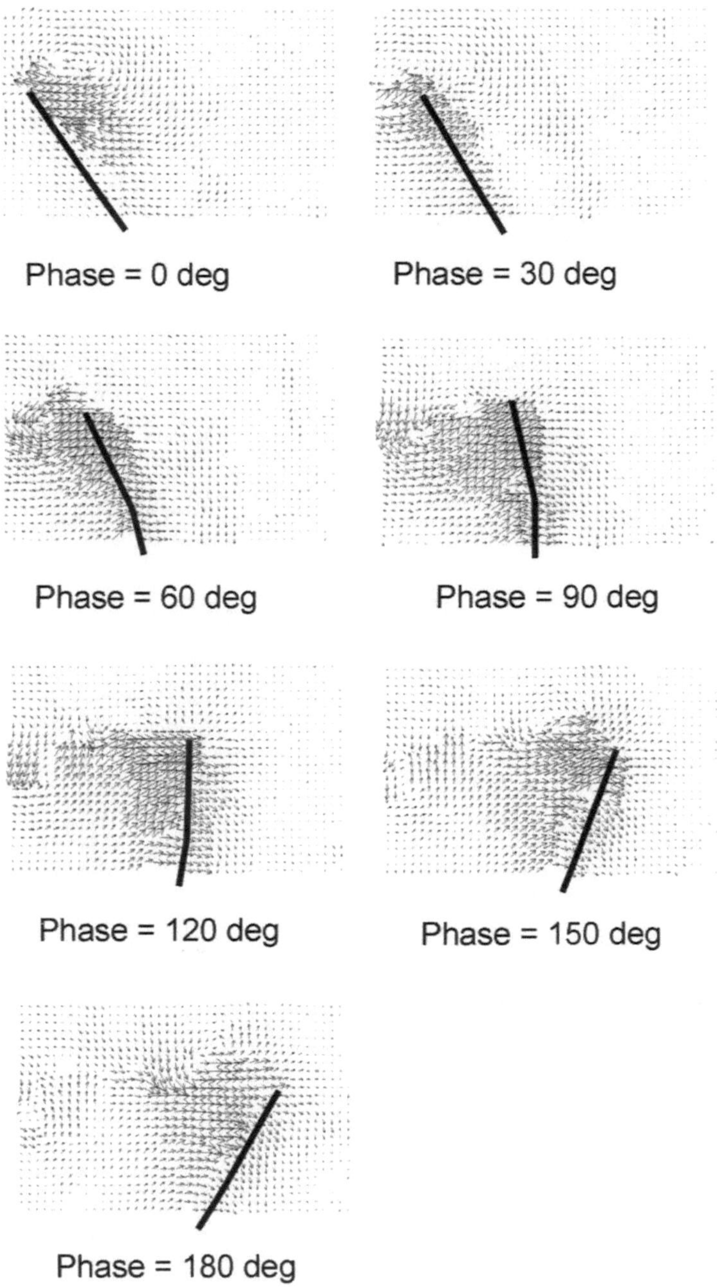

Fig.7. Change of fin shape and flow in one movement cycle ($\underline{H = 0\ (0\mathrm{mm})}$ static condition, $T = 3$ s, $\theta_{max} = 30$ deg).

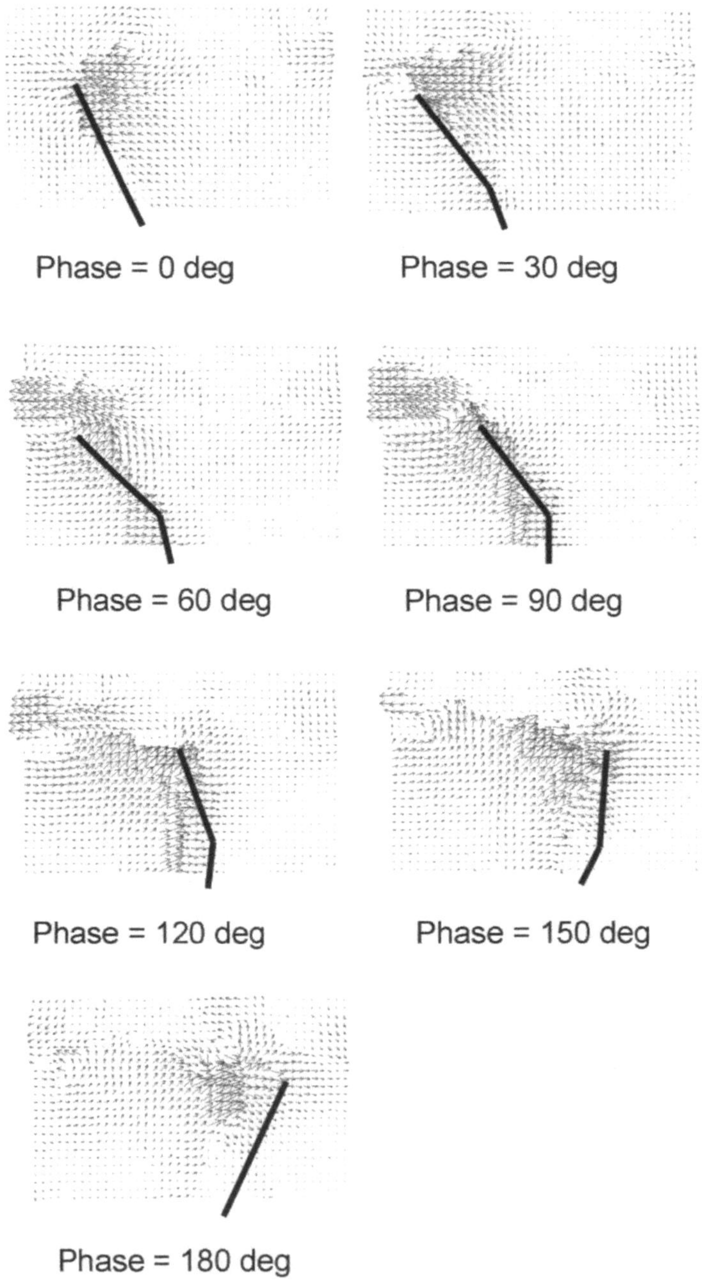

Fig. 8. Change of fin shape and flow in one movement cycle ($H = L$ (20 mm) static condition, $T = 3$ s, $\theta_{max} = 30$ deg).

Phase = 150 deg (phase for the minimum thrust force): Area of high flow velocity on the left side of fin is still greater, but the negative x-directional thrust force component would be greater from the fin position. Phase = 180 deg: Flow velocity is reduced. The negative x-directional thrust force component would be reduced from the fin position.

***H* = *L* (minimum apparent stiffness)**
Phase = 60 deg: The flow velocity on the left side of fin is smaller than that at the same phase when $H = 0$. Since the fin is bent, the positive thrust force on the fin is smaller than that when $H = 0$. Phase = 90 deg (phase for the maximum thrust force): Positive x-directional thrust force is the maximum by the bending position of fin. But the area of high flow velocity on the left side of the fin is smaller than that at the same phase when H=0. Phase = 150 deg: the positive x-directional thrust force would be reduced by the fin position. Phase = 180 deg (phase for the minimum thrust force): the negative thrust force component would be generated by the fin position, but since the flow is smaller than that when $H = 0$, magnitude of negative x-directional thrust force is smaller than that when $H = 0$.

3.2 Dynamic Condition of Effective Spring Length

3.2.1 Thrust Force

After the experiments under the static condition of effective length of the spring H, we controlled the dynamic change of effective length during motion. Figure 9 shows the change of H and x-directional thrust force in one movement cycle. H is controlled to obtain greater thrust force. For the phases of 0 - 90 deg and 180 - 270 deg, H was set at 0 mm because the thrust forces in these phases are greater when $H = 0$ mm (static condition of H, shown in Fig. 6). For the phases of 90-180 deg and 270-360 deg, H was set at 20 mm because the thrust forces in these phases are greater when $H = 20$ mm (static condition of H, shown in Fig. 6). In Fig. 9, x-directional thrust forces during the phases of 0-90 deg and 180-270 deg are the closer to those when $H = 0$ mm shown in Fig. 6, and x-directional thrust forces during the phases of 90 - 180 deg and 270 - 360 deg are closer to those when $H = 20$ mm shown in Fig. 6. Thus the average thrust force with the dynamic change of H ($F_{xavg} = 0.424$ N) is greater than that under the static condition of H_{opt} ($F_{xavg} = 0.416$ N).

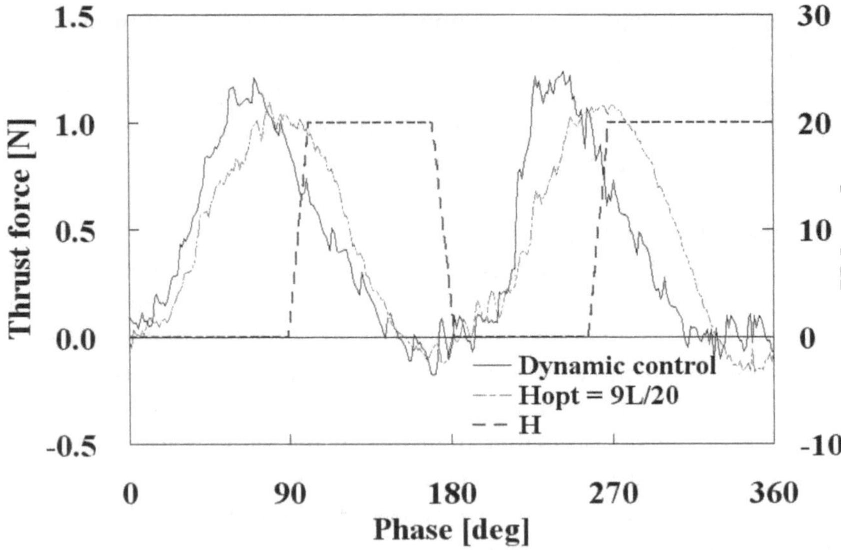

Fig. 9. Change of effective length of spring H and variation of x-directional thrust force in one movement cycle ($T = 3$ s, $\theta_{max} = 30$ deg) . Thrust force for $H_{opt} = 9L/20$ (Static condition) is added for comparison.

3.2.2 Fin Shape and Flow Around the Fin

Figure 10 shows the change in fin shape and flow around the fin in one movement cycle. The effective spring length H is changed during motion (dynamic condition of H). Bending of fin and flow fields around the fin during phases 0 - 90 deg and 90 - 180 deg are closer to those during phases 0 - 90 deg when $H = 0$ and phases 90 - 180 deg when $H = L$, respectively (shown in Figs. 7 and 8). This result is consistent with that of thrust force in one movement cycle described section 3.2.1.

4 Conclusions

This study was conducted to develop a propulsion mechanism in fluid, using a fin with a dynamic variable-effective-length spring. We discussed the influence of thrust force on the effective length of the spring under static and dynamic conditions. The following results were obtained.

(1) Bending of the fin increases with an increase in effective length of the spring H.

(2) Flow velocity difference between the both sides of fin is influenced on the thrust force.

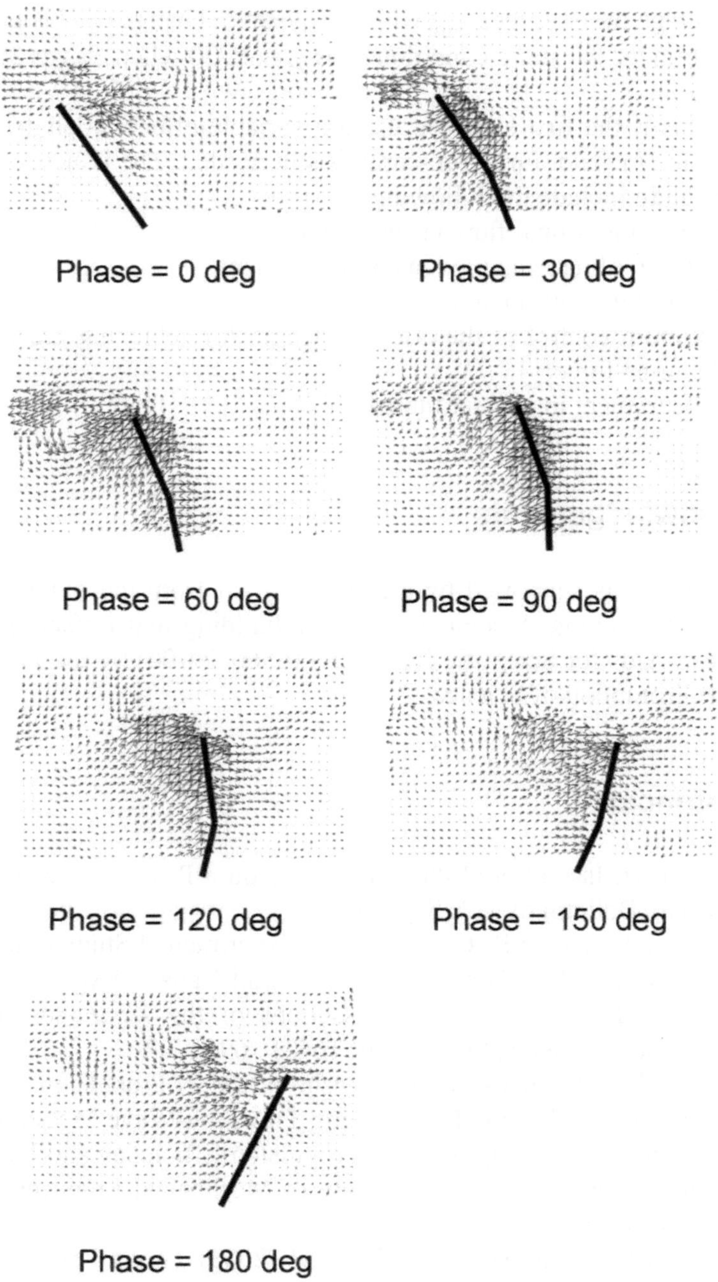

Fig. 10. Change of fin shape and flow in one movement cycle (H:dynamic conditon (0-20 mm), $T = 3$ s, $\theta_{max} = 30$ deg).

(3) Bending of the fin reduces negative thrust force.
(4) Average thrust force F_{xavg} with the dynamic change of H is greater than that under the static condition of $H_{opt} = 9L/20$ when θ_{max} is 30 deg (T = 3 s).

Since this is the first study on using a variable-effective-length spring for propulsion in fluid, further research is still needed. The issues to be tackled in the future include the following:
(1) Three dimensional flow visualization
(2) Addition of heaving movement mechanism,
(3) Discussion of efficiency,
(4) Optimum control of the effective length for achieving greater and stable thrust force,
(5) Measurement in a water tunnel.

Acknowledgements

This work was supported by a research grant from the Fundamental Research Developing Association for Shipbuilding and Offshore, and by a Grant-in-Aids for Scientific Research (16560224-00) by the Japan Society for the Promotion of Science.

References

Morikawa H, Isshiki N (1980) The Study on a Propulsion System by Fin Stroke. Bulletin of M.E.S.J., 8 (1)

Nakashima H, Tokuo K, Ono K (2000) Experimental Study of a Two-Joint Dolphinlike Propulsion Mechanism (2nd Report, Experiment of Self-Propelled Large Robot). Transactions of the Japan Society of Mechanical Engineers, Series C, 66 (643) : 695–702

Watanabe M, Muramatsu K Kobayashi N (2002) Propulsion Performance of an Aquatic Mobile Robot Using Traveling-Wave Motion of a Flexible Fin (Relationship between Propulsion Efficiency and Flow Pattern. Transactions of the Japan Society of Mechanical Engineers, Series C, 68 (665): 188–196

Kobayashi S, Mashima T, Morikawa H (2003) Thrust Force Characteristics of Propulsion Mechanism in Fluid Using Variable-Bending-Stiffness Fin Modeled on Ciliary Movement. JSME International Journal, Series C, 46 (4): 1340–1345

Chapter 22

Elastic Pectoral Fin Actuators for Biomimetic Underwater Vehicles

Naomi Kato [1]
Yoshito Ando[2], Ariyoshi Tomokazu[1], Hiroyoshi Suzuki[1],
Koichi Suzumori[3], Takefumi Kanda[3], Satoshi Endo[3]
[1]Department of Naval Architecture and Ocean Engineering, Graduate School of Engineering, Osaka University, 2-1 Yamadaoka, Suita, Osaka 565-0871, Japan
[2]Mitsubishi Heavy Industries, LTD.
[3]Department of Mechanical and System Engineering, Graduate School of Natural Science and Technology, Okayama University, 1-1 Naka 3-chome, Tsushima, Okayama 700-8530 Japan

Summary. This paper describes the development of two new types of pectoral fins made of elastic materials. The fins were designed to have flexibility and multifunctionality for use not only as propulsive devices in biomimetic underwater vehicles but in other applications such as grippers, thus avoiding the environmental damage often caused by rigid fins. We developed an actively controlled pneumatic fin and a passively controlled flexible fin.

The former is composed of silicone with a build-in FMA (flexible micro-actuator) and generates propulsive force by means of a lifting motion driven by the FMA. The latter type of flexible fin is formed of silicone rubber and generates propulsive force by means of a dragging motion. We experimentally verified the elastic pectoral fins' ability to generate propulsive force and used FEM to analyze the fins' behavior.

Key words. Biomimetic Underwater Vehicle, Mechanical Pectoral Fin, Elastic, Pneumatic Actuator, Flexible Micro Actuator, Passive Flexible Fin, Finite Element Method

1 Introduction

Lindsey(1978) classified swimming modes of fish into body and/or caudal fin (BCF) mode, and median and/or paired fin (MPF) mode, according to portion and parts in fish body as the propulsive structure for thrust generation. Furthermore, BCF mode and MPF mode were classified into undulatory motion involving the passage of a wave along the propulsive structure, and oscillatory motion controlled by fin rays such as a fan with several degrees of freedom without exhibiting a wavy motion. Pectoral fin propulsion has classically been divided into two categories which represent biomechanical extremes in the use of appendages for propulsion: drag-based and lift-based mechanisms of thrust production (Blake 1979 and Webb 1973).

In drag-based propulsion the pectoral fins move mainly along a horizontal stroke plane in a rowing mode of propulsion. The fins usually form a high angle with the main axis. The fins generate thrust force while the fins are retracted posteriorly (power stroke) if the velocity of fin retraction is greater than the velocity of water movement. During the recovery (protraction) stroke, the fin is feathered so that drag on the fin is reduced and the fin is brought forward. There is no positive thrust generated during the recovery stroke.

In lift-based propulsion the pectoral fins move mainly along a vertical stroke plane in a flapping mode of propulsion with a low angle of attack to the direction of forward progression. The fins generate thrust force during both upstroke and downstroke.

Focusing on the oscillatory motion of pectoral fin in MPF mode, we have studied its biomimesis to develop a mechanical pectoral fin for precisely maneuvering underwater vehicles and developed a biomimetic underwater vehicle equipped with 2 pairs of mechanical pectoral fins (Kato et al. 2004, Kato 2005,Kato et al. 2005, 2006). The mechanical pectoral fin uses a rigid plate and is capable of producing both the lift-based and the drag-based swimming modes.

On the other hand, several types of flexible fin for MPF mode have been developed so far, for instance, Davis et al.(1998) using flexible hydraulic tubes, Toda et al.(2004) using a pair of undulating side fins consisting of flapping rods, Sandberg et al.(2006) using a set of flapping rods producing flexible curvature of pectoral fin and Ichikizaki et al. (2006) for flexible pectoral fins.

This paper deals with the development of new pectoral fins made of elastic materials for use as not only a propulsive device, but also as grippers, allowing the system to avoid damaging the environment in such as a way a did the rigid fins of our previous mechanical pectoral fin device. In light of these concerns, we developed two kinds of elastic propulsive devices: an actively controlled pneumatic fin and a passively controlled flexible fin. These fins are referred to as the "Active Pneumatic Actuator Fin" and the "Passive Flexible Fin", respectively. This paper describes these two new elastic pectoral fins' actuators, focusing on the following areas:

(1) The development of the Active Pneumatic Actuator Fin and the Passive Flexible Fin,
(2) Verification tests for estimating the propulsive forces generated by these new elastic fins,
(3) Finite Element Method (FEM) analyses of the behavior of the two new elastic fins.

2 Mechanical Pectoral Fin Device

In this study, we improved the mechanical pectoral fin's device producing rowing, feathering, and flapping motions (Fig.1) by installing air bulb controllers inside it for the Active Pneumatic Actuator as shown in Figure 2. Rowing, feathering, and flapping motions are characterized by back-and-forth motions, twisting motions, and up-and-down motions, respectively.

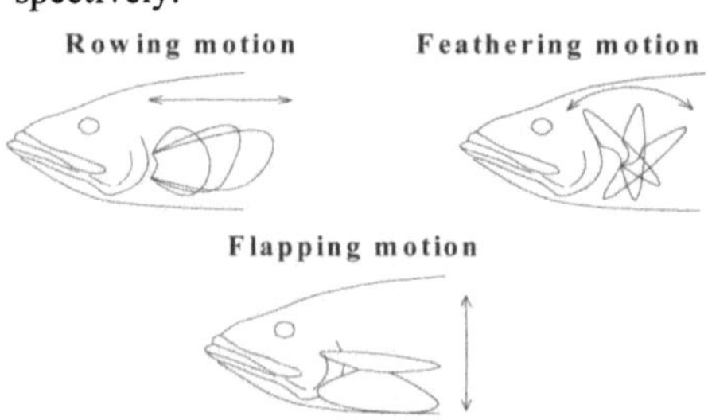

Fig.1. Basic motion of pectoral fin

Fig.2. Improved mechanical pectoral fin device

3 Active Pneumatic Actuator Fin

3.1 FMA (Flexible Micro Actuator)

The Active Pneumatic Actuator Fin consists of three Flexible Micro Actuators (FMAs). FMAs made of silicon rubber with fiber reinforcement were previously developed by one of the authors (Suzumori et al. 1989,1990,1991). The FMA has three pressure chambers inside a rubber tube as shown in Figure 3, and motions with three degrees of freedom are realized by controlling the pressure in these chambers independently. The distinctive features of the FMA are:

1. Ease of miniaturization because of its simple structure,

2. High ratio of output to weight,
3. Its possible use as a component of a manipulator.

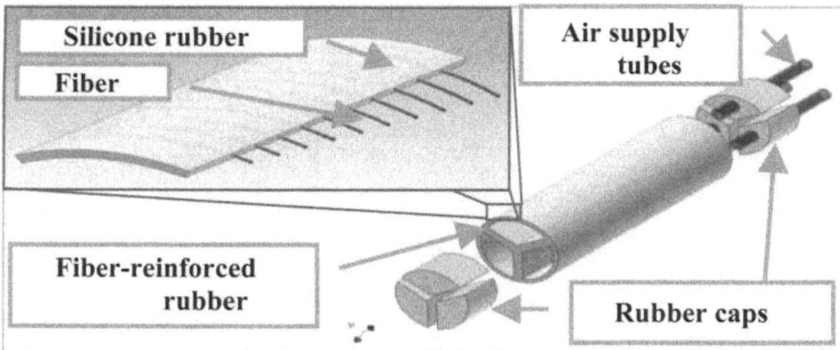

Fig.3. Structure of FMA

3.2 Propulsive principle

The Active Pneumatic Actuator Fin, which can produce a feathering motion, requires only a flapping motion by the mechanical pectoral fin device in order to generate propulsive forces using a lift-based swimming mode with small angles of attack against uniform flow, whereas the rigid mechanical pectoral fin needs both flapping and feathering motions in order to perform lift-based swimming mode. To produce a feathering motion, the Active Pneumatic Actuator Fin uses three build-in FMAs.

3.3 Active Pneumatic Actuator Fin

We developed a prototype of the Active Pneumatic Actuator Fin and performed verification tests on it. However, the driving displacement of the fin was small. In addition, the analog valves necessary to drive FMAs inside the fin cannot be installed in biomimetic underwater vehicles, because these valves require a relatively large space. To solve these problems, fin shape and pressurization method were improved as shown below.

(A) Shape of the Active Pneumatic Actuator Fin

To allow the Active Pneumatic Actuator Fin a larger range of motion, the fin shapes are designed as shown in Figs.4 and 5 (Endo, et al., 2005b). Their span is 130mm and width 113mm. The web between the FMAs is made of silicon rubber and is crinkled. The thickness of the web is 1.5mm. The diameter of the pressurized section of the FMA is $\Phi 18$. Figure 5 shows the Active Pneumatic Actuator Fin with FMAs.

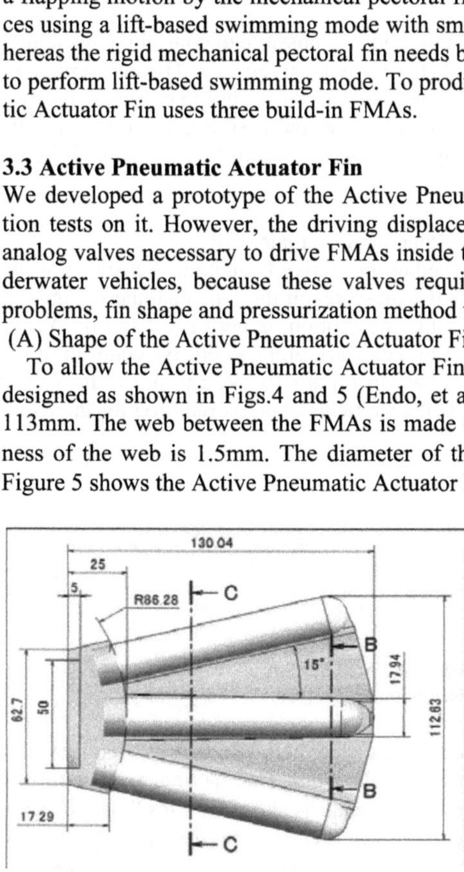

Fig.4. Dimensions of Active Pneumatic Actuator Fin [mm]

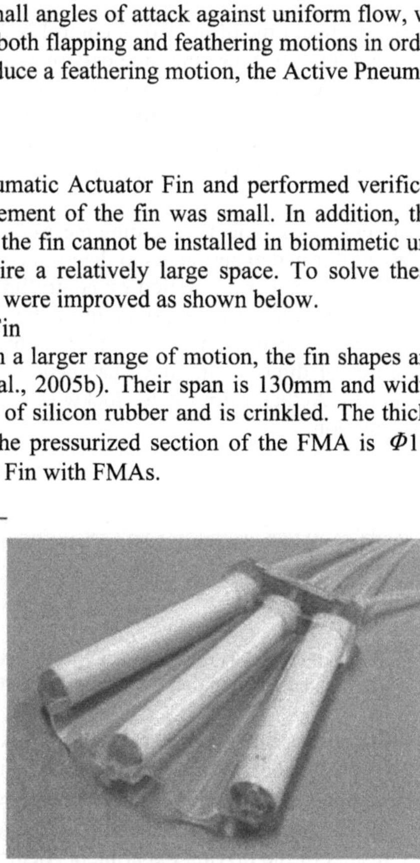

Fig.5. Active Pneumatic Actuator Fin

(B) PWM controlled digital valves

Small digital valves instead of analog valves were installed in the mechanical pectoral fin device. These valves are driven with the PWM control in order to realize air pressure control like analog control.

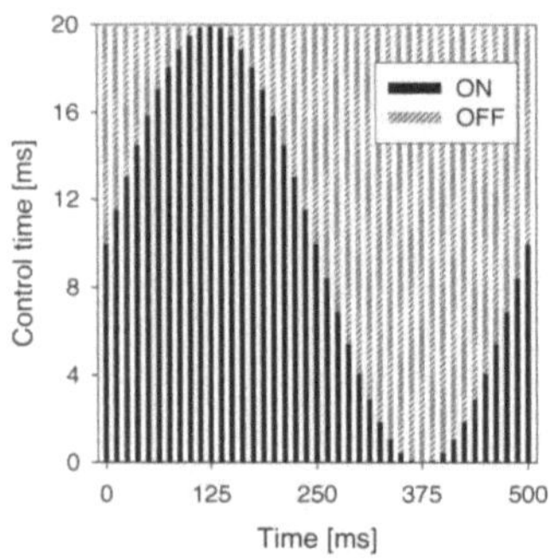

Fig.6. Control time of PWM driving

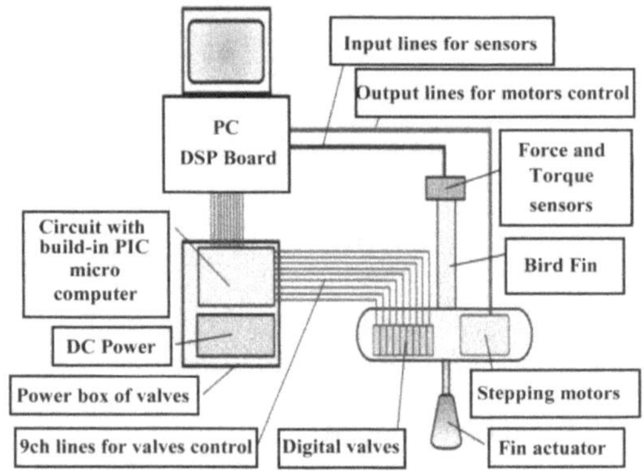

Fig.7. Control system of air pressure

The cyclical time of the PWM control is 20ms, and it is possible to control the resolution of 1ms. Figure 6 shows the time pattern at a 2 Hz of driving frequency. The air pressure of the FMA is controlled by the ratio of ON- and OFF-time as shown in Fig.6. Figure 7 shows the control system. Nine control codes of the output from the DSP board are sent to PIC microcomputers. Based on these codes, the digital valves are controlled by the PWM control program in PIC microcomputer.

(C) Driving experiments regarding the Active Pneumatic Actuator Fin

We performed experiments to determine whether the Active Pneumatic Actuator Fin can produce the feathering motion by controlling the pressures in the nine chambers. Figure 8 shows an example of the principle of the feathering motion as performed by the Active Pneumatic Actuator Fin. The filled chambers by color are pressurized. By controlling the pressures of three of the nine chambers as shown in Figure 10, we can make the Active Pneumatic Actuator Fin rotate clockwise as well as anticlockwise as required for the feathering motion.

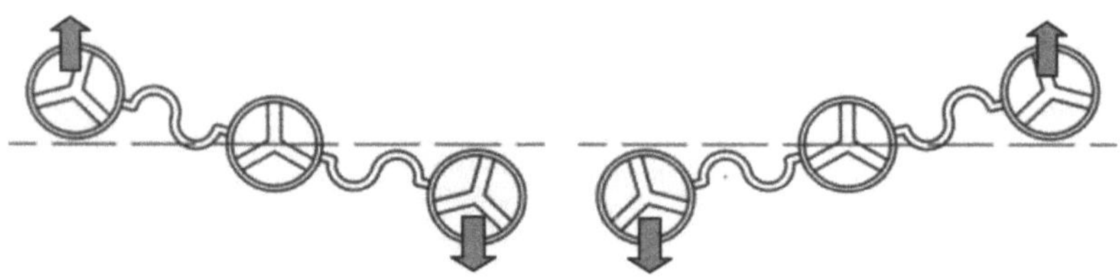

Fig.8. Principle of Feathering Motion by Active Pneumatic Actuator Fin (Left; clockwise rotation, Right; anticlockwise rotation)

In these experiments, the maximum pressure in the chambers was 0.2MPa. To make the Active Pneumatic Actuator Fin perform a more precise feathering motion, we pressurized the chambers with the phase difference of π at both ends as shown in Figs.9 and 10 without pressurizing the center chambers. Figure 11 shows a view of the driving experiments. The results of the driving experiments demonstrate that the Active Pneumatic Actuator Fin can successfully perform a feathering motion.

Fig.9. Control numbers of FMA chambers

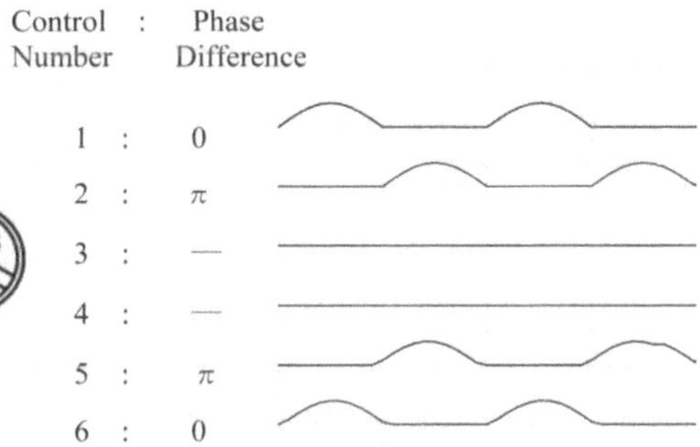

Fig.10. Pressure variations with time in each chamber

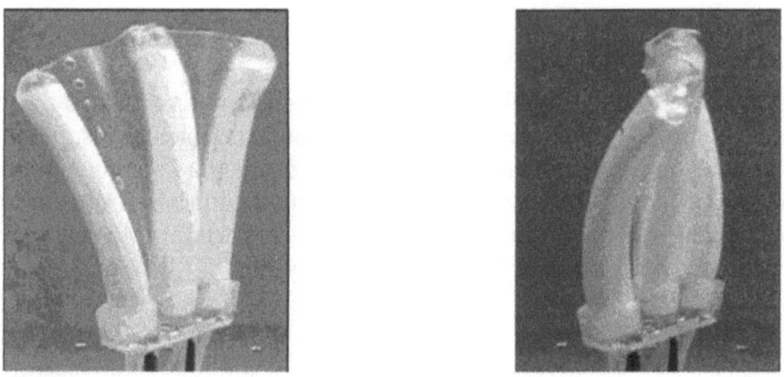

Fig.11. View of driving experiments

3.5 Propulsive performance by the Active Pneumatic Actuator Fin in a uniform flow

We performed experiments to verify that the Active Pneumatic Actuator Fin can generate propulsive forces in a uniform flow using the Towing Tank at Osaka University, which measures 100 meters in length, 7.8 meters in width, and 4.35 meters in depth.

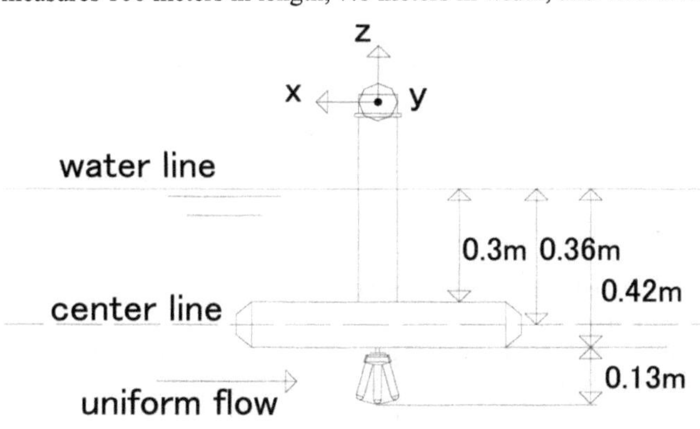

Fig.12. View of experiment at the Towing Tank

The mechanical pectoral fin device shown in Figure 2 was submerged at a depth of 0.36 meters at the center line of the fuselage as shown in Fig.12. A 6-components force sensor was attached to the top of the strut to measure the hydrodynamic forces. The hydrodynamic forces acting on the fin only were estimated by subtracting the forces acting on the device without the fin from the total forces acting on the device with the fin. The uniform flow was set at 0.16m/s, and the frequency of the flapping motion produced by the mechanical pectoral fin device and the feathering motion produced by the Active Pneumatic Actuator Fin was set at 1Hz to produce lift-based swimming motion. The flapping motion is described as follows:

$$\phi_{FL} = \phi_{FLA} \cdot \cos\left(\omega_{fin} \cdot t + \Delta\phi_{FL}\right) \quad (1)$$

where ω_{fin} denotes the angular velocity of the fin, t and the time, Φ_{FLA} the amplitude of the flapping motion, and $\Delta\Phi_{FL}$ the phase difference between the flapping motion and the pressure at No.1 and No.6 FMA chambers in Fig.9. We set Φ_{FLA} as 40.0 ° ,referring to the optimum parameters of the mechanical pectoral fin device using a rigid fin for the lift-based swimming mode (Kato et al. 2004). $\Delta\Phi_{FL}$ was changed every 30.0 ° within the range from –150 ° to 180 ° . The non-dimensional propulsive force coefficient,C_x , is defined by the following equation:

$$C_x = F_x / \left(0.5 \cdot \rho \cdot U^2 \cdot S\right) \quad (2)$$

where F_x denotes the propulsive force, ρ the fluid density, U the uniform flow velocity, and S the fin's surface area. Figure 13 shows the comparison of the experimental results between combination of flapping motion and feathering motion, and only the flapping motion. We can see that the fin can generate thrust over a region by using the lift-based swimming mode consisting of the flapping motion produced by the mechanical pectoral fin device and the feathering motion produced by the Active Pneumatic Actuator Fin.

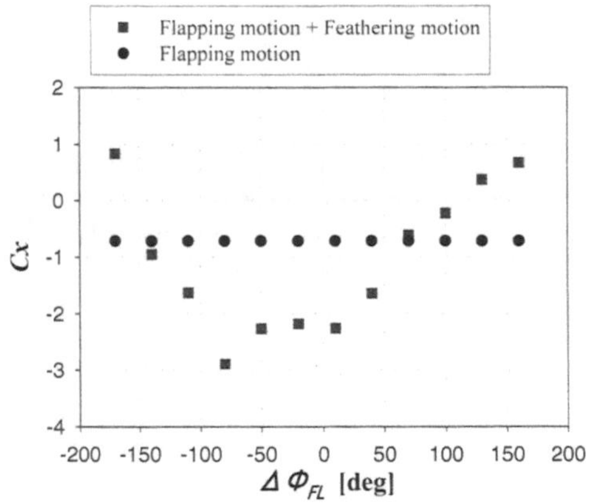

Fig.13. Comparisons of the experimental results in uniform flow

3.6 Finite element analysis of the behavior of individual FMA

(A) Hyper elastic Material of FMA

We used FMA made of silicon rubber. The typical properties of silicon rubber include

1. The ability to remain elastic even while undergoing large deformation,
2. A highly nonlinear relationship between load and extension,
3. Near incompressibility.

These characteristics can be generally defined as hyperelastic. It is appropriate for modeling the mechanical behavior of elastomers such as silicon rubber of which the FMA and the web are made. Hyperelasticity is defined as a material that has an elastic potential function W. A stress tensor is derived by differentiating W with respect to the conjugate strain tensor as

$$S_{ij} = \frac{\partial W}{\partial E_{ij}} \tag{3}$$

where the following definitions can be applied:

$$\mathbf{E} = (\mathbf{C} - \mathbf{I})/2 \tag{4}$$

where **E** denotes the Green-Lagrange strain tensor, ***C*** the right Cauchy-Green deformation tensor, and ***S*** the second Piola-Kirchhoff stress tensor. The elastic potential function W is defined as the functions of the ratios of expansion and contraction, λ_1, λ_2, and λ_3, in three dimensions.

$$W = f(\lambda_1, \lambda_2, \lambda_3) \tag{5}$$

In this analysis, we use the Ogden model for W since this model can represent the behavior of slightly compressible rubber. The elastic potential function W in the Ogden model is defined as

$$W = \sum_{n=1}^{N} \frac{\mu_n}{\alpha_n} \left[J^{-\alpha_n/3} \left(\lambda_1^{\alpha_n} + \lambda_2^{\alpha_n} + \lambda_3^{\alpha_n} \right) - 3 \right] + 4.5K \left(J^{1/3} - 1 \right)^2 \tag{6}$$

where μ_n and α_n are material constants, K is the initial bulk modulus, and J is the volumetric ratio, defined by

$$J = \lambda_1 \lambda_2 \lambda_3 \tag{7}$$

The number N of the terms expressed in the Ogden model was set as 3 in this study.

(B) FE Analysis of the Flexible Micro Actuator

We computationally analyzed the behaviors of the individual FMA when each chamber is pressurized at 0.2 MPa using FEM software MSC.Marc. To determine the Ogden constants in Eq. (7), uniaxial tensile tests were performed using specimens made of the same materials as the FMA and the web. Optimizing the Ogden constants derived by the tensile tests, we used FEM to analyze the behavior of the individual FMA as shown in Fig. 14. Figure 15 shows the FE model of the FMA.

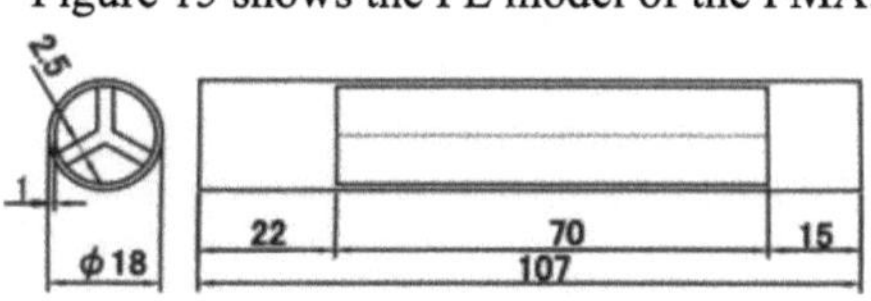

Fig.14. Dimensions of FMA [unit: mm]

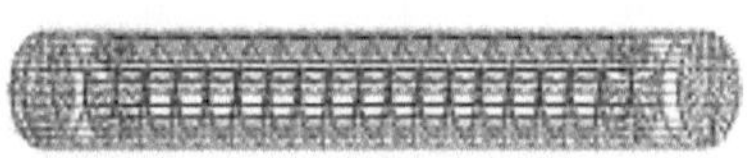

Fig.15. FE model of FMA

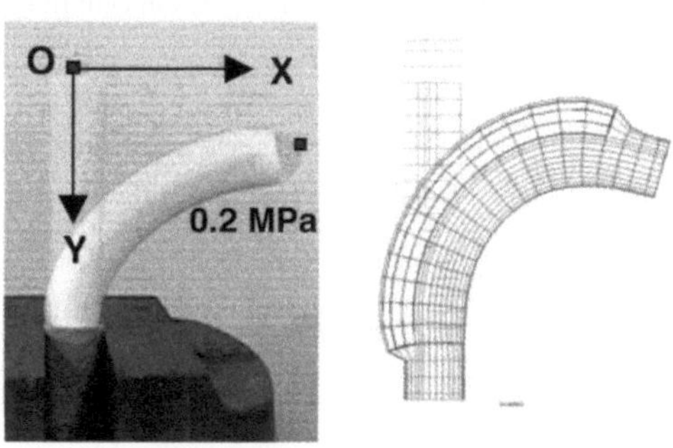

Fig.16. Behavior of Image Analysis and FEM Analysis (Left: Measured Image, Right: FE Analysis)

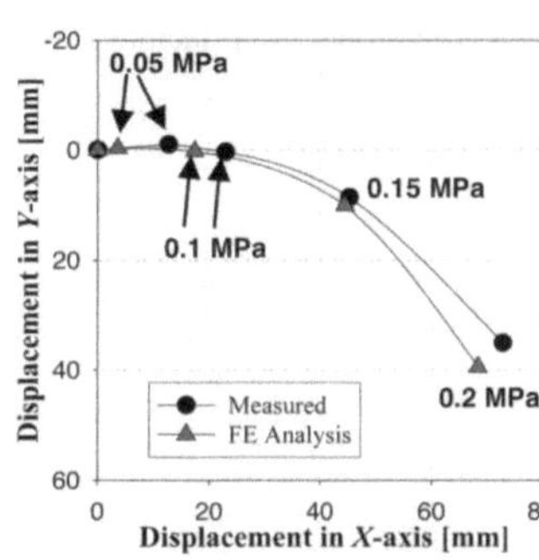

Fig.17. Comparison of displacements in X and Y-axes of FMA

Figure 16 shows the behaviors of the FMA at 0.2MPa by FE and image analyses. By defining the X-Y axes as shown in Fig.16, we compared the components of the displacements at the top of FMA in the X-Y axes as determined by FE analysis with those obtained by the image analysis as shown in Fig.17. We can see that the behavior of the FMA by FE analysis expresses well the experimentally measured behavior at each pressure level.

4 Passive Flexible Fin

4.1 Propulsive principle

Passive flexible fin performs a drag-based swimming mode consisting of only a rowing motion as a way to produce propulsive forces. The rowing motion is divided into two strokes: a power stroke and a recovery stroke. The power stroke is defined as the motion from the anterior position to the posterior position. The recovery stroke is defined as the reverse motion of the power stroke. It is ideal to produce propulsive forces during the power stroke and to minimize drag forces during the recovery stroke in order to produce the maximum propulsive forces during one cycle (Endo et al. 2005a). It is necessary for the fin to change its shape depending on the moving phase in order to change its drag coefficient. We developed a fin that is made of silicone rubber containing grooves in order to allow the fin to change its structural stiffness. The grooves fabricated on the fin make the fin harder during the power stroke, while softer during the recovery stroke. This working principle is shown in Fig.18.

The outline shape of the fin is shown in Fig.19. In optimizing those two strokes in the passive flexible fin, two types of groove pattern were designed and tested: a level groove as shown in Fig. 20 and a vertical groove as shown in Fig.21.

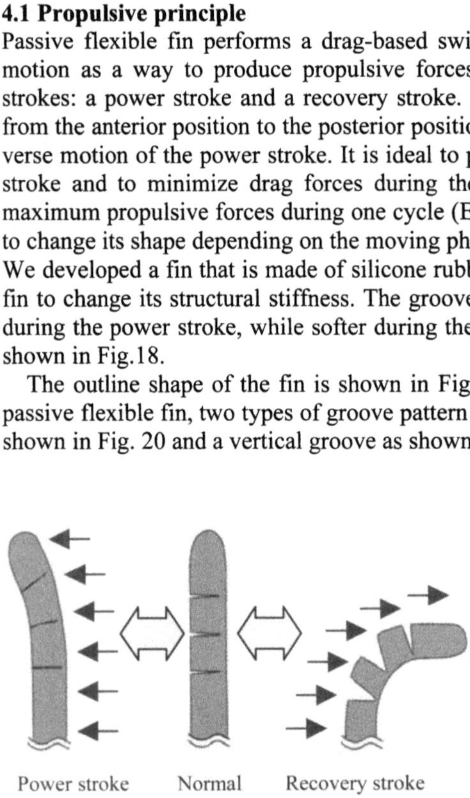

Fig.18. The propulsive principle

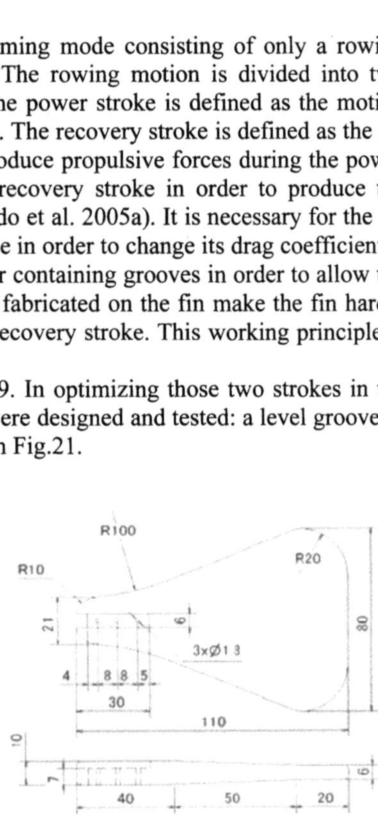

Fig.19. Dimensions of Passive Flexible Fin Unit [mm]

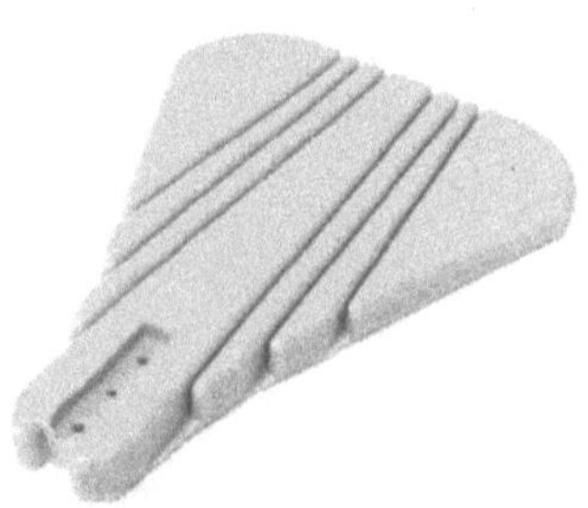

Fig.20. Level grooved passive fin

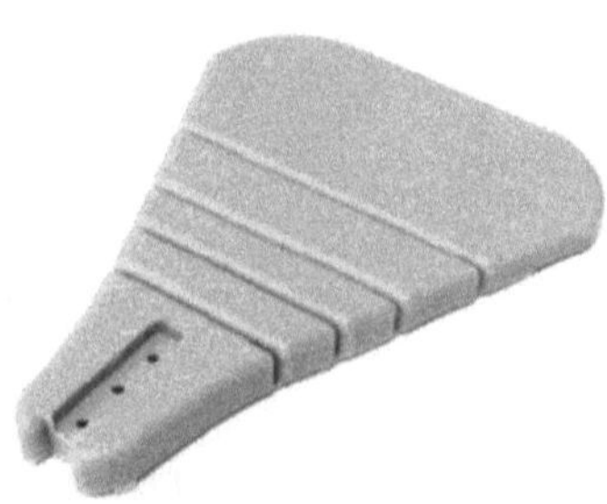

Fig.21. Vertically grooved passive fin

4.2 Propulsive performance of the Passive Flexible Fin in still water

We experimentally measured the propulsive forces of the Passive Flexible Fin by performing a simple rowing motion in still water. In these experiments, the amplitude of the rowing motion was set at 30 degrees, and the frequency of the rowing motion was set at 1.0Hz. As well, the measurement time of the rowing motion was set at 10s. The non-dimensional propulsive force coefficient, C_{x0} is defined by the following equation:

$$C_{x0} = Fx/(0.5\cdot\rho\cdot(\omega_{fin}\cdot c)^2\cdot S) \tag{8}$$

where F_x denotes the propulsive force, ρ the fluid density, ω_{fin} the angular velocity of the fin, c the maximum chord of the fin, and S the surface area of the fin.

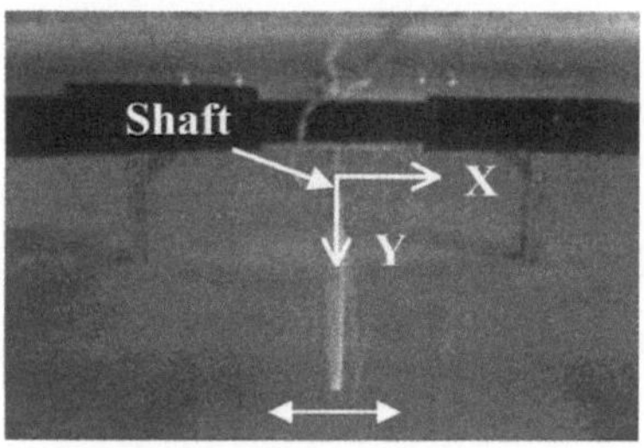

Fig.22. View of the experiments

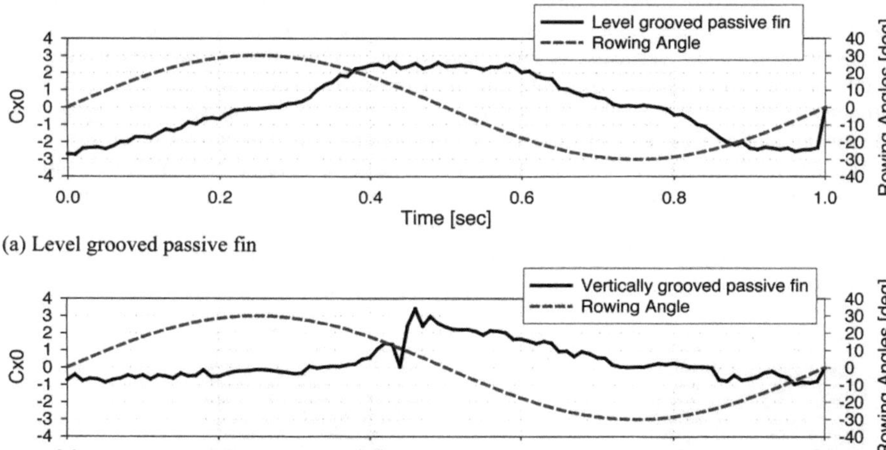

(a) Level grooved passive fin

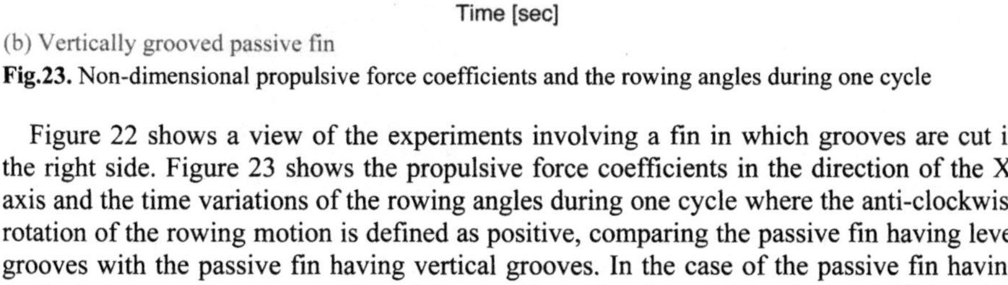

(b) Vertically grooved passive fin

Fig.23. Non-dimensional propulsive force coefficients and the rowing angles during one cycle

Figure 22 shows a view of the experiments involving a fin in which grooves are cut in the right side. Figure 23 shows the propulsive force coefficients in the direction of the X-axis and the time variations of the rowing angles during one cycle where the anti-clockwise rotation of the rowing motion is defined as positive, comparing the passive fin having level grooves with the passive fin having vertical grooves. In the case of the passive fin having vertical grooves, the average value of the non-dimensional propulsive force coefficient during one cycle is 0.306. In the case of the passive fin having level grooves, this value is 0.031.

We can see from these results that the vertical grooves on the fin's surface contribute more effectively than do the level grooves to the generation of propulsive forces during the power stroke and to the decrease of drag forces during the recovery stroke.

4.3 Dynamical finite element analysis of Passive Flexible Fin behaviors

(A) Hyper-elastic Material of Passive Flexible

In this analysis, we used the Mooney-Rivlin model as W since this model can represent the behavior of incompressible isotropic rubber. The elastic potential function W in the Mooney-Rivlin model (the *James-Green-Simpson* model) is defined as

$$W = C_{10}(I_1-3)+C_{01}(I_2-3)+C_{11}(I_1-3)(I_2-3)+ \\ C_{20}(I_1-3)^2+C_{30}(I_1-3)^2 \qquad (9)$$

where c_{ij} are material constants, and the principal invariants I_i are a function of λ_i , defined by

$$\begin{cases} I_1 = \lambda_1^2+\lambda_2^2+\lambda_3^2 \\ I_2 = \lambda_1^2\lambda_2^2+\lambda_2^2\lambda_3^2+\lambda_3^2\lambda_1^2 \\ I_3 = \lambda_1^2\lambda_2^2\lambda_3^2 = 0 \end{cases} \qquad (10)$$

(B) FE analysis of Passive Flexible Fin

We used FEM to analyze the dynamical behavior of the vertically grooved passive fin, which can effectively generate thrust in air. To determine the Mooney-Rivlin constants in Eq. (10), uniaxial tensile tests were carried out using specimens composed of the same materials as those of the verticaliy grooved passive fin. Optimizing the Mooney-Rivlin constants derived from tensile tests, we analyzed the dynamical behavior of the vertically grooved passive fin as it produced a rowing motion in air. In this analysis, the amplitude of the rowing motion was set at 30 degrees, and the period of the motion was set at 0.5s.

We also analyzed the rowing motion of the vertically grooved passive fin in air through image analysis at an interval of 0.03s. We compared the motion obtained by the image analysis with that determined by FE analysis. Figure 24 shows a comparison of the results of the image analysis and the FE analysis where the X-Y coordinate is defined. The tip of the fin was analyzed.

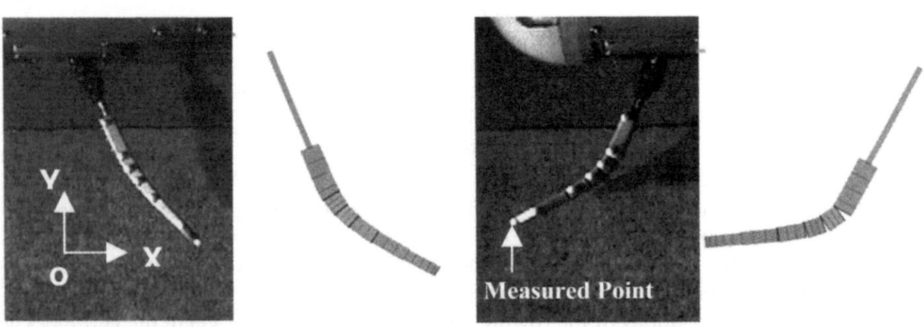

(a) Comparison of maximum displacement (b) Comparison of minimum displacement

Fig.24. Comparisons of displacement in X-axis (Left: Measured Image, Right: FE Analysis)

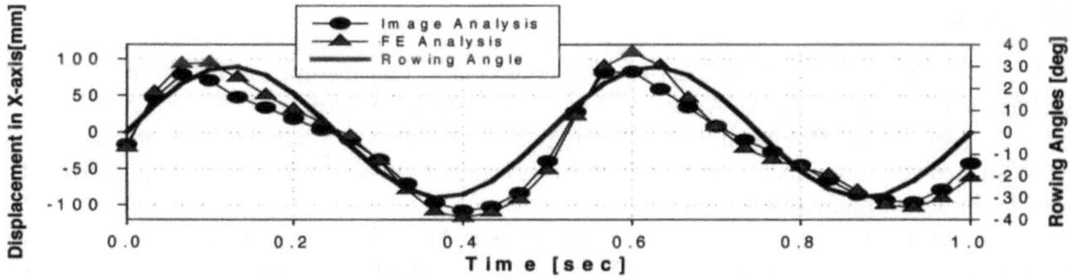

Fig.25. Comparison of the displacement in X-axis

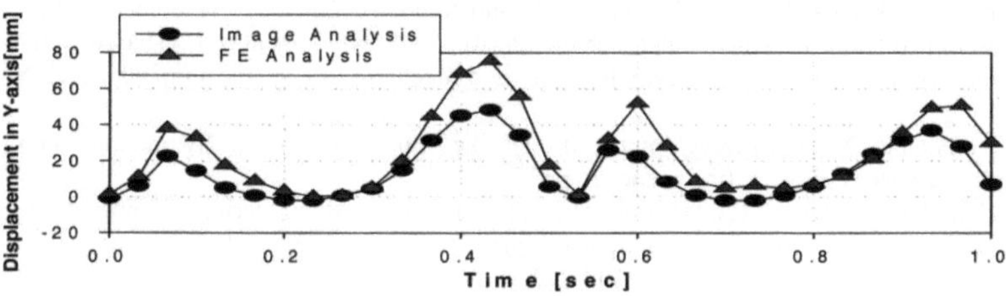

Fig.26. Comparison of the displacement in Y-axis

Figure 25 shows a comparison of the displacements in the X-axis between the results of FE analysis and image analysis. Figure 26 shows a comparison of the results on the Y-axis. We can see that FE analysis can express dynamic behavior quantitatively. However, we can see that there is a difference in displacement peaks on the Y-axis between the FE analysis results and those of the experiments. These phenomena may be attributed to the high-nonlinear characteristics of the elastomers.

5 Conclusion

In the present study, we discussed the development of two new types of elastic pectoral fin with the following results:

1. The Active Pneumatic Actuator Fin with built-in FMAs was found to work successfully through the employment of digital valves controlled by PIC microcomputers
2. The Active Pneumatic Actuator Fin can generate propulsive force by using a lift-based swimming motion.
3. The highly non-linear behavior of a single FMA was analyzed.
4. The vertical grooves on one side of a Passive Flexible Fin were found to contribute to the generation of propulsive force during the power stroke and to the decrease of drag force during the recovery stroke.
5. FE analysis fairly well expresses the dynamic behavior of the vertically grooved passive fin as experimentally measured quantitatively by image analysis.

In our future work, we will use a combination of FE analysis and Computational Fluid Dynamics (CFD) analysis to investigate the behavior of both fins more precisely in order to simulate their performance when employed on an underwater vehicle.

6 Acknowledgements

This research was funded for three years beginning in 2005 by the Ministry of Education, Culture, Sports, Science and Technology, Japan, (Grant No. 16360435) as the project title "Application of flexible Microactuator for an underwater vehicle".

7 References

Blake, R.W. (1979) The Mechanics of Labriform Locomotion. I. Labriform Locomotion in the Angelfish(Pterophyllum eimekei): An Analysis of the Power Stroke, J. Exp. Biol., 82, pp.255-271

Davies, J.B.C., Lane, D.M., Robinson, G.C., O'Brien, D.J., Pickett, M., Sfakiotakis, M. and Deacon, B. (1998) Subsea Applications of Continuum Robots, Proc. of 1998 International Symp. on Underwater Technology, pp.363-369

Endo, S., Suzumori, K., Kanda, T., Kato, N., Suzuki, H., and Ando, Y. (2005a) Development of Flexible Pectoral Fin Actuator for Underwater Vehicle (1st Report, Development of Drag-based Passive Pectoral Fin), Trans. JSME, (in Japanese), Robotics and Mechatronics Symposium2005, 1A1-N-069

Endo, S., Suzumori, K., Kanda, T., Kato, N., Suzuki, H., and Ando, Y. (2005b) Development of Flexible Pectoral Fin Actuator for Underwater Vehicle (2nd Report, Development of Lift-based Active Pectoral Fin), Trans. JSME, (in Japanese), Robotics and Mechatronics Symposium2005, 1A1-N-070

Ichikizaki, T., Yamamoto, Y.(2006) Development of Fish-like Robot with Various Swimming Functions, Proc. of JASNAOE No.3, (in Japanese), pp27-30

Kato, N., Liu, H. and Morikawa, H. (2004) Biology-Inspired Precision Maneuvering of Underwater Vehicles in "Bio-mechanisms of Swimming and Flying," edited by N.Kato, J. Ayers and H.Morikawa, Springer-Verlag, Tokyo, pp.111-125

Kato, N.(2005) Controllers for Biomimetic Marine Vehicles, Applied Mechanics Reviews, Transactions of the ASME, Vol.58, No.4, pp.238-252

Kato, N, Liu, H and Morikawa, H (2005) Biology-Inspired Precision Maneuvering of Underwater Vehicles (Part 3), Int. J. Offshore and Polar Eng., Vol.15, No.2, pp.81-87

Kato, N., Ando, Y., Shigetomi, T., and Katayama, T. (2006) Biology-Inspired Precision Maneuvering of Underwater Vehicles (Part 4), International Journal of Offshore and Polar Engineering,Vol.16, No.3, p.p. 195-201

Lindsey, C. C. (1978). Form, function and locomotory habits in fish. In Fish Physiology, vol. 7 (ed. W. S. Hoar and D. J. Randall), pp.1–100. New York: Academic Press

Suzumori, K., (1989) Flexible Microactuator (1st Report, Static Characteristics of 3 DOF Actuator), Trans. JSME, (in Japanese), C55, pp.2547-2552

Suzumori, K., (1990) Flexible Microactuator (2nd Report, Dynamic Characteristics of 3 DOF Actuator) Trans. JSME, (in Japanese), C56, pp.1887-1892.

Suzumori, K., and Tanaka, H., (1991).Flexible Microactuator (3rd Report, An Application to a Multi-finderd Robot Hand), Trans. JSME, (in Japanese), C57, pp.1261-1266.

Toda, Y., Suzuki, T., Uto, S. and Tanaka, N. (2004) Fundamental Study of a Fishlike Body with Two Undulating Side-Fins," in Bio-mechanisms of Swimming and Flying, edited by N.Kato, J. Ayers and H. Morikawa, Springer-Verlag, Tokyo, pp.93-110

Sandberg, W.C. and Ramamurti,R. (2006) 3-D Unsteady Computations of Flapping Flight in Insects, Fish, and Unmanned Vehicles, Proc. of ISAMBEC (CD-ROM)

Webb, P.W. (1973) Kinematics of Pectoral Fin Propulsion in Cymatogaster Aggregate, J. Exp. Biol., 59, pp.697-710

Chapter23

Design, Development, and Testing of Flapping Fins with Actively Controlled Curvature for an Unmanned Underwater Vehicle

John Palmisano[1], Jason Geder[2], Ravi Ramamurti[2], Kerr-Jia Liu[3], Jonah Jonah Cohen[1], Tewodros Mengesha[3], Jawad Naciri[1], William Sandberg[2], and Banahalli Ratna[1]

[1] Center for Biomolecular Science and Engineering, Naval Research Laboratory, Washington, DC
[2] Laboratory for Computational Physics and Fluid Dynamics, Naval Research Laboratory, Washington, DC
[3] Department of Mechanical and Aerospace Engineering, George Washington University, Washington, DC

Summary. This paper describes the design, construction, and testing of a biomimetic pectoral (side) fin with actively controlled curvature for UUV propulsion. It also describes the development of a test UUV and the design of a fin control system for vertical plane motion. A 3D unsteady computational fluid dynamics (CFD) analysis has been carried out to computationally optimize the fin design including a full study of the primary design parameters. The fin has been constructed and it can reproduce any specified deformation time-history. The full dynamics of the proposed vehicle have been modeled and the forces produced by the flapping fins computed. Finally, the stability of motion in the vertical plane has been analyzed and a control system has been designed.

Key words. Biomimetic pectoral fin, UUV, unsteady CFD, PID control, adaptive curvature

1 Introduction

Unmanned underwater vehicles (UUVs) have proven very useful in a wide range of applications including inspection, surveillance and exploration. Current technology allows UUVs to excel at complex operations from deep-sea diving to high-speed and long distance traversal. However, one weakness of this proven technology is low-speed, high maneuverability operations that are required for near-shore and littoral zone missions. A UUV designed for high performance in this niche area is essential to accomplish inspection and surveillance in such challenging environments.

To confront the issue of low-speed maneuverability in the presence of currents and near-shore obstacles, flapping fin mechanisms have been studied to understand how certain aquatic animals achieve their high levels of controllability. Blake (1979) determined that labriform locomotion, using pectoral fin oscillation with very little body undulation, is more efficient and allows for better maneuverability in low-speed operations than carangiform locomotion, using body and caudal fin oscillation. This result is important to UUV applications because artificial fins based on those of labriform swimmers can be mounted on a rigid hull without sacrificing internal volume for complex body actuation mechanisms. Kato (2000) developed a UUV based on the small-mouth bass with rigid flapping fins. Licht et al. (2004) presented a design for an underwater vehicle and investigated the issue of vehicle control, again using rigid flapping foils. Kato et al. (2002) and Ando et al. (2006) have developed both lift-based and drag-based deformable pectoral fins for use on UUVs. None of these fins, however, provide the precise control of surface curvature time-history that we have determined to be necessary to meet our performance goals for hovering and low-speed maneuvering.

Walker and Westneat (1997) experimentally studied fin kinematics of a class of lift-based labriform swimmers, specifically the bird wrasse, whose lift-based pectoral fin force production was a good match with our performance objectives. Ramamurti and Sandberg (2002) computationally studied force production by the bird wrasse and obtained good agreement with the results of Walker and Westneat (1997). We applied this validated computational method to the design and development of a biomimetic pectoral fin propulsor with actively controlled curvature.

We have chosen a two-fin test vehicle design in order to demonstrate deforming fin force production and vertical plane control in an underwater environment. We present a model of the dynamics of the vehicle for steady, level flight. The stability of motion in the vertical plane is analyzed, and vehicle geometries and linear control design techniques are considered.

2 Fin Design

The fin design begins with an in-depth analysis of the common bird wrasse pectoral fin, placing emphasis on fin kinematics, fluid dynamics, and anatomy. We identified dominant parameters and computationally investigated their effects on force production. The results of our parametric study are presented to aid in future pectoral fin design. We then present the design and rationale of the physically constructed device that can produce the required fin kinematics. Lastly, the experimental measurements from our flapping fin tests are used to verify that the flapping fin is deforming in time as desired and to determine the actuator power required.

2.1 CFD Parameterization

To understand the effects of fin kinematics on propulsive force production, we used a 3D unsteady computational fluid dynamics code, FEFLO, which is an unstructured grid-based unsteady Navier-Stokes solver with automatic adaptive remeshing (Löhner 1988). Already successfully demonstrated by Ramamurti et al. (1996) and by Ramamurti and Sandberg (2001), this computational capability was extremely important in designing the pectoral fin. Not only did it help us rule out ineffective designs such as a rigid flapping fin (Ramamurti and Sandberg 2004), but also in testing design variations in search for performance gains. In this manner, we identified four major independent parameters that can be varied to change the force generation time-history during a stroke cycle: fin surface curvature, root angle of attack, bulk rotation angle, and frequency.

Fin surface curvature is defined as the shape of the fin with respect to time. There are 14 fin rays (also referred to as ribs) in the bird wrasse pectoral fin, each contributing to the fin curvature kinematics. For simplicity of design, manufacture, actuation and control, it is ideal to have the fewest possible number of ribs. But for more effective fin propulsion, it is ideal to maximize the number of ribs as more control points result in a smoother fit to desired fin curvature time-histories. Using CFD it was determined that 5 ribs resulted in a good balance between effective curvature and our specific mechanical implementation constraints (Ramamurti and Sandberg 2006). The overall rib and fin dimensions are scaled up from the bird wrasse pectoral fin to match our operational UUV requirements (Palmisano et al. (2007)). By imitating bird wrasse kinematics observed by Walker and Westneat (1997), we have determined an effective fin tip deflection time profile to define fin curvature. The curvature of each rib is defined by rib geometry and changes as a function of tip deflection. Due to a small deflection angle, rib 3 passively actuates between ribs 2 and 4.

The angle of attack, α, is the angle of the fin root with respect to the horizontal axis (Fig. 1). In the bird wrasse pectoral fin, this angle varies throughout the stroke cycle. However, a CFD analysis based on our simplified design and specific curvature time-history showed no thrust benefit to varying α over time. Instead, a fixed α of 20°, as seen in Fig. 1, was found to maximize forward thrust. This is consistent with the morphology of common labrid fish, as fast fish tend to have an α closer to 20° while slow fish have an α closer to 70° (Wainwright et al. 2002).

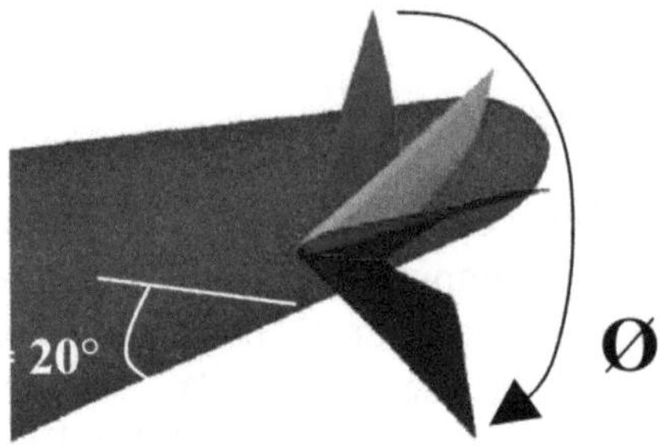

Fig. 1. α, Ø, and fin curvature (at 1 Hz)

The bulk rotation angle, Ø, is the peak-to-peak amplitude of a stroke (Fig. 1). In the bird wrasse, this angle is about 74°. However, we found that as the bulk rotation angle increases, the thrust will also increase. We chose a nominal peak-to-peak amplitude of 114° for our design, but this angle can be increased or decreased in flight by a controller for various operational conditions. An increase in bulk rotation angle also results in an increase of sinusoidal lift amplitude – an effect potentially degrading to UUV stability.

The fin flapping frequency was varied between 0.3Hz and 3.3Hz to study its effects on propulsion. It has been found that the upstroke thrust production increases as the square of the frequency ratio (Ramamurti and Sandberg 2006). A frequency of 1.2 Hz was chosen as a compromise between effective fin operation and mechanical limitations, although a higher frequency would be more desirable. Similar to varying the bulk rotation angle, changing the frequency can also be used as a vehicle control parameter.

2.2 Construction

Using the CFD results, and inspiration from the biological construction of the bird wrasse, we proceeded to construct the device. When encountering implementation conflicts, it was common to refer back to CFD to understand how a modification would affect propulsion. There are four main construction challenges: to design an actively deforming rib that we can exert full control over, to make a flexible skin connecting the ribs to serve as the fin surface area, to select small, lightweight, feedback controllable actuators that can independently bend each rib, and to devise a method for bulk rotation actuation.

Our fin design starts with the structural analysis, optimization, and construction of each of the five artificial ribs. Our fin rib is a compliant structure, designed to bend at specific base-to-tip deflection angles when subjected to translational forces at the base. The fin is composed of five of these independent ribs, and when actuated in particular patterns, their independent deflections are combined to create fin curvature as in Fig. 2.

Fig. 2. Independently bending ribs forming pectoral fin curvature

The topology of the compliant ribs was designed by using a structural-optimization based synthesis approach (Trease et al. 2003) and FEA with the ultimate objective to ensure the ribs could achieve the required deflection under fluid pressure when subjected to linear actuation at the base. The five designed ribs, each individually optimized based on different required lengths and curvatures, were built from ABS using a 3D printer. Details of the rib design are outlined by Palmisano et al. (2007).

A flexible skin of thickness .4mm ± .1mm is added to create a surface area between the individual ribs (Palmisano et al. 2007). This skin must easily stretch under rib deflection, yet not sag under fluid pressure. The skin was tested for chemical stability in sea water with no detectable degradation over a one month period. Skin thickness optimization was done experimentally and verified through FEA. Thick skin was shown to contribute significantly to required input force, while skin made too thin would tear. The empty space inside the fin is allowed to fill with water in order to equalize hydrostatic pressure and reduce skin sagging.

Servo motors are used as feedback controllable actuators for driving the ribs and bulk rotation (Fig. 3). Control in this manner is straightforward, simple to implement, and very affordable. All fin components were also made from ABS by 3D printing.

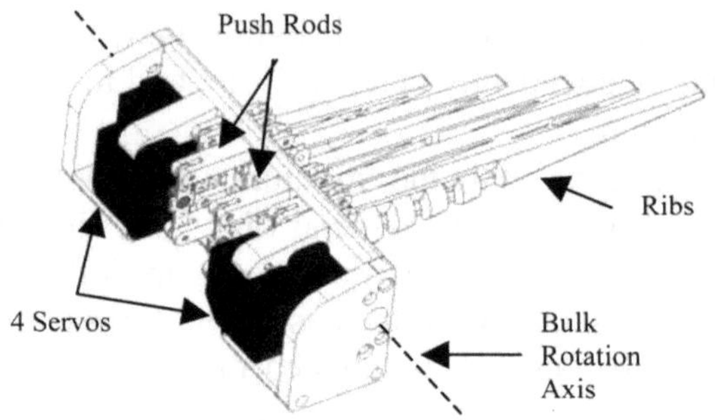

Fig. 3. Isometric view of fin system

All servos are controlled by the PIC16F877 microcontroller with a read-through list of pre-defined rotation angles and timing data points. Seven selected servo rotation angles for each of the four actuated ribs and the bulk motion are used during a complete stroke cycle. These angles are determined by manually measuring servo angle vs. rib curvature for each individual rib. The same microcontroller simultaneously collects all sensor data.

2.3 Testing and Results

Fin kinematics, generated forces, and power are measured from the fin. All measured data is for still water at a prescribed bulk rotation angle of 114°, an angle of attack of 0°, and a cycle frequency of 1.2 Hz. Because testing occurs in still water, the 0° angle of attack can be rotated in post processing calculations.

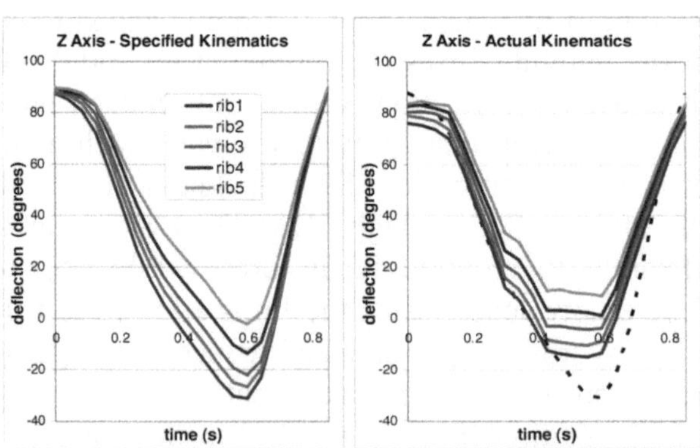

Fig. 4. (a) Specified CFD rib tip angle + bulk rotation angle (1.2Hz). (b) Actual experimental camera measurements (1.2Hz).

To verify that the actual experimental kinematics matched the specified kinematics, we traced the 3D location of each rib tip across an entire cycle. Two high speed digital video cameras angled at 90° to each other synchronously recorded images of the flapping fin at a predefined time step. Using image analysis software, matching rib tip coordinates selected from these image sequences were transformed into 3D coordinates (Geder and Sandberg 2007). Fig. 4 demonstrates good agreement between desired rib tip plus bulk rotation angles vs. actual experimental angles.

The measured thrust and lift shown in Fig. 5 are 0.028N and 0.031N, respectively, while the CFD thrust and lift recomputed with experimental kinematics are 0.0262N and -0.0289N (Palmisano et al. 2007). The expected differences in the force time histories between the experiment and CFD result from fluid-structure interaction and ingestion of air due to free surface interaction not present in CFD. Submerged vehicle testing will provide an environment more conducive to measuring the fin force production and propulsion efficiency.

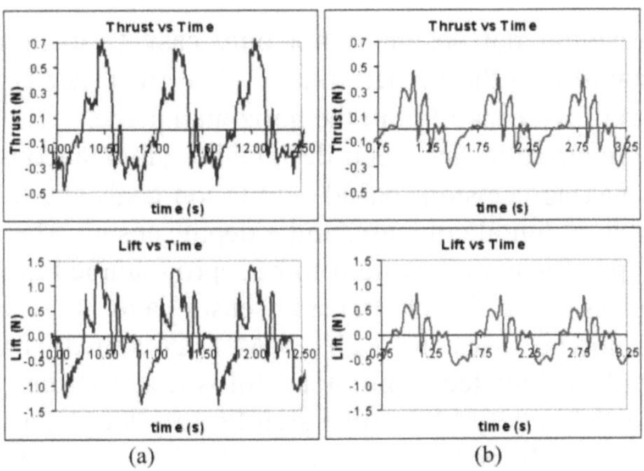

Fig. 5. (a) Measured forces over 3 cycles for experimental kinematics. (b) CFD computed forces over 3 cycles for experimental kinematics.

The total mean power consumption of all the servos was determined experimentally to be 9.81W, while control electronics consumed an additional 0.3W (Palmisano et al. 2007). This information was used for battery sizing, as described in the following section.

3 Vehicle Design

The goal of the test vehicle is to provide a platform which carries a set of two actively controlled curvature fins in order to demonstrate thrust production and vertical plane controllability in an underwater test environment. The flooded hull design measures 1.3" high, 13" long and 7" wide (Fig. 6). It has an out of water weight of 2.2lb, and is slightly negatively buoyant in water. The center of mass is located near the center of buoyancy. Key hull design parameters were manufacturability, vertical plane stability, and drag reduction.

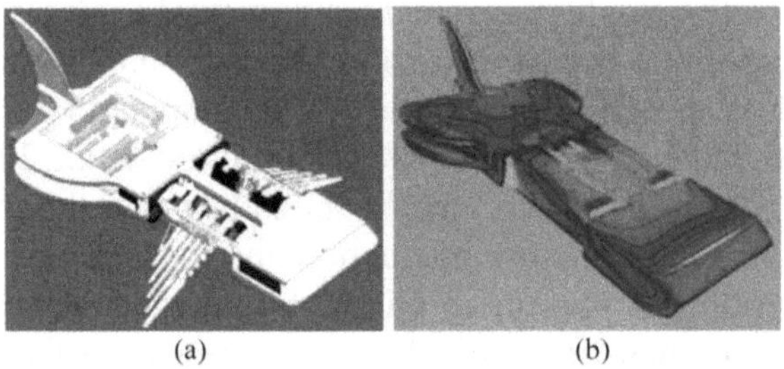

Fig. 6. (a) UUV design with two fins. (b) CFD drag analysis on UUV body

The horizontal flat design is not only easy to manufacture but also improves stability in the vertical plane. A vertical tail is incorporated in the design to provide stability in the horizontal plane.

The microcontroller has been designed to simultaneously control both fins, while making decisions based on mission goals and feedback from an accelerometer, a digital compass, and a depth sensor. The microcontroller has bootstrap software, allowing it to be reprogrammed in-vitro, as well as having wireless capabilities to transmit sensor data.

The prototype uses a 7.2V NiMH 2600mAh battery with an experimentally calculated continuous full-speed runtime of 50 minutes. Actual expected battery life can be significantly longer, as the vehicle can go into various drift, position holding, and efficiency modes that consume much less energy.

4 Vehicle Dynamics and Control System Design

With the vehicle design finalized, we then modeled the dynamics of the system in order to investigate stability issues and determine the necessary control coefficients. For experimental verification of the modeled dynamics, we will quantitatively measure displacement, velocity, rotation, and acceleration in each plane by use of both on-board and off-board sensors. Propulsive forces can then be back-calculated using experimental data as input to a more sophisticated, transient trajectory computation (Alin et al. 2007).

4.1 Equations of Motion

An undersea vehicle in motion is a complicated dynamic system. The vehicle has both rigid and elastic motions including control surface deflections and twisting of hydrodynamic surfaces. The elastic motion is observed in the bending and twisting of the controlled curvature fins. Isolating these fins, the rest of the body is treated as rigid, and as such, simplifications can be made by removing elastic terms from the equations of motion. We further reduce the equations by placing the body-fixed coordinate frame at the center of buoyancy and assuming symmetry about the x-z plane.

4.2 Evaluation of Forces and Moments

The forces and moments acting on the vehicle can be represented by the sum of hydrodynamic, hydrostatic and pectoral fin forces and moments. Added mass is computed using Taylor series expansion and simplified geometry (Jakuba 2003). An exact solution of the hydrodynamic force and moment terms requires solving the Navier-Stokes equations with boundary conditions.

Because of the complex nature of the actively controlled curvature pectoral fins and the vorticity dynamics caused by the periodic flapping motion, 3-D unsteady CFD computations are needed to determine the force time-histories for an accurate UUV controls analysis. We derived all necessary force and moment coefficients and integrated them into the equations of motion, providing a fully described six degree-of-freedom system.

4.3 Control System Design

To verify the thrust and lift performance of the fins, straight line motion of a vehicle is first considered. In this mode of motion, sway, yaw and roll (Y, N, L) can be assumed zero leaving us with 3DOF motion in the vertical plane characterized by surge, heave and pitch (X, Z, M). A linearized set of vertical plane equations (Eq. 4.1) about steady, level flight allows use of classical control techniques and provides an adequate representation of the system in the neighborhood of this flight condition (Fig. 7c). Steady level flight is characterized by a nominal surge velocity and by zero values for heave and pitch velocities, and pitch angle. Similar linearizations are carried out for other maneuvers including climbing and diving. Later control systems will incorporate nonlinear control techniques as the need for maneuverability across all flight conditions arises.

$$
\begin{aligned}
& m(\dot{u} + z_G \dot{q}) = \\
& \qquad X_{u|u|} u_0 |u| + X_{\dot{u}} \dot{u} + X_{uw} u_0 |w| - (W - B)\theta + X_{pec} \\
& m(\dot{w} - u_0 q + x_G \dot{q}) = \\
& \qquad Z_{uw} u_0 w + Z_{uq} u_0 q + Z_{\dot{w}} \dot{w} + Z_{\dot{q}} \dot{q} + (W - B) + Z_{pec} \\
& I_y \dot{q} - m x_G (\dot{w} - u_0 q) + m z_G \dot{u} = \\
& \qquad M_{uw} u_0 w + M_{uq} u_0 q + M_{\dot{w}} \dot{w} + M_{\dot{q}} \dot{q} - W z_G \theta - W x_G + M_{pec}
\end{aligned}
\tag{4.1}
$$

Representing the system from Eq. 4.1 in matrix form allows us to determine stability by implementing an open-loop Bode analysis. Fig. 7a shows that the system is stable in pitch and heave velocities without any form of closed loop control. However, to navigate to specified vehicle waypoints properly, feedback control is necessary.

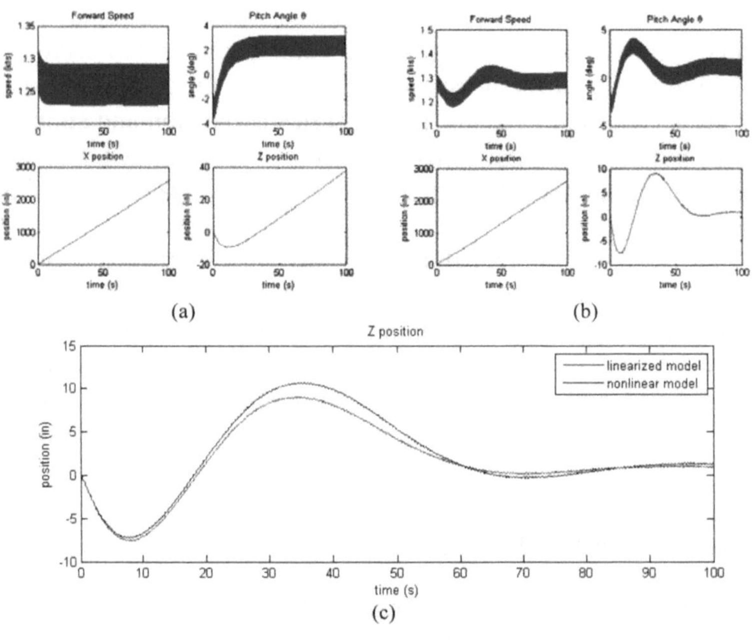

Fig. 7. (a) Open-loop vehicle response to fin operating at 2 Hz. (b) Closed-loop vehicle response to control of depth by way of fin frequency. (c) Closed-loop vehicle response comparison between a full nonlinear model of the system and a model linearized about steady, level flight.

Adding feedback control to the system, as shown in Fig. 8, provides the means for the UUV to respond to commands or a changing environment within specified performance parameters. Proportional-integral-derivative (PID) control has been implemented as a simple and effective means of controlling our vehicle for the desired testing. When control over a wider range of operating conditions is required in future testing, more robust, nonlinear control schemes will be implemented. Controllable parameters for the vehicle include pectoral fin flapping frequency and stroke amplitude. Altering these variables produces an increase or decrease in thrust and lift.

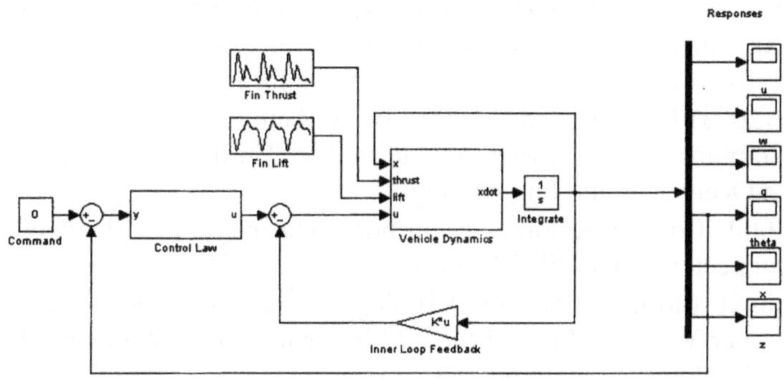

Fig. 8. Block diagram of vehicle feedback control system

As an example, the responses in Fig. 7b are a result of using depth as the feedback control variable with fin flapping frequency as the control input. In this simulation, we have discretized the control input to only change every 3 seconds. This allows the fins to operate at a single frequency over 4-5 cycles preserving the integrity of each stroke. With PID control, we have successfully stabilized depth at a commanded value.

5 Conclusions

We have described the design, development, and testing of actively controlled curvature pectoral fins for use on a UUV and demonstrated the achievement of a desired curvature time-history. We also have described the prototype vehicle that has been designed to incorporate these fins. The 3-D unsteady CFD analysis used to assist in the fin design and development processes, as well as for the vertical plane controller design, has also been described.

References

Alin N, Fureby C, Svennberg SU, Sandberg WC, Ramamurti R, Bensow RE (2007) Large Eddy Simulation of the Transient Flow Around A Submarine During A Maneuver. 45th AIAA Aerospace Sciences Meeting and Exhibit, AIAA-2007-1454, Reno, NV

Ando Y, Kato N, Suzuki H, Ariyoshi T, Suzumori K, Kanda T, Endo S (2006) Elastic Pectoral Fin Actuators for Biomimetic Underwater Vehicles. Proc of the 16th International Offshore and Polar Engineering Conference, pp 260-267

Blake RW (1979) The mechanics of labriform motion I. Labriform locomotion in the angelfish (*pterophyllum eimekei*): An analysis of the power stroke. J Exp Biol 82:255-271

Geder J, Sandberg WC (2007) Multi-Camera, High-Speed Imaging System for Flapping Fin Kinematics Measurement. Naval Research Laboratory Memorandum Report NRL/MR 6401-07. Submitted for publication January 2007

Jakuba M (2003) Modeling and Control of an Autonomous Underwater Vehicle with Combined Foil/Thruster Actuators. SM Thesis, WHOI-MIT Joint Program in Oceanographic Engineering

Kato N (2000) Control performance in horizontal plane of fish robot with mechanical pectoral fins. IEEE J Ocean Eng 25:121-129

Kato N, Liu H, Morikawa H (2002) Biology-Inspired Precision Maneuvering of Underwater Vehicles. Proc of the 12th International Offshore and Polar Engineering Conference, vol 2, pp 269-276

Licht S, Polidoro V, Flores M, Hover FS, Triantafyllou MS (2004) Design and Projected Performance of a Flapping Foil AUV. IEEE J Ocean Eng 29:786-794

Löhner R (1988) An adaptive finite element solver for transient problems with moving bodies. Computers & Structures 30:303-317

Palmisano J, Ramamurti R, Lu K, Cohen J, Sandberg W, Ratna B (2007) Design of a Biomimetic Controlled-Curvature Robotic Pectoral Fin. IEEE International Conference on Robotics and Automation, Rome, IT

Ramamurti R, Sandberg WC (2001) Simulation of Flow About Flapping Airfoils Using A Finite Element Incompressible Flow Solver, AIAA J., 39 (2):253-260

Ramamurti R, Sandberg WC (2002) Fluid dynamics of flapping aquatic flight in the bird wrasse: three-dimensional unsteady computations with fin deformation. J Exp Biol 205:2997-3008

Ramamurti R, Sandberg WC (2004) The influence of fin rigidity and gusts on force production in fishes and insects: a computational study, 42nd AIAA Aerospace Sciences Meeting, AIAA 2004-404, Reno, NV

Ramamurti R, Sandberg WC (2006) Computational Fluid Dynamics Study for Optimization of a Fin Design. 24th AIAA Applied Aerodynamics Conference, AIAA-2006-3658, San Francisco, CA

Ramamurti R, Löhner R, Sandberg WC (1996) Computation of unsteady flow past a tuna with caudal fin oscillation. In: Rahman M, Brebbia CA (Eds) Advances in Fluid Mechanics Vol. 9. Computational Mechanics Publications, Southampton, UK, pp 169-178

Trease BP, Lu KJ, Kota S (2003) Biomimetic Compliant System for Smart Actuator-Driven Aquatic Propulsion: Preliminary Results. ASME Int. Mechanical Eng. Congress & Exposition, IMECE2003-41446

Wainwright P, Bellwood D, Westneat M (2002) Ecomorphology of Locomotion in Labrid Fishes. Environmental Biology of Fishes 65:47-62

Walker JA, Westneat MW (1997) Labriform Propulsion in Fishes: Kinematics of Flapping Aquatic Flight in the Bird Wrasse, *Gomphosus Varius* (Labridae). J Exp Biol 200:1549-1569

Chapter 24

Controlling Biomimetic Underwater Robots With Electronic Nervous Systems

Joseph Ayers [1] and Nikolai Rulkov [2,3]

[1] Dept of Biology and Marine Science Center, Northeastern University, East Point, Nahant, MA 01908, USA
[2] Information Systems Laboratories, San Diego CA 92121, USA
[3] Institute for Nonlinear Science, UCSD, La Jolla, CA 92093, USA

Summary. We are developing biomimetic robots based on neurobiological model systems, the lobster and the lamprey. Existing implementations of these robots are based on finite state machine based controllers that instantiate a set of finite state machines based on the organizational units of the animal model nervous systems. These state machines include leg or body axis central pattern generators (CPGs) that generate leg movements or undulations, postural pattern generators that control compensatory appendages and/or adaptive sensors and sensory integration networks that process sensor information. The use of neuron models instead of finite state systems allows one to replicate in great detail the real behavior of the neurobiological system (a network) and, thanks to spiking nature of the models, provides a link between the control functions and the experimental measurements from the animal. The key feature of these models is that because they are based on capturing of nonlinear dynamical behavior of neurons rather than neuronal conductance models, they are simpler, can operate in real time and are thus suitable for robotic control applications.

Key Words. Biomimicry, Lobster, Lamprey, Electronic neuron, Robot

1 Introduction

Animals adapt with impunity to complex environments that challenge the most sophisticated engineered systems. A standard approach in robotics is to control systems with rule-based finite state machines (Brooks 1991). Animals, however, control their behavior with networks of neurons and simple animals have correspondingly simple nervous systems. The dynamical principles organizing the neuron networks that control animal behavior are efficient, robust and flexible. How they achieve this is still not clear even for the cases of relatively small biological networks where all connections are established (Selverston, Rabinovich et al. 2000). The main difficulty in the achieving a complete picture and identifying the fundamental mechanisms for dynamical control of behavior is that the network needs to be studied while it operates under closed loop conditions.

A recent trend in the analysis of animal behavior has involved the construction biomimetic robots that include animal like sensors, actuators and neural circuit controllers and analysis of their dynamics during interaction with the environment

(Webb 2001). Development of such robots allows the investigator to instantiate hypotheses, but what is presently missing is the capability to instantiate hypothetical networks from neurons and synapses and test their operation in a vehicle. We are developing electronic nervous systems (ENS) composed of analog and computed neurons and synapses for biomimetic robots based on neurobiological model systems, the lobster and the lamprey (Ayers 2004).
Existing implementations of these robots are based on finite state machine based controllers that

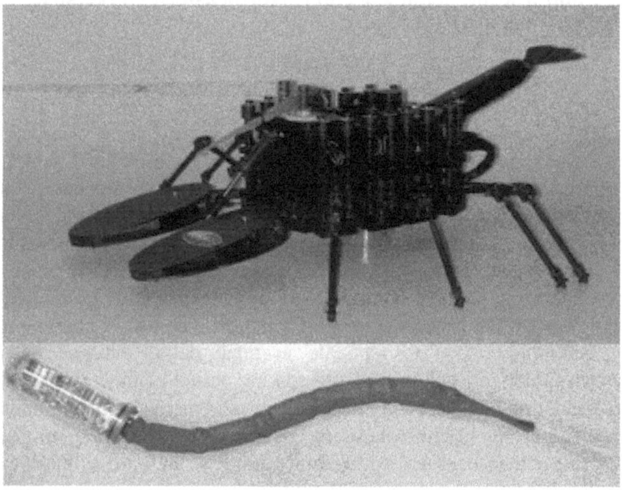

Fig. 1. Upper Panel: The lobster-based ambulatory robot. Lower Panel: The lamprey-based undulatory robot.

instantiate a set of finite state machines based on the organizational principles of the animal model nervous systems (Ayers and Witting 2007). These state machines include leg or body axis central pattern generators (CPGs) that generate leg movements or undulations, postural pattern generators that control compensatory appendages and/or adaptive sensors and sensory integration networks that process sensor information. A hierarchical command and coordination level that sequences behavior makes choices and establishes the inter-segmental coordination pattern of the locomotory CPGs, modulates the state machines (Ayers 2004). The state machines implement connectivity and dynamical rules of the system elements. For example the leg CPGs are based on a synaptic network containing neuronal oscillator, pattern generation and recruiter levels. The central commands have 2-5 states and recruited by input from exteroceptive sensors. Although these systems are quite effective in laboratory conditions their effectiveness in unpredictable field environments degrades.

In this manuscript, we report progress on replacing the finite state machine controllers with electronic nervous systems formed of synaptic networks of electronic neurons (Rulkov 2002). The use of neuron models instead of finite state systems allows one to replicate in great detail the real behavior of the neurobiological system (a network) and, thanks to spiking nature of the models, provides a link between the control functions and the experimental measurements from the animal. Now we can not only suggest and implement a control mechanism but also compare its operation with real data from the models.

2 Electronic Neuron Central Pattern Generators

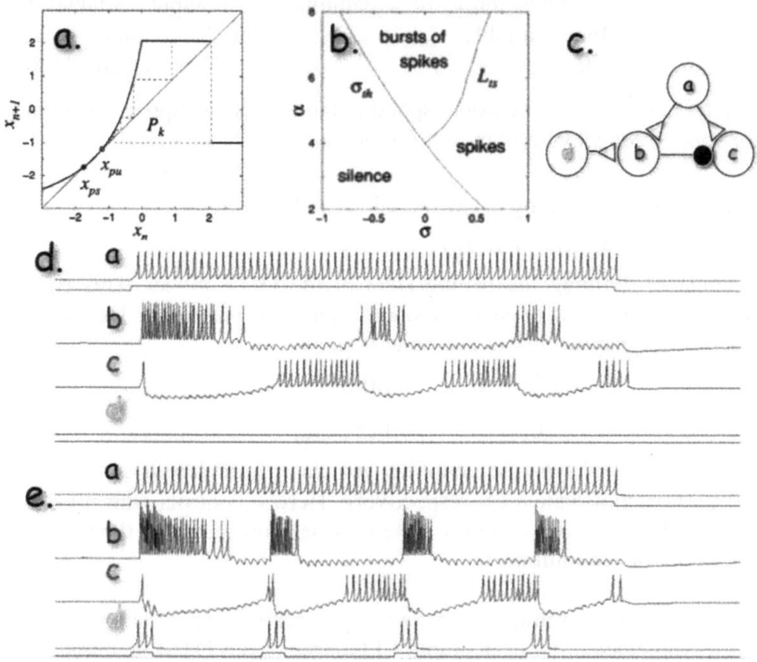

Fig. 2. Discrete Time Map (DTM) Based Neurons. a. Return Map of the function (see text). b. Bifurcation diagram for different ranges of ⟨ and ſ. c. Synaptic network modeled in e & f. Cell b is configured in the bursting regime. Cell c is configured in the spiking regime. d. Effect of long time constant synaptic inputs from a (a command neuron) on b & c. e. Entrainment of the bursting rhythm evoked in b & c by pulsed short time constant synaptic input from d (a coordinating neuron).

A two-dimensional discrete time map that describes spiking and spiking-bursting behavior of a neuron model (Rulkov 2002) can be written as follows

$$x_{n+1} = f_\alpha(x_n, x_{n-1}, y_n + \beta_n),$$
$$y_{n+1} = y_n - \mu(x_n + 1) + \mu\sigma + \mu\sigma_n,$$

where x_n is the fast and y_n is the slow (due to 0<*m*.<<1) dynamical variables. The nonlinear function is written in the following form

$$f_\alpha(x_n, x_{n-1}, u) = \begin{cases} \alpha/(1-x_n)+u, & x_n \le 0, \\ \alpha+u, & 0 < x_n < \alpha+u \text{ and } x_{n-1} \le 0, \\ -1, & x_n \ge \alpha+u \text{ or } x_{n-1} > 0, \end{cases}$$

where the third argument u stands for y_n or a combination of input variables that depend on the model type. Input variables b_n and s_n incorporate the action of synaptic inputs I^{syn} and can be written as $\beta_n = \beta_E I_{syn}$, $\sigma_n = \sigma_E(I_{syn})$, where b_E is a constant that controls how quickly neuron respond to the input and supports dynamical mechanisms for spike frequency deceleration for DC pulses of current.

A simplified one-dimensional analog of the model is can be written as

$$x_{n+1} \;=\; f_\alpha(x_n, x_{n-1}, b^{rs} + \beta^e I_n),$$

where b^{rs} is a constant defining the resting state of the DMT model. This type of model does not allow the onset of self-sustained regime of bursting. The simplest map-based model for a synaptic current can be written as

$$I_{n+1}^{syn} = \gamma I_n^{syn} - \begin{cases} g_{syn}(x_n^{post} - x_{rp}), & \text{spike}_{pre}, \\ 0, & \text{otherwise}, \end{cases}$$

where g_{syn} is the strength of synaptic coupling, indexes *pre* and *post* stand for the presynaptic and postsynaptic variables, respectively. Here *g* controls the relaxation rate of the synaps (0 £ *g* < 1) and x_{rp} defines the reversal potential and, therefore, the type of synapse: excitatory or inhibitory.

Chemical-based synapse models allow the generation of realistic neuronal circuits. Fig. 2d-f illustrates the implementation of a simple CPG circuit with DTM neurons. DTM synapses can be configured with different time constants to mediate slow (neuromodulatory command) or fast effects (perturbation) on other neurons. In this example, the command neuron a, turns on the oscillation in b and c through long time constant synapses. Neuron d (here a coordinating neuron) is able to perturb and entrain the rhythm in b and c due to short time constant synapses.

The key feature of these models is that because they are based on capturing of nonlinear dynamical behavior of neurons rather than neuronal conductance processes. Therefore, they are simpler, can operate in real time and are thus suitable for robotic control applications. Using these circuits we have been able to instantiate the same control models implemented with the finite state machines. Unlike the state machine models the EN models are capable of a high degree of modulation of intrinsic variability including chaos, ensuring a high degree of adaptability as well as a rapid return to robust stability (Selverston, Rabinovich et al. 2000).

3 Higher Order Control

In the intact animal, a CPG is only one component of the neural machinery underlying locomotion. Descending command units from the brain can modify the movements of single limbs while the CPG is operating so as to provide steering and velocity control (Ayers 2000). In addition, limb proprioceptive reflexes also operate concurrently with the CPG drive to adjust the motor pattern to the immediate environment. Exteroceptive reflexes, result from feedback extrinsic to the organism and act upon sets of CPGs through command and/or modulatory interneuronal systems. Exteroceptive reflexes mediate taxes and kineses, responses to gravity, surge and optical flow, etc., and can both trigger behavioral acts as well modulate them on an ongoing basis (Kennedy 1977). Animals fuse layers of exteroceptive reflexes to mediate orientation, navigation and investigation. For example, many animals have a sense of direction relative to the earths magnetic field (Lohmann, Pentcheff et al. 1995). Orientation to water currents are fundamental to rheotaxis (Moore and Grills 1999) and chemotaxis (Hanna, Grasso et al. 1999).

The command networks that orchestrate behavior in the brain consist of sensory interneurons, low level commands, organizing commands, pattern generators and command logic. Sensory interneurons will process input from the different exteroceptive sensors (compass, antennae, inclinometers, bump detectors) to realize releasers for different behaviors as well as to modulate ongoing behavior. We are developing a hierarchy of commands. Low-level commands act on specific task groups such as the antennae, legs, claws and abdomens. Command organizers act on sets of low level commands to mediate complete behavioral acts. Command pattern generators pattern sequences of the activity of command organizers and terminate goal-achieving sequences. Behavioral choices are mediated by synaptic interactions between these different elements (Edwards 1991). Command logic is the prioritizing connectivity that mediates behavioral choice and hierarchy (Kovac and Davis 1977; Takahata 1998).

We have performed detailed modeling of the exteroceptive control of rheotaxic behavior mediated by antennal sensors and have developed a hypothetical network that will mediate rheotaxic behavior in flow in surge. Surge and flow can be detected by strain gauges embedded in the antenna that are bent by flow as well as the optical flow of suspended particulates. Two layers of interneurons that distinguish lateral versus axial flow and activate walking command systems to mediate rotational and yawing components of rheotaxis process the sensory input.

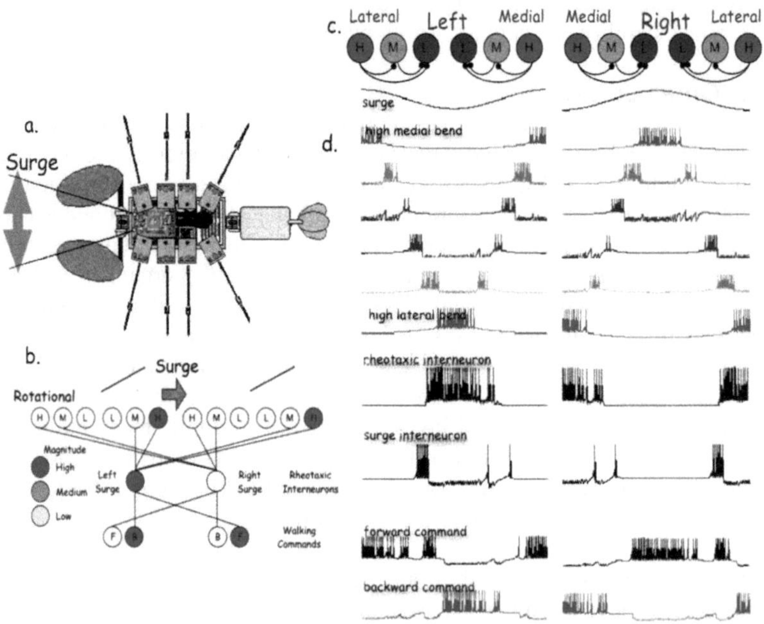

Fig. 3. LabView™ Discrete time map neuron simulations of rheotaxic responses to surge. a. Orientation of antennae and orientation of surge. b. The circuit connecting antennal bend afferents, with sensory interneurons (surge) and walking command neurons (F: forward; B: backward). The upper trace shows the flow velocity for lateral flow. c. The bend afferents are connected by lateral inhibition such that the high bend (H) inhibits the medium (M) and low (L) bend and the medium bend inhibits the low bend. d. rotational responses to lateral surge with the antenna held forward as indicated in a. The 2nd through 7th traces indicate the responses of low, medium and high bend afferents (color coded). The 8th and 9th traces indicate the responses of rotational and surge interneurons. The 10th and 11th traces indicate the responses of forward and backward command neurons

Fig. 3 illustrates a simple network to mediate the rotational component of rheotaxic behavior. The antennae we have developed each have 6 labeled lines that transduce three degrees of bending to the left and right. If the antennae are projected forward (Fig. 3a) and flow occurs from the left, the left antennae would be bent medially and right antennae will be bent to the laterally (Fig. 3). A left rheotaxis interneuron would respond to this and activate the left backward walking command and the right forward command to initiate a rheotaxic rotational turn to the left, into the flow by activating the backward command on the left side and the forward command on the right side.

The suite of exteroceptive reflexes will involve additional layers similar to this to mediate heading control (sensor: compass), translation and angular rotation (sensor optical flow vision sensor and micro engineered machine (MEMs) accelerometer), balance (sensor: MEMs inclinometers) and collisions (sensors: antennae and bump detectors). We have been able to implement a sense of direction by coupling the sine and cosine signals from a Hall-effect compass to a network of DTM sensory neurons. This process involves coupling the output of the analog sensor to the bias current of the DTM neuron. Lateral inhibition (Fig. 3c) is then used to range fractionate the output. By this means, we can endow the robot with a neuronal-based sense of direction.

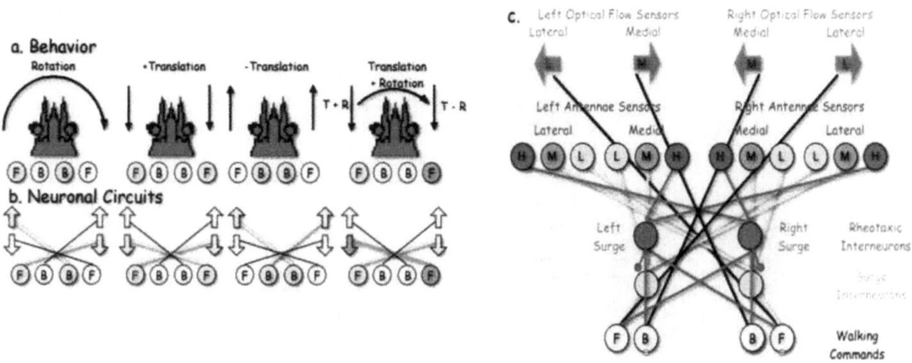

Fig. 4. Sensory fusion of hydrodynamic and optical flow reflexes. **a.** Behavioral responses to combinations of angular rotation and translational optical flow. **b.** Neuronal circuits mediating the exteroceptive reflexes. The arrows indicate unidirectional optical flow sensitive sensors. Circles represent forward (F) and backward (B) walking commands **c.** Network fusing optical flow sensors and antennal bending sensors to mediate rheotaxis.

We are developing layered sets of exteroceptive reflexes that in some cases will fuse two or more modalities. For example, a Ladybug™ optical flow sensor (CentEye, Inc.) can be used to instantiate bilateral unidirectional flow sensitive neurons that frequency code the velocity of unidirectional motion in the same fashion as crustaceans (Wiersma and Yamaguchi 1966). Crustaceans exhibit four basic compensatory responses responses to the angular and translational components of optical flow (Fig. 4a). Like the hydrodynamic flow example described above, these optical flow layer responses can be instantiated by simple networks connecting sensors with locomotory command neurons. Fig. 4b. This optical flow layer can fuse with the corresponding hydrodynamic flow layer to provide more robust modulation (Fig. 4c).

4 Undulatory Robots

We are also developing an ENS for the control of swimming in an undulatory robot based on the biomechanics and neuronal control mechanisms of the sea lamprey Fig. 1. The propulsive system of the undulatory robot (Fig. 1) is a polyurethane strip that is flexed by five bilateral pairs of shape memory alloy (Nitinol) actuators. We found that five actuator pairs projecting from 25 to 65% of the body axis were adequate to generate lampriform swimming movements. Alternating bilateral flexions propagate down the body axis with the same duration as the undulatory period, eliminating vortices and producing highly efficient swimming. Climbing and diving are mediated by a SMA actuated pivot between the hull and body axis (Fig. 1) that allows the generation of low-pressure areas above and below the hull. Turning is mediated by biasing the amplitude of flexions to the two sides. Current sensors include a pitch inclinometer, compass and a depth gauge. Power is provided by 10-nickel metal hydride batteries providing a mission duration of about 1 ½ hours.

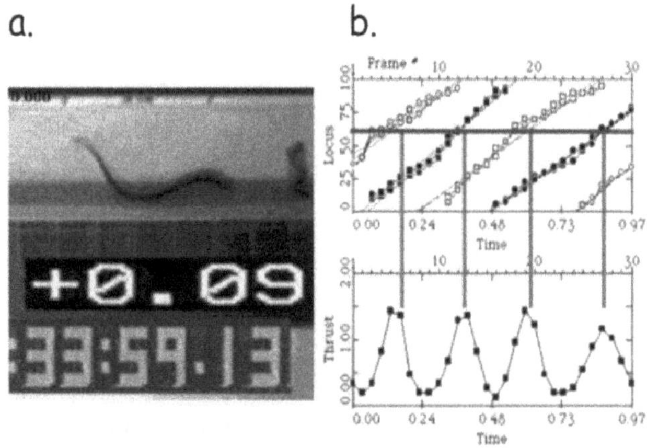

Fig. 5. Kinematic Analysis of lamprey swimming. a. Experimental assessment of thrust production during lamprey swimming. The specimen is tethered to a strain gauge by a thread cemented to the skin at 25% of body length. The numbers represent the instantaneous thrust and time code. b. Correlation of the position of lateral flexions (expressed as % of body length) in each frame of the movie with instantaneous thrust (from the video meter). The undulations form propagating flexion waves. Thrust peaks, when the corresponding flexion wave reaches about 65% of body length.

To quantitatively analyze the undulatory behavior of the lamprey, we employed a computer program uses stop-frame analysis of video images to establish the dynamics of body flexion's (Ayers 1989). Filmed sequences of swimming were then digitized and analyzed on a frame-by-frame basis (Ayers, Carpenter et al. 1983). Kinematic analysis of body curvature demonstrates swimming behavior is organized into lateral flexion waves that propagate either from nose to tail or tail to nose. During swimming, the propagation time of flexion waves down the body axis is equal to the period so specimens always maintain an S-shape during swimming.

We developed multimedia techniques to correlate the electrophysiology of behaving animals with their kinematics (Ayers 1990). The electromyograms or CNS recordings were recorded on the audio tracks of a videocassette recorder, the behavior in the

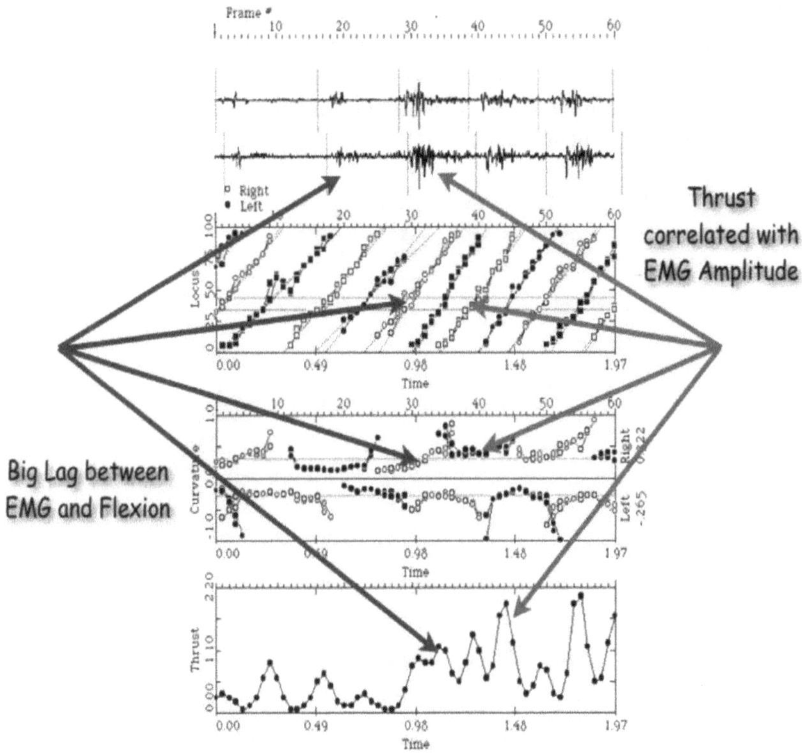

Fig. 6. Multimedia analysis of lamprey swimming. The top trace is the video frame number. The second two traces are electromyographic recordings from myotomes on the right side of the body. The third trace is a graph of the locus of flexions sorted into flexion waves. The horizontal dashed lines in this figure indicate the locus of the electromyographic electrodes. The third trace is the curvature of the lateral flexions sorted into flexion waves. The bottom trace is the instantaneous swimming thrust. The vertical dashed lines in the second and third traces indicate when the flexion wave passes over the recording electrode

video signal while slower parameters like thrust were embedded in the video signal using a video meter. After digitization, the data could be combined in displays that directly correlate electrophysiology with kinematics and thrust generation (Fig. 6). We found that there was considerable variation in output on a cycle-by-cycle basis, particularly in response to vibrational stimulation of the aquarium. The most striking finding of these experiments is the long phase lag between the electromyographic excitation of the muscle and the mechanical propagation of the flexion wave past the recording site. Secondly there is considerable cycle-by-cycle variation in the amplitude of thrust and this was correlated with the amplitude of the corresponding electromyographic burst. Surprisingly, amplitude could vary profoundly while timing of the undulations was quite stable. This finding implies that descending modulatory influences can bypass the neuronal oscillator to directly recruit muscle force.

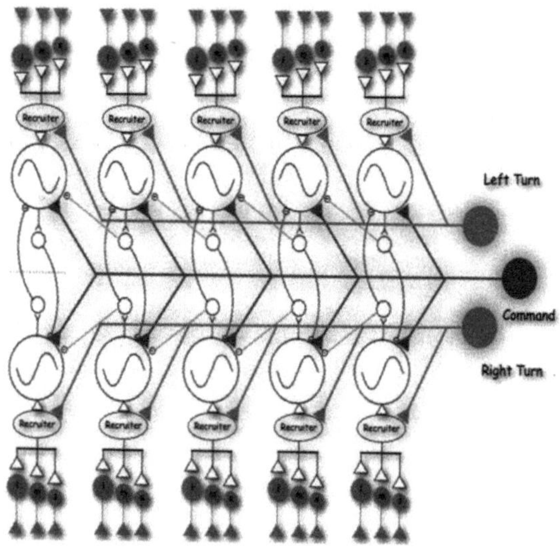

Fig. 7. CNS model for the undulatory robot. Undulations are generated by 5 segmental bilaterally paired CPGs that drive SMA actuators. The CPGs activate recruiters that activate motor neurons that provide pulse width duty cycle modulation of the actuators. Larger motor neurons produce longer duration pulses. The CPGs are coordinated among themselves by coordinating neurons that inhibit the contra-lateral CPG and the posterior CPG to generate a propagating wave of motor neuron activity. The CPGs are brought into activation by a descending modulatory command that provides bias current to the CPG elements. Bilateral paired modulatory interneurons activate the recruiters of the same side to mediate turning.

Variations in period and curvature of flexion waves differentiate lamprey behavior acts. Thrust occurs when a propulsive wave travels backward over the body at a velocity greater than the speed of forward swimming. Kinematic analysis demonstrates lamprey generate a peak of thrust as each flexion wave passes the cloaca (Fig. 8). The thrust peaks at approximately sixty percent of the body and there are two peaks of thrust for every swimming cycle.

Based on these observations, we have developed network model to replace the state machine for the control of the undulatory robot shown in Fig. 7. This network consists of a concatenated bilateral set of CPGs that activate the motor neurons for each SMA actuator. The CPGs are brought in to operation by descending command neurons that provide bias current to the CPG neurons through long time constant excitatory synapses. Coordinating neurons mediate the intersegmental propagation of flexion waves down the body so that propagation time equals the cycle period. Recruiting networks mediate the size principle of motor neuron recruitment and gradation of muscular force. Amplitude modulatory interneurons bypass the CPG to directly modulate the recruiters to mediate turning. Modulatory interneurons for turning, pitch, and roll are modulated by sensor input from the compass, pitch and roll inclinometers. Descending commands that alter the bias of excitation in anterior vs. posterior segments will be used to reverse the direction of propagation of the flexion waves to allow backing up following collisions sensed by bump sensors.

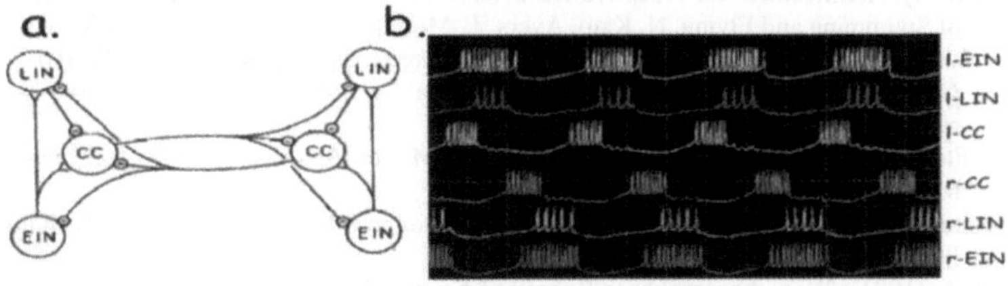

Fig. 8. a. CPG Network for the generation of the undulatory motor program. LIN: Lateral Interneuron, CC: CC Interneuron, EIN: Excitatory Interneuron. Adapted from (Buchanan and McPherson 1995). **b.** Swimming motor program generated by the synaptic network indicated in Fig. 8a

The details of the CPG networks are shown in Fig. 8a. The swimming motor program is generated by three populations of interneurons connected by excitatory and inhibitory synapses (Buchanan 1996). The excitatory interneurons appear to be the primary organizers. Lateral inhibitory interneurons shorten the burst in CC (contralateral, caudal projecting) interneurons that contribute to bilateral and inter-segmental coordination. Fig 8b indicates that output of this network instantiated with DTM neurons. Note the alternating discharge of the EINs on the two sides. This CPG forms the basis of our undulation controller. The motor neurons activate the SMA actuators. We use a comparator to gate a pulse from the EN motor neuron action potential. This pulse is, in turn, used to gate the power transistor applying current to the SMA actuator. Pulse width duty cycle modulation can be achieved by both increasing the frequency of motor neuron discharge as well as modulating the width of the motor neuron action potential (Aldrich, Getting et al. 1979).

5 Conclusions

It is feasible to construct electronic nervous systems for the control of biomimetic robots. Motor pattern generating circuits of electronic neurons and synapses can generate adaptive motor programs by modulation of intrinsic variability. ENs can proportionally control artificial muscle using simple interfaces based on pulse width duty cycle modulation by neuronal action potentials. Use of sensors that generate neuron-like pulse trains will allow both proprioceptive and exteroceptive modulation of motor output using the same reflexes of the animal models. Such architectures can lead to the generation of relatively cheap and highly adaptive biomimetic robots.

References

Aldrich, R., P. A. Getting, et al. (1979). "Mechanism of frequency-dependent broadening of molluscan neuron soma spikes." J. Physiol. **291**: 531-544.

Ayers, J. (1989). Recovery of oscillator function following spinal regeneration in the Sea Lamprey. Neuronal and Cellular Oscillators. J. W. Jacklet, Marcel Dekker Inc.: 371-406.

Ayers, J. (2000). Finite State Analysis of Behavior and the Development of Underwater Robots. Artificial Ethology. O. Holland, MacFarland, D. Oxford, Oxford University Press.

Ayers, J. (2004). Architectures for Adaptive Behavior in Biomimetic Underwater Robots. Bio-mechanisms of Swimming and Flying. N. Kato, Ayers, J., Morikawa, H. Tokyo, Springer-Verlag: 171-187.

Ayers, J. (2004). "Underwater walking." Arthropod Structure & Development **33**(3): 347-360.

Ayers, J., G. A. Carpenter, et al. (1983). "Which behavior does the lamprey central motor program mediate?" Science **221**: 1312-1314.

Ayers, J., Fletcher, G. (1990). "Color Segmentation and Motion Analysis of Biological Image Data on the Macintosh II." Advanced Imaging **5**(39-42).

Ayers, J. and J. Witting (2007). "Biomimetic Approaches to the Control of Underwater Walking Machines." Phil. Trans. R. Soc. Lond. A **365**: 273-295.

Brooks, R. A. (1991). "New Approaches to Robotics." Science **253**: 1227-1232.

Buchanan, J. T. (1996). "Lamprey spinal interneurons and their roles in swimming activity." Brain Behav Evol **48**(5): 287-96.

Buchanan, J. T. and D. R. McPherson (1995). "The neuronal network for locomotion in lamprey spinal cord: evidence for the involvement of commissural interneurons." J Physiol Paris **89**(4-6): 221-33.

Edwards, D. H. (1991). "Mutual inhibition among neural command systems as a possible mechanism for behavioral choice in crayfish." J Neurosci **11**(5): 1210-23.

Hanna, J. P., F. W. Grasso, et al. (1999). "Temporal Correlation Between Sensor Pairs in Different Plume Positions: A Study of Concentration Information Available to the American Lobster, Homarus americanus, During Chemotaxis." Biol Bull **197**(2): 250-251.

Kennedy, D., Davis, W. J. (1977). Organization of Invertebrate Motor Systems. Handbook of Physiology. S. R. K. Geiger, E. R., Brookhart, J. M., Montcastle, V. B. Bethesda, MD, American Physiological Society. **Section 1, Part 2:** 1023-1087.

Kovac, M. P. and W. J. Davis (1977). "Behavioral Choice: Neural mechanisms in Pleurobranchaea." Science **198**: 632-34.

Lohmann, K., N. Pentcheff, et al. (1995). "Magnetic orientation of spiny lobsters in the ocean: experiments with undersea coil systems." J Exp Biol **198**(10): 2041-2048.

Moore, P. A. and J. L. Grills (1999). "Chemical orientation to food by the crayfish Orconectes rusticus: influence of hydrodynamics." Anim Behav **58**(5): 953-963.

Rulkov, N. F. (2002). "Modeling of spiking-bursting neural behavior using two-dimensional map." Phys Rev E **65**: 041922.

Selverston, A. I., M. I. Rabinovich, et al. (2000). "Reliable circuits from irregular neurons: a dynamical approach to understanding central pattern generators." J Physiol Paris **94**(5-6): 357-74.

Takahata, M. (1998). "Behavioral hierarchy in crayfish: Multiple gate control of sensori-motor pathways." Neuroscience Research - Supplement(22): S54.

Webb, B. (2001). "Can robots make good models of biological behaviour?" Behav Brain Sci **24**(6): 1033-50; discussion 1050-94.

Wiersma, C. A. and T. Yamaguchi (1966). "The neuronal components of the optic nerve of the crayfish as studied by single unit analysis." J Comp Neurol **128**(3): 333-58.

Chapter 25

Micro-energy Converter Using Insect Wings

Seiichi Sudo[1], Kazumasa Nashimoto[1], Koji Tsuyuki[2], Tetsuya Yano[1], and Muneo Futamura[1]

[1] Faculty of Systems Science and Technology, Akita Prefectural University, 84-4 Ebinokuchi, Tsuchiya, Yurihonjo, Akita 015-0055, Japan
[2] Department of Mechanical Systems and Design Engineering, Iwaki Meisei University, 5-5-1 Iino, Chuohdai, Iwaki, Fukushima 970-8551, Japan

Summary. This paper is concerned with the development of novel type of micro wind energy converter using insect wings. A series of experimental studies are made on the function of wing apparatus of flying insects. In the first place, the structural properties of insect wings are studied through measurements of certain morphological parameters. The scanning electron microscopic observation of an insect wing shows the morphological characteristics and kinematic characteristics of a fly. In the second place, the novel type of micro wind energy converter is constructed. The micro wind energy converter is composed of the permanent magnet (NdFeB magnet), magnetic fluid (kerosene-based magnetic fluid ferricolloid HC-50), coil, and insect wings. The power generation characteristics of the micro wind energy converter using fly wings are examined. The airfoils of energy converter are Diptera insect wings. It is confirmed that the output signals reach hundreds of mV. The power generation characteristics of the micro wind energy converter with ABS resin wings are also examined. The size of artificial wing is almost equal to the insect wing. It is found that the micro wind energy converter using insect wings shows higher output voltage compared with artificial wings.

Key words. Insect Wing, Wing Morphology, Micro Wind Energy Converter, Permanent Magnet, Magnetic Fluid

1 Introduction

There has been increasing interest recently in micromachines and microrobots. With the critical development in micromachining technology and Micro Electro Mechanical System (MEMS) technology, the applications of miniaturized and ultraminiaturized components and devices have been rapidly increasing (Esashi 2002; Higuchi 2003). Therefore, micromachines and microrobots might become realistic in a few years. Actually, several microrobots have been proposed. There is biomimetic approach in one of several approaches (Shimoyama 1994). Biomimetics can essentially be defined as the practice of 'reverse engineering' ideas and concepts from nature and implementing them in a field of technology. Authors also proposed the micro magnetic swimming mechanism propelled by the alternating magnetic field in a water pipe (Sudo et al. 2003). This swimming machine imitates bacterium behavior. In general, lifetime that the microrobot can be driven with the battery is reduced with a decrease in the robot size. The energy that can be stored in its battery is proportional to its volume. Therefore, energy to drive microrobots will be supplied from an internal battery and the external energy source (Shimoyama 1996). On the other hand, the living things have been evolving for a long time naturally and selectively. The locomotive organs of animals are the products of two very potent optimizing processes, the processes of evolution by natural selection and learning by experience (Alexander 1984). Extensive investigations on the locomotive mechanisms of a great many animals such as birds, fish, flying insects and others have been conducted and reported (Azuma 1992). Authors have also studied detailed wing morphology of same types of insects with a scanning electron microscope, and analyzed the free flight of a wasp with a three dimensional motion analysis system (Sudo et al. 1999, 2000, 2005). Those studies revealed that insect wings have superior aerodynamic characteristics.

In this paper, novel type of micro wind energy converter using insect wings is proposed. In the first place, the surface roughness of Diptera insect wing is measured by a three-dimensional, optical shape measuring system, and insect wing is observed using a scanning electron microscope. Secondly, the micro wind energy converter using Diptera insect wings is constructed, and then the power generation characteristics of the micro wind energy converter are examined. It is revealed that the output voltage from the micro energy converter reaches hundreds of mV.

2 Experimental Apparatus and Procedures

2.1 Observation of Insect Wings

The surface shape of insect wings was measured by the three-dimensional, optical shape measuring system. This measuring system is composed of an automatic focus microscope with a laser beam driving servomechanism and the high accuracy XY-stage. In this experiment, test insect wings are Diptera wings. The test insect wing was severed from the insect body before the measurement and mounted on the XY-stage. When the laser beam impinges on the surface of the test wing, a diffused or scattered reflection occurs. The scattered light reflection is then focused through an objective lens on a unique semiconductor sensor. The output signal from the photodetector gives the position of the measured surface relative to the gage probe. A three-dimensional measurement of the wing was made by scanning of the XY-stage. A schematic diagram of the measuring device is shown in Fig. 1. Microscopic observations of test insect wing were conducted using a scanning electron microscope. The sample wings were coated with gold by the ion sputtering method for conductivity. The test insects were collected from fields in Yurihonjo, Japan.

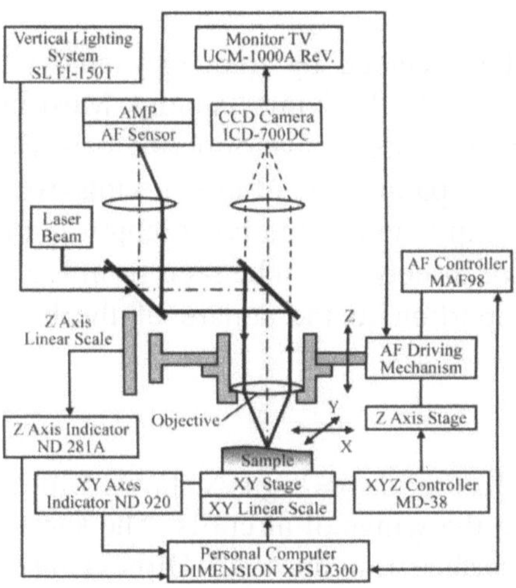

Fig. 1. Block diagram of experimental apparatus for surface shape measurement

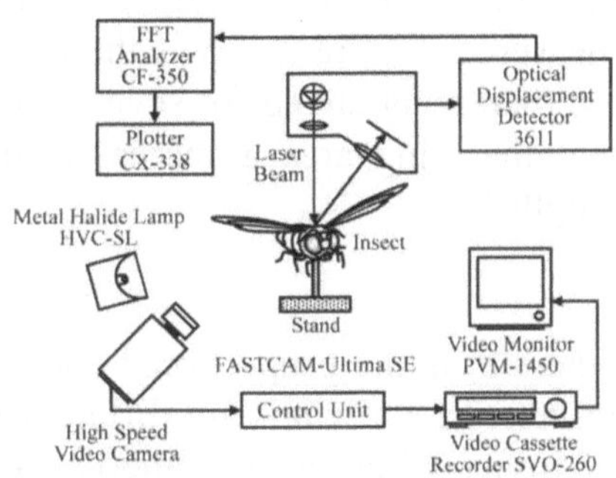

Fig. 2. Block diagram of experimental apparatus for flapping analysis

2.2 Analysis of Insect Flapping and Measurement of Skeleton Vibration

Figure 2 shows a schematic diagram of the experimental apparatus for flapping analysis of insects. Experiment on insect flapping was conducted with a high speed video camera system. The experimental system is composed of a high speed video camera, video cassette recorder, video monitor, and a personal computer. In the experiment, a live insect was stuck on the wooden needle with the adhesive as shown in Fig. 2. The flapping behavior of insect is recorded by the high speed camera. A series of frames of flapping behavior were analyzed by the personal computer. After the flapping behavior of the insect was recorded, the displacement of the exo-skeleton vibration was measured by the optical displacement detector system. The displacement signals were analyzed by the dual channel fast Fourier transform (FFT) analyzer. The flapping frequency of test insect was decided by frequency analysis.

3 Flapping Characteristics of Diptera Insects

3.1 Wing of Diptera Insects

The flapping frequency of test insect is decided by frequency analysis. Winged insects have the superior ability to fly by wing flapping. Most insects have two pairs of wings, but Diptera insects have only one pair. Before the experiments, some morphological parameters of insect wings were measured. Figure 3 shows the relationship between the body length L and the total wing area S for several kinds of insects. It can be seen from Fig. 3 that the wing area S increases in proportion to the square of the body length L.

$$S = c_1 L^2 \tag{1}$$

where c_1 is the proportional constant. In Fig. 3, the sign Δ shows Diptera insects. Insect wings are different from the wings of aircrafts. The surface of insect wings is not smooth. The surface roughness of Diptera insect wing was measured by the three-dimensional, optical shape measuring system. In the measurement, only the upper surface of the insect wing was measured. Figure 4 shows the three-dimensional display of measurement result for the right wing of the gadfly *Tabanus rufidens*. Most insect wings are composed of veins and membranes that are identical on the upper and under surfaces. The ups and downs shown in Fig. 4 formed along the veins, and the difference of elevation is especially remarkable between the longitudinal veins. The corrugated wing acts as several functions for the insect flight (Sudo et al. 1999).

In this study, microscopic observations of wing surface of Diptera insects were also conducted using a scanning electron microscope. Figure 5 shows an example of the scanning electron micrograph of wing surface of the robber fly *Promachus yesonicus* Bigot. Robber flies are common, swift-flying predators. They have three ocelli and a beard face of distinct setae around the stout and pointed proboscis. They feed on other flies, especially deer flies and horse flies, as well as flying ants, small bees, true bugs, grasshoppers, butterflies, and moths. They are extremely versatile flier. It can be seen from Fig. 5 that Diptera wing is clothed in minute hairs. Minute hairs are characteristics for insects with a high wing beat frequency, and most of hairs on the surface of wing are tilted. These hairs fulfill the function as the drag reduction riblets (Sudo et al. 2000).

In this paper, novel type of micro wing energy converter using Diptera wings is proposed. Diptera wings are suitable as the air foils of micro wind energy converter for the above-mentioned reasons.

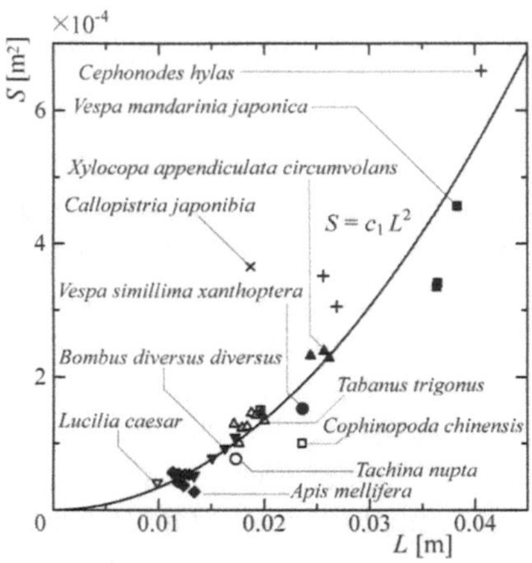

Fig. 3. Relationship between the body length and the total wing area of insects

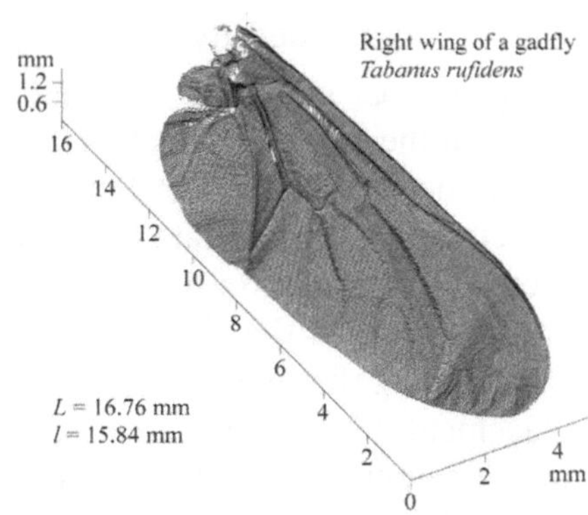

Fig. 4. Three-dimensional description of the surface shape measurement for the wing of gadfly, *Tabanus rufidens*

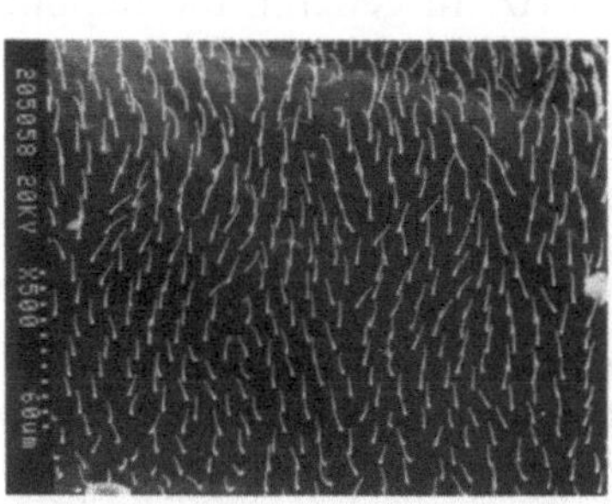

Fig. 5. Scanning electron micrograph of the wing of robber fly, *Promachus yesonicus* Bigot

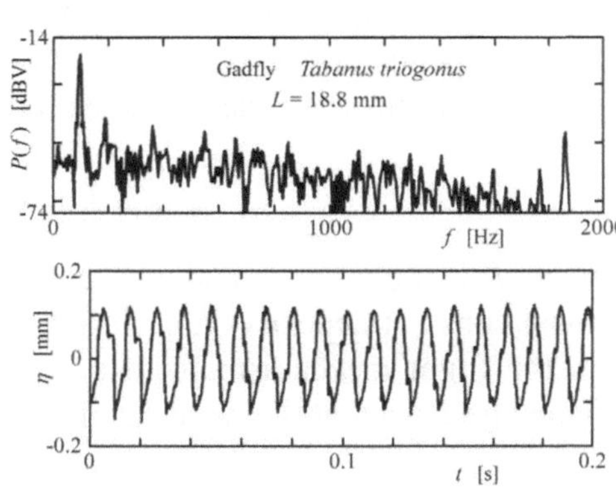

Fig. 6. Extrinsic skeleton vibration during the flapping motion of the gadfly, *Tabanus triogonus*

3.2 Flapping Frequency of Gadfly

Generally, winged insects produce aerodynamic force by flapping their wings and fly in the air (Brodsky 1994). The airflow over a wing generates forces of lift and drag which combine to create a resultant force. Lift is a component normal to the wing, and drag is a component parallel to the wing. Lift F_L and Drag F_D are described as follows;

$$F_L = \frac{1}{2} C_L \rho U_0^2 S_d \tag{2}$$

$$F_D = \frac{1}{2} C_D \rho U_0^2 S_d \tag{3}$$

where ρ is the density of air, U_0 is the velocity of airflow, and S_d is the wing area. C_L and C_D are lift and drag coefficients. These coefficients depend on the angle of attack. The velocity at each place of the wing is different, because the insect flaps around the wing root. The fact shows that insect wing is appropriate for the airfoil of micro energy converter. The velocity U_0 depends on the size and the flapping frequency of insect wing, and it relates to the characteristics of micro energy converter. Therefore, the flapping frequency of the gadfly was measured by the optical displacement detector system as shown in Fig. 2. Figure 6 shows an example of the output signal from the optical displacement detector. The lower figure in Fig. 6 corresponds to the extrinsic skeleton vibration during the flapping motion of the gadfly, *Tabanus triogonus*. The upper figure shows the power spectrum of the extrinsic skeleton vibration. The power spectrum has a sharp peak. The dominant frequency agrees with the flapping frequency of the gadfly. It was observed that the flapping frequency of gadfly depends on the size, and f_i = 80 ~ 100 Hz. In general, the flapping frequency of insects is described as follows;

$$f_i = K m^{-\frac{1}{6}} \tag{4}$$

where K is the proportional constant and m is the insect mass. The proportional constant depends on each order of flying insects, that is, it depends on the flight mechanism for the generation of aerodynamic forces. The flapping velocity at the wing tip of insect can be obtained from Eq. (5).

$$U = \omega l \tag{5}$$

where ω is the angular flapping frequency, l is the wing length of insect. The flapping velocity of the gadfly is $U = 9.4$ m/s for $f_i = 100$ Hz and $l = 15$ mm. Therefore, it is necessary to use the micro energy converter using gadfly wings in the area of the airflow velocity under $U_0 = 9.4$ m/s.

4 Caracteristics of Micro Wind Energy Converter

4.1 Composition and Principle

The micro energy converter which converted the airflow energy into electric energy was constructed in this paper. The micro energy converter is composed of the Diptera insect wings, NdFeB permanent magnet, magnetic fluid, and the coil. The length of gadfly wing was about 15 mm. Three insect wings were bounded to the shaft with the angle of $2\pi/3$ as shown in Fig. 7. The permanent magnet is columnar of 5 mm outside diameter and 5 mm height. Table 1 shows the physical properties of NdFeB permanent magnet used in the experiment, and Table 2 shows the magnetic properties of the magnetic material. It can be seen from Table 2 that the permanent magnet has excellent magnetic characteristics. Magnetic fluid was adsorbed by the permanent magnet. Table 3 shows the physical properties of kerosene-based magnetic fluid (ferricolloid HC-50) used in the experiment. In general, the permanent magnet encompassed with magnetic fluid is suspended in the fluid (Rosensweig 1985). This kind of phenomenon is called "self-levitation". The phenomenon of self-levitation occurs when a permanent magnet is immersed in magnetic fluid. Figure 8 shows a short columnar magnet stably suspending itself in magnetic fluid. In Fig. 8, V_m is the volume of magnetic fluid, and g is the gravitational acceleration. The permanent magnet, which it nearly six times as dense as the fluid, is seen hovering above the acrylic plastic base. Self-levitation of a permanent magnet in magnetic fluid is attributed to the magnetic force;

$$F_m = -V\mu_0 M_m \Delta H_m \tag{6}$$

where V is the volume of magnet, μ_0 is the permeability of free space, M_m is the intensity of magnetization, and H_m is the magnetic field. The shaft that bonded three insect wings was bounded to the permanent magnet. When the insect wings receive the air flow, self-levitating magnet in magnetic fluid can rotate very easily, because the magnetic fluid plays the role of excellent lubrication. The energy converter has the coil as shown in Fig.

7. Faraday's law shows that a current can be produced by a changing magnetic field. A current is set up in the circuit as long as there is relative motion (rotational motion of the magnet) between the magnet and the coil. The electromotive force (emf) induced in a circuit is directly proportional to the time rate of change of magnetic flux through the circuit. It can be written as follows;

$$V_i = -\frac{\partial \Phi}{\partial t} \tag{7}$$

where Φ is the magnetic flux threading the circuit, which can be expressed as follows;

$$\Phi = \int_S B_n ds \tag{8}$$

where B_n is the normal component of magnetic flux density, and ds is small area. The permanent magnet used in the experiment has higher magnetic flux density, even if it is small. If the circuit is a coil consisting of n loops all of the same area and if the flux threads all loops, the induced emf is

$$V_i = -n\frac{\partial \Phi}{\partial t} \tag{9}$$

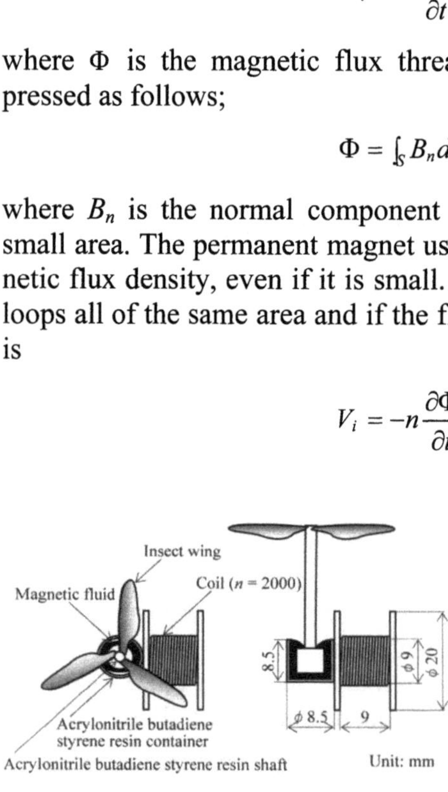

Fig. 7. Outline diagram of micro wind energy converter with insect wings

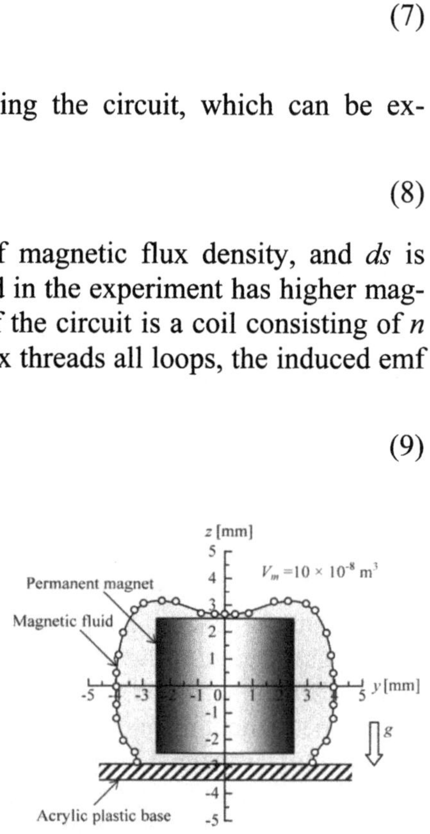

Fig. 8. Free surface shape of magnetic fluid encapsulated permanent magnet

Table 1. Physical properties of permanent magnet

Permanent magnet	$Nd_2Fe_{14}B$
Temperature coefficient	0.12 %/°C
Density	7300-7500 kg/m^3
Curie temperature	310 °C
Vickers hardness	HV 500-600

Table 2. Magnetic characteristics of permanent magnet

Residual magnetic flux density	*Br* (T)	1.62-1.33
Coercive force	*bHC* (kA/m)	859-970
Coercive force	*iHC* (kA/m)	>955
Maximum energy product	(*BH*)max (kJ/m^3)	302-334

Table 3. Physical properties of magnetic fluid

Magnetic fluid	Ferricolloid HC-50
Density	1250 kg/m^3
Viscosity	9 mPa s
Saturation magnetization	32 kA/m
Surface tension	0.0277 N/m

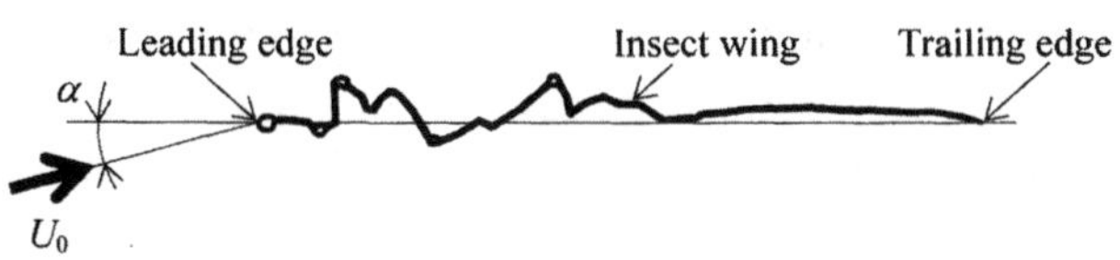

Fig. 9. Definition of the angle of attack α for the airflow U_0

The installation angle of the insect wing to the shaft influences the rotational characteristics (lift characteristics) of the micro energy converter. In this experiment, the insect wing was bounded to the shaft for the airflow at the angle of attack α. In general, the resultant of two forces of lift and drag depends on the angle of attack α. The definition of the angle of attack α for the insect wing is shown in Fig. 9.

The micro energy converter with artificial wings made of the acrylonitrile-butadiene-styrene (ABS) resin was also produced. The artificial wing with the same shape as the insect wing was produced by Dimension 3D Printer. In the experiment, two converters with real insect wings and artificial insect wings were compared for power generation characteristics.

4.2 Power Generation Characteristics

First the power generation characteristics of the micro wind energy converter with insect wings were examined. Secondly the power generation characteristics of the micro energy converter with ABS resin wings were examined. Thirdly, two converters were compared for power generation characteristics.

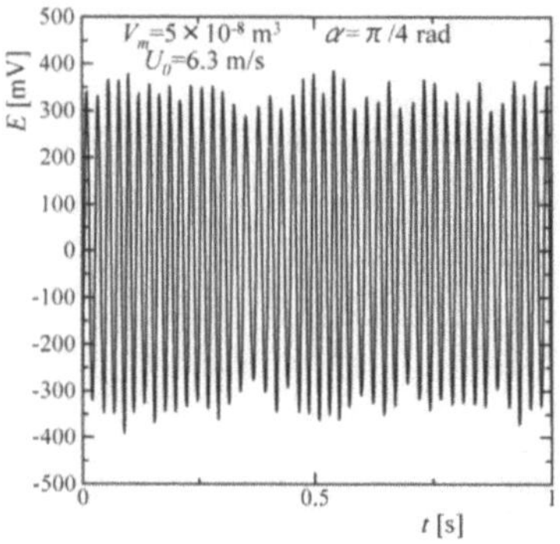

Fig. 10. Output signal from the micro energy converter with insect wings

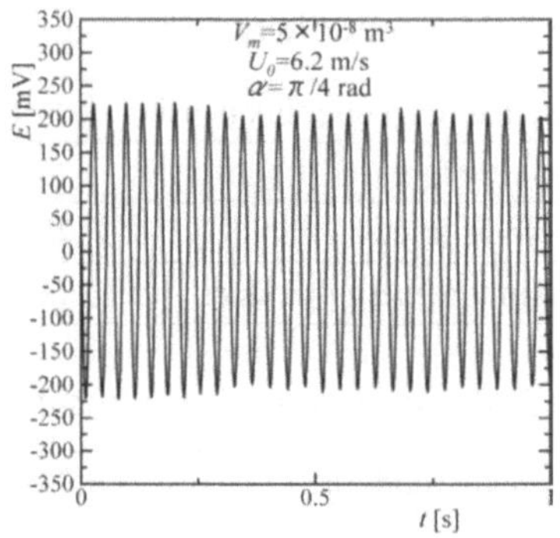

Fig. 11. Output signal from the micro energy converter with artificial wings

Figure 10 shows the output signal from the energy converter with Diptera insect wings subjected to the airflow of U_0 = 6.3 m/s. The angle of attack α in Fig. 10 is the installation angle of the insect wing to the shaft. In a general aircraft wing, the lift-drag ratio depends on the angle of attack. As the angle of attack rises, lift initially increases faster than drag so the lift-drag ratio is a maximum at an angle of attack of about 5 degrees. Thereafter, drag increases rapidly and, at an angle of attack of about 15 degrees, the aerofoil stalls. This effect is explained by a separation of the flow on the upper surface of the aerofoil (Wegener 1996). However, the insect wings do not stall during their flight, even if insects flap at higher angle of attack. The effective angle of attack of insect wing in the micro energy converter varies according to rotation of the runners. These facts show the insect wings are suitable for the micro wind energy converter at lower Reynolds number flow. It can be seen from Fig. 10 that the output voltage E exceeds 200 mV. The output signal vibrates to the positive and negative. This vibration corresponds to the rotation of the magnet. As before, when the magnet is rotated toward the stationary conducting loop, a current is induced. The variation of the amplitude in the output signal is observed in Fig. 10. This phenomenon occurs due to no completely same size of the insect wings. The different aerodynamic moments M act on each insect wing.

$$M = \frac{1}{2}\rho U_0^2 C_m S_d c \tag{10}$$

where C_m is the pitching moment coefficient, and c is the chord length. The imbalance of the angular moment generates the variation of the amplitude.

Figure 11 shows the output signal from the energy converter with ABS resin artificial wings. The signal shows a steady output compared with the insect wings. Three artificial wings are the complete same sizes. Therefore, there is no imbalance of the angular moment on the micro energy converter. However, the output voltage is lower compared with the insect wing converter. This fact shows that the lift generation function of Diptera wing is excellent. Figure 12 shows the comparison of the frequencies of the output signals. The frequencies for the micro energy converter with insect wings show higher values at the constant airflow velocity. Faraday's law of induction shows that the emf induced in a circuit is directly proportional to the time rate of change of magnetic flux through the circuit. Therefore, the micro energy converter with insect wings has excellent characteristics for power generation. This fact shows the importance of minute morphology of the corrugated wing with minute hairs.

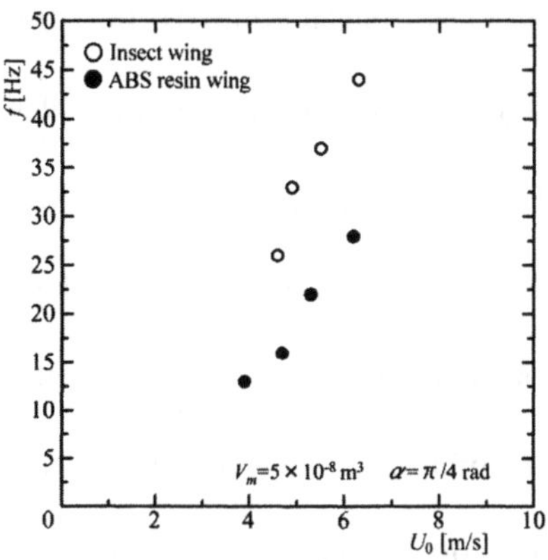

Fig. 12. Uniform airflow velocity U_0 versus frequency of output signal f

5 Conclusions

Flapping characteristics of Diptera insects were studied using the optical displacement detector system, and the structural design of Diptera insect wings was studied using the three-dimensional optical shape measuring system and the scanning electron microscope. Furthermore, the micro wind energy converter using Diptera insect wings were constructed, and the power generation characteristics of the energy converter were examined. The results obtained are summarized as follows;

1. The flapping frequency of gadfly *Tabanus triogonus* depends on their size, and it is in the area from about 80 Hz to 100 Hz.
2. Diptera wings are corrugated and clothed in minute hairs. Minute hairs are tilted.

3. The micro wind energy converter composed of Diptera insect wings, permanent magnet, magnetic fluid, and coil outputs the voltage of hundreds of mV.
4. The frequency of the output signal from the micro wind energy converter increases with the velocity of uniform airflow.
5. The frequencies for the micro energy converter with Diptera wings show higher values at the constant airflow velocity compared with the converter with ABS resin artificial wings.

References

Alexander RM (1984) The Gails of Bipedal and Quadrupedal Animals. Int J Rob Rese 3:49–59

Azuma A (1992) The Biokinetics of Flying and Swimming. Springer-Verlag, New York

Brodsky AK (1994) The Evolution of Insect Fight. Oxford Science Publications, New York

Esashi M (2002) Micromachines. Sngyo Gijutsu Service Center, Tokyo (in Japanese)

Higuchi T (2003) Micromachine Technologies. Sangyo Gijutsu Service Center, Tokyo (in Japanese)

Rosensweig RE (1985) Ferrohydrodynamics. Cambridge University Press, New York

Shimoyama I (1994) Artificial Insect. J Jpn Soc Mech Engn 97:60–63 (in Japanese)

Shimoyama I (1996) Scale Effects in Microrobots. J Robotic Soc Japan 14:1106–1108 (in Japanese)

Sudo S, Tsuyuki K, Ikohagi T, Ohta F, Shida S, Tani J (1999) A Study on the Wing Structure and Flapping Behavior of a Dragonfly. JSME Int J 42:721–729

Sudo S, Tsuyuki K, Tani J (2000) Wing Morphology of Some Insects. JSME Int J 43:895–900

Sudo S, Tsuyuki K, Ito Y, Tani J (2001) The Wing Apparatus and Flapping Behavbior of Hymenoptera. JSME Int J 44:1103–1110

Sudo S, Segawa S, Honda T (2003) Frequency Dependence of Locomotive Velocity of Magnetic Micro Machine. J JSEM 3:235–241 (in Japanese)

Sudo S, Takaki Y, Hashiguchi Y, Nishiyama H (2005) Magnetic Fluid Devices for Driving Micro Machines. JSME Int J 48:464–470

Sudo S, Tsuyuki K, Kanno K (2005) Wing Characteristics and Flapping Behavior of Flying Insects. Exp Mech 45:550–555

Wegener PP (1996) What Makes Airplanes Fly?. Springer-Verlag, New York

Chapter 26

Clapping-wing Micro Air Vehicle of Insect Size

Yoshiyuki Kawamura[1], Satoshi Souda[1], Satoshi Nishimoto[1],
and Charles P. Ellington[2]

[1] Department of Intelligent Mechanical Engineering
Faculty of Engineering, Fukuoka Institute of Technology, Japan

[2] Department of Zoology, University of Cambridge, UK

Summary. The efficiencies for generating the axial thrust and the lift of a flapping wing micro air vehicle (MAV) were increased by clapping the wings together, using the "Weis Fogh clap and fling mechanism" from studies on the aerodynamics of the insect flight. Making use of this effect, we have succeeded in the development and the flight of an insect-size clapping wing MAV. The span and the weight were 10cm and 2.3g, respectively, which are very close to those of the hawkmoth (*Manduca sexta*), and amongst the smallest and the lightest ever built. The flight characteristics of the clapping MAV were similar to those of the hawkmoth. These results are the first step towards MAVs that mimic insects at lower Reynolds numbers.

Key words. MAV, Micro Air Vehicle, Flapping, Clapping, Insect, Hawkmoth

1 Introduction

Since the success in powered flight by the Wright brothers in 1903, human beings have mainly been developing flight technologies in order to obtain faster and larger aircraft, characterized as air vehicles with large Reynolds numbers. Recently, owing to advances in production technologies relating to the semiconductor industry, especially to those of cell phones and personal computers, essential technologies for making a micro air vehicle (MAV) have advanced dramatically, such as small and lightweight powerful micro motors, micro computers and batteries with large capacity.

In parallel with this progress, the reason why insects can create large lift forces has been studied (Weis-Fogh 1973). The lift cannot be explained by conventional aerodynamic theory applied to fixed wings, and the generation of a spiral leading edge vortex at low Reynolds numbers was found to play an important role (Ellington 1996, Johansson 2003, Dickinson 1999 and Fry 2003),. Numerical calculations (Isogai 1999, Lian 2005) as well as theoretical analyses (Ellington 1999) have been performed to investigate the possibility for developing flapping MAVs of insect size.

Recently, we have succeeded in the fabrication and the flight of an insect size flapping MAV, making use of "Weis Fogh clap and fling mechanism" (Weis-Fogh

1973). It has a span of 10 cm and a weight of 2.3 g, which are very close to the values of the hawkmoth (*Manduca sexta*) (Willmott 1997A, Willmott 1997B), which has been studied extensively. The flight characteristics of this flapping MAV were also compared with those of the hawkmoth.

2 Flight with high angle of attack

Fig. 1 shows the experimental facilities for measuring vertical force applied on the wings as a function of the angle of attack of the body for various flapping frequencies and wind velocities. The total span of the machine is 15 cm, and the flapping motion is driven by a DC power supply. The diameter of the exit of the wind tunnel is 25 cm. The vertical force force, which is perpendicular to the wind velocity, was measured by an electric balance with a resolution of 0.01 g. Flapping frequency was changed from 0 Hz to 20 Hz and was measured optically without contact. The flapping angle was ± 23 degrees. Angle of attack was set from 0 degree to 40 degrees. The wind velocity was changed from 0 m/s to 5m/s.

Fig. 2 shows the vertical force for various angles of attack and frequencies. Fig. 2(A) shows the lift for a flapping frequency of 0 Hz. In this case, the flapping angle was fixed to be 0 degrees, and the wing angle of attack of 20 degrees, the lift drops suddenly for all velocities, which is considered to be the stall. Fig. 2 (B) and (C) show the vertical force applied on the wings for flapping frequencies of 10 Hz and 20 Hz, respectively, where stall does not appear up to the angle of attack of 40 degrees.

These experimental results show that the flapping motion of the wings depresses the stall up to large angles of attack, keeping the lift coefficient high and leading to the possibility of quasi-hovering flight of a flapping MAV with large angle of attack.

Fig. 1. Experimental setup for measuring the vertical force at various angles of attack of the body and flapping frequencies.

(A)

(B)

(C)

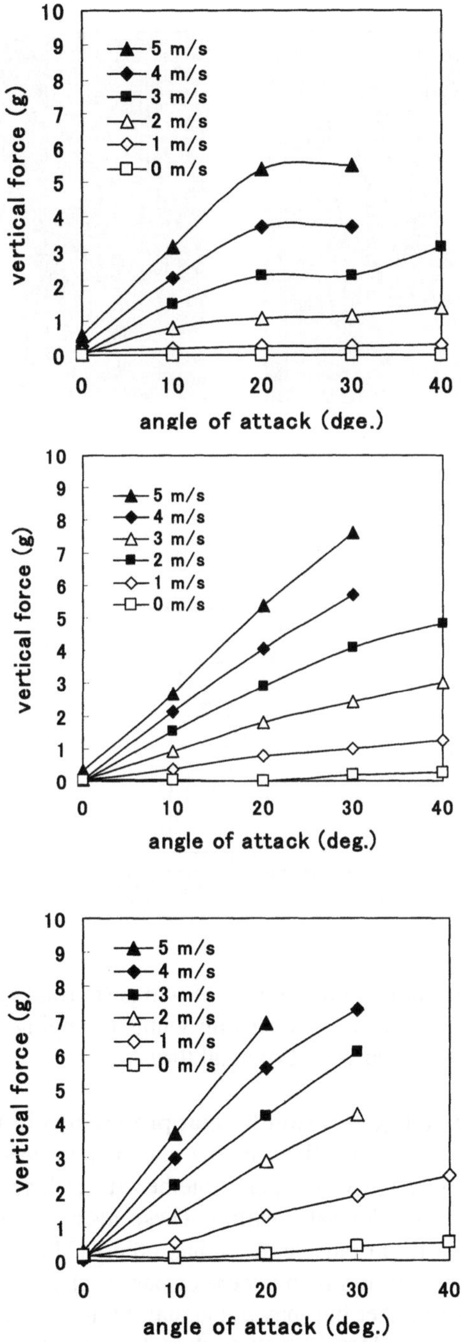

Fig. 2. Vertical force of the flapping wings as a function of angle of attack for various wind velocities. Flapping frequencies of (A), (B) and (C) are 0 Hz, 10 Hz and 20 Hz, respectively.

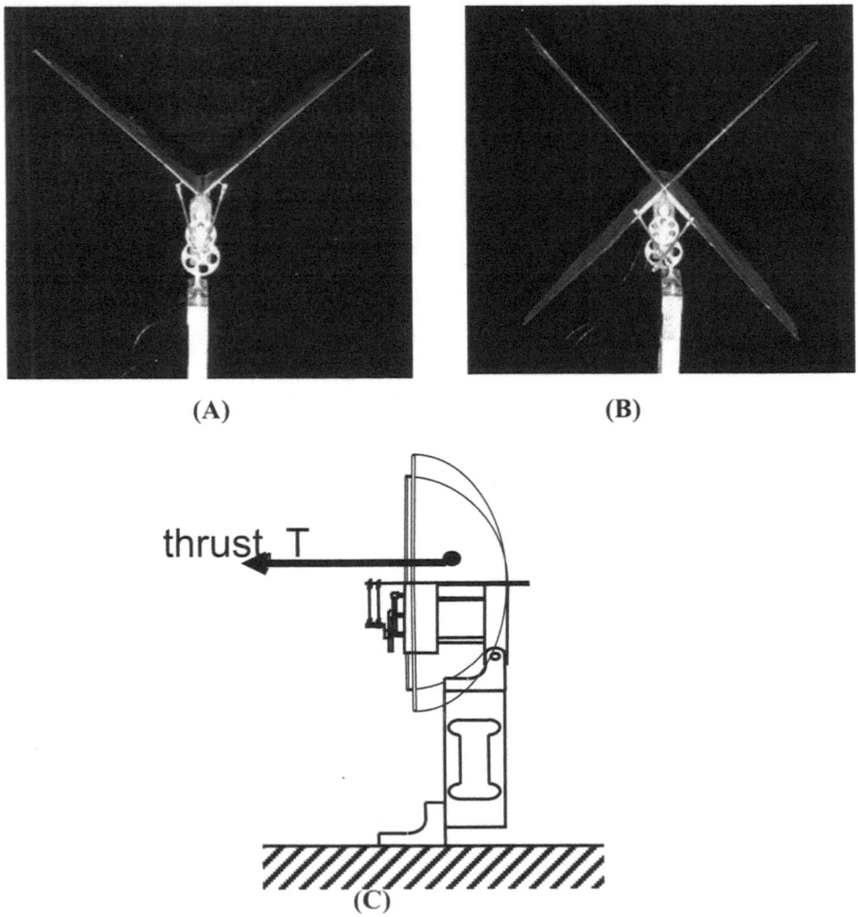

Fig. 3. (A) Flapping machine with one pair of wings (normal type). **(B)** Flapping machine with two pairs of X-like intersecting wings (clapping type). **(C)** Experimental apparatus for the measurement of axial thrust.

3 Improvement in thrust generation by clapping wings in hovering flight

Some types of insects (e.g. butterflies, moths and tiny insects) use a "clap and fling" wing motion, where the wings clap together at the highest position of the stroke and sometimes at the lowest position. These insects produce, on average, 25% more force per unit flight muscle mass than others (Marden 1987). The most common variation of this motion is the "near or partial clap and fling", where the wings come very close together but may not actually touch, or do so only in distal regions. Hawkmoths such as *Manduca sexta* show this variation.

We investigated the possibility of increasing force by introducing a clapping motion of the wings. We made two types of flapping machine for comparison.

One has two normal wings as shown in Fig. 3 (A), and the other has four X-like wings which are composed of two pairs of wings intersecting with each other as shown in Fig. 3(B). "Delfly", which is Remote control flapping MAV having a span of 35 cm and the weight of 17 g, used this type of wings, and has been succeeded in realizing vibration-free flight of the ornithopter at Delft University of Technology in Nederland.

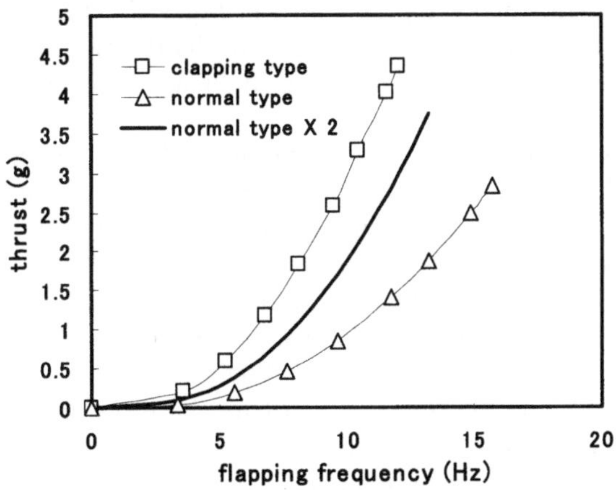

Fig. 4. Comparison of the axial thrust produced by the normal type (open triangles) and the clapping type (open squares) models. The thick solid line shows the doubled values of those for the normal type. The thrust for clapping is about 40% higher than the doubled values for the normal type.

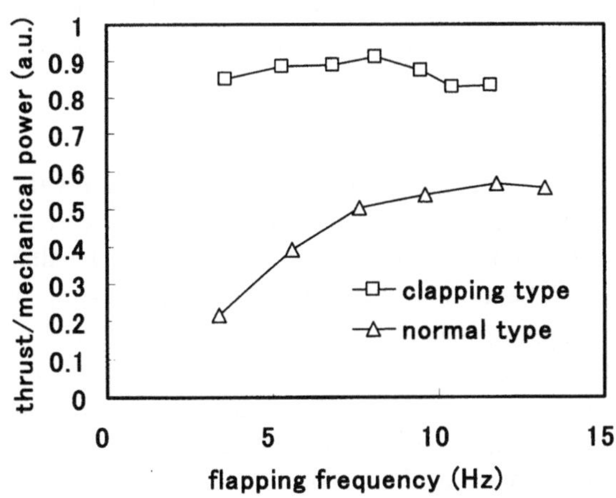

Fig. 5. Ratio of thrust to mechanical power as a function of the flapping frequency for the clapping (open squares) and normal (open triangles) types of machines.

For both types, the span and maximum chord were 20 cm and 5 cm, respectively. For the normal type, the flapping angle was ± 25 degrees, which corresponds to a stroke angle of 50 degrees, although for the clapping type, the flapping angle was ± 50 degrees, which corresponds to a stroke angle of 100 degrees. Whereas the clapping does not occur in the case of the normal type, it occurs once a cycle at both sides of the wings in the other. Each wing is comprised of a carbon rod driven by a DC motor through a linkage, and a thin plastic film attached to the carbon rod. The experimental setup is shown in Fig.3 C. The experiments were performed in still air to measure the axial thrust of the flapping wings, which provides the essential vertical force for hovering flight when the body axis becomes vertical.

In order to extract the net increase in axial thrust due to the "clapping effect", compensating for the difference in the total number of wings between these two models, the thrust produced by the clapping model was compared with double the value produced by the normal model. As is shown in Fig. 4, the thrust produced by the former is about 40% larger than twice the thrust produced by two wings on the normal model. The increase in the thrust can be considered to be due to the "Weis Fogh clap and fling mechanism" (Weis-Fogh, 1973).

In order to investigate the figure of merit for the thrust generation, we analyzed the ratio of the thrust to the product of the flapping frequency and the current of the DC motor. The flapping frequency is proportional to the angular velocity of the DC motor, and the current is proportional to the output torque of the motor. Therefore, their product is proportional to the mechanical power generated by the motor, and the ratio of the thrust to this product represents the figure of merit for the thrust generation. (The product of the voltage applied to the motor and the current was not used to calculate the mechanical power, because the motor efficiency changes depending on the operating condition of the motor.) Fig.5 shows that the experimental results for clapping wings have figure of merits 2 to 3 times higher than those for normal wings, which is qualitatively consistent with the results for real insects (Marden 1987).

4 Improvement in lift generation by clapping wings in level flight

The figure of merit for level flight was also investigated by realizing the force balance in level flight for both types of flapping machine. The resultant vertical force (Lift$+$ Thrust $\times \sin\theta$) and the resultant horizontal force (Drag $+$ Thrust $\times \cos\theta$) acting on the flapping machine were measured by a pair of load cells, as shown in Fig. 6. The velocity of the wind tunnel was adjusted so that the horizontal component of axial thrust balanced the drag, making the resultant horizontal force zero. By this adjustment, the force balance for level flight at the flight velocity equivalent to the velocity of the wind tunnel could be realized. The flapping frequency was 10 Hz, and the angle of attack of the body ranged from 0 degree to 40 degrees.

The wind velocities to realize level flight are shown in Fig.7. Over the total range of the angle of attack, the wind velocities for clapping wings were lower than those for normal wings, meaning that the clapping machine flies at a slower velocity than the normal wing motion.

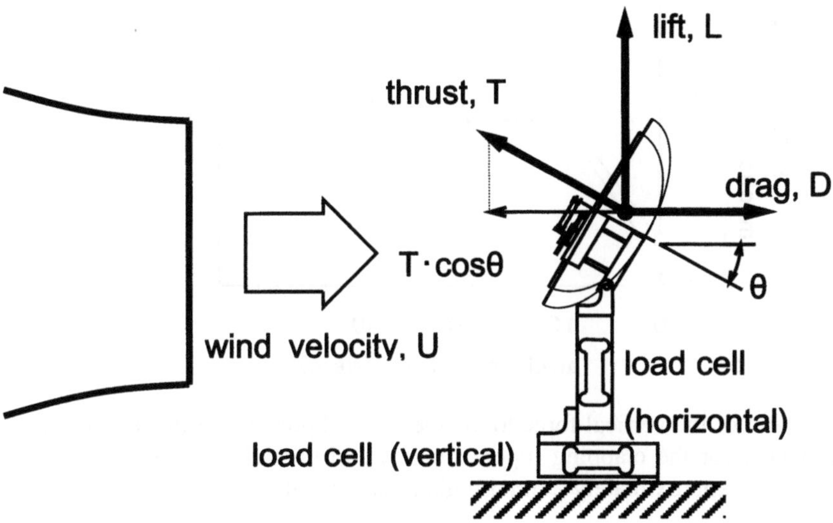

Fig. 6. Experimental setup for realizing the force balance of flight finding the wind velocity where the resultant horizontal force on the flapping machine is zero.

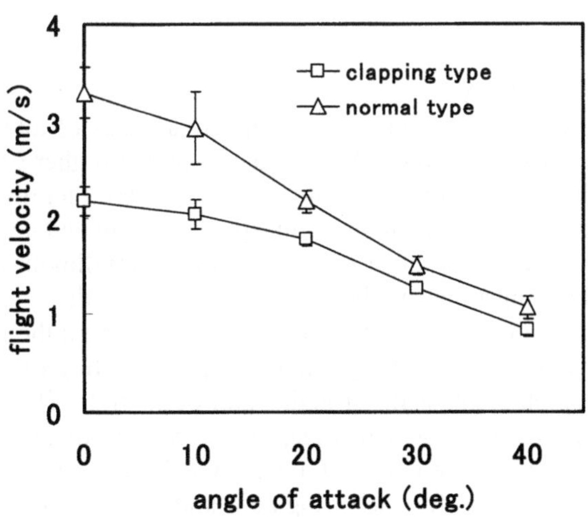

Fig. 7. Wind velocities corresponding to level flight with zero resultant horizontal force as a function of angle of attack for the clapping (open squares) and the normal (open triangles) types of wings.

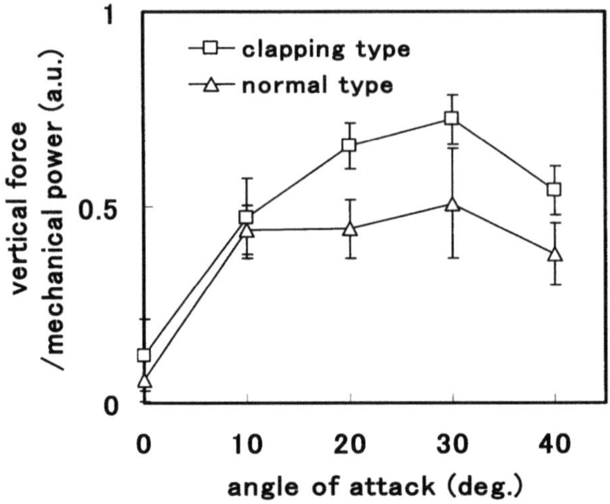

Fig. 8. Ratio of the vertical force to the mechanical power as a function of the angle of attack for the clapping (open squares) and normal (open triangle) wings. The former is approximately 2 times higher than the latter.

Fig. 8 shows the ratio of the resultant vertical force to the product of the flapping frequency and the current of the DC motor as a function of the angle of attack for the both types of wings. For angles of attack higher than 20 degrees, the ratios for clapping wings were larger than those for normal wings by about 1.5 times.

5 Flight of clapping wing MAV

The results described in sections 3 and 4 show the advantages of intersecting X-like wings for an insect-size flapping MAV in order to obtain higher vertical force and thrust given the limited output power of a DC motor. We therefore designed and fabricated a clapping wing MAV with a wingspan of 10 cm and a weigh of 2.3 g, which are almost the same as those of the hawkmoth (Willmott 1997A, and Willmott 1997B), and amongst the smallest and lightest ever built.

Fig. 9 shows the clapping wing MAV that we have developed; the side view is shown in Fig. 10A. We used intersecting X-like wings, consisting of two pairs of wings driven by a common crank shaft through a pair of conrods. The linkage was designed so that these two pairs of wings clap at the bottom of the stroke (Fig. 10B) and at the peak of the stroke (Fig. 10C): i.e. two claps in each stroke cycle. The stroke angle was 50 degrees for each wing pair, and therefore 100 degrees for the combined wing motion. The clapping wing MAV has hori zontal and vertical tail fins like conventional airplanes. The motor current and the rudder are controlled remotely by an infrared transceiver system. A typical value of the flapping frequency necessary for hovering was about 35 Hz.

Fig.9. Clapping wing MAV of insect size with a span of 10 cm and a weight of 2.3 g.

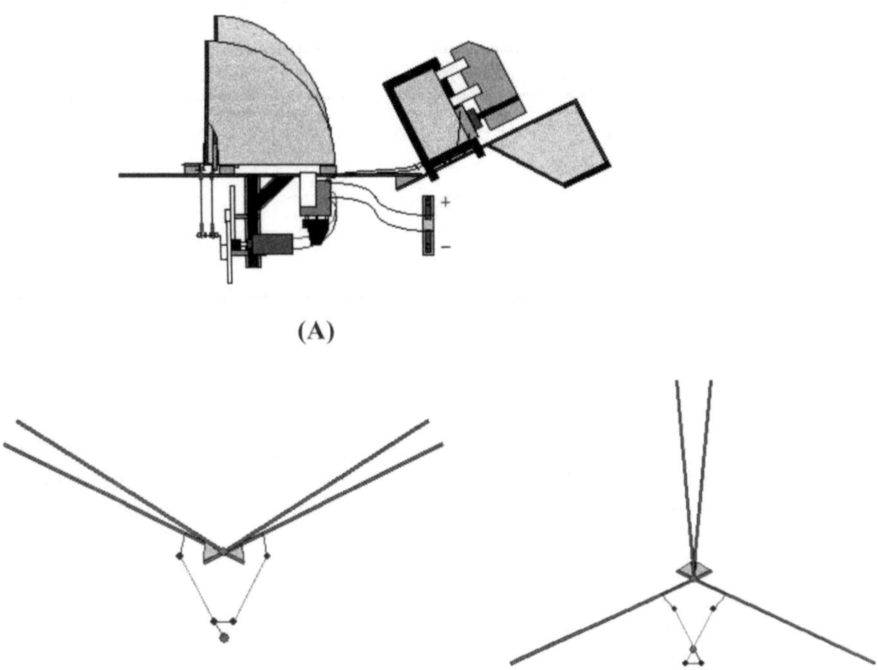

Fig. 10 (A) Side view of the clapping wing MAV of insect size. **(B)** Front view when it claps at the side position of the stroke. **(C)** Front view when it claps at the highest position of the flapping stroke.

Table 1. Comparisons of the specifications between the wing clapping MAV and the Hawkmoth.

	wing clapping MAV	Hawkmoth [15][16]
wingspan	10 cm	10 cm
aspect ratio	4	4.8
weight	2.3 g	1.8 g
flapping frequency	35 Hz	25 Hz
stroke amplitude	100 degrees*	110 degrees**
number of wings	4 sheets	2 sheets

(*: total stroke angle generated be two wings clapping with each other, **: averaged value between hovering and level flight)

Table 2. Comparisons of the flight characteristics between the wing clapping MAV and the hawkmoth. (Figures in the table represent the relative increases or decreases measured in case of the hovering flight to the values measured in case of the level flight.)

	wing clapping MAV	Hawkmoth [15][16]
stroke plane angle α	30 degrees down	35degrees down
flapping frequency	5 % increase	no change
stroke amplitude	mechanically fixed	15% increase

The flight tests were performed in still air inside a gymnasium. The position of the center of gravity of the clapping wing MAV was adjusted by changing the position of the battery. A tendency towards level flight was observed when the center of gravity was moved forward, and towards hovering when it was moved rearward,. For hovering flight, the flapping frequency was increased slightly above that for level flight.

The flapping motion of the wings made it possible for the MAV to move on a rough surface, such as a thick carpet, and even to take off from it. This is an additional advantage of flapping wings in comparison to fixed wing designs.

6 Comparisons with the hawkmoth

The specifications of the clapping wing MAV and hawkmoth are compared in Table 1. There are many similarities between these two except for the number of wings.

Whereas our clapping wing MAV changes the flight mode between level-flight and hovering by moving the position of the center of gravity, hawkmoths change it by shifting the wing motion forward and rearward (Willmott,1997A). When changing from level flight to hovering, the stroke plane angle decreases about 30 degrees for the MAV, while it decreases 35 degrees for the Hawkmoth (Willmott,1997A). The stroke amplitude of our clapping wing MAV does not change, because the wings are driven by a fixed mechanical linkage, while that of the hawkmoth increases by 15 % (Willmott,1997A). However, the increase in the flapping frequency of the MAV was about 5%, whereas the hawkmoth does not show a significant change (Willmott,1997A). (It is difficult, in general, for insects to change their flapping frequency because the wings are driven by a resonance mechanism relating to the body construction.) These comparisons in flight characteristics are summarized in Table 2.

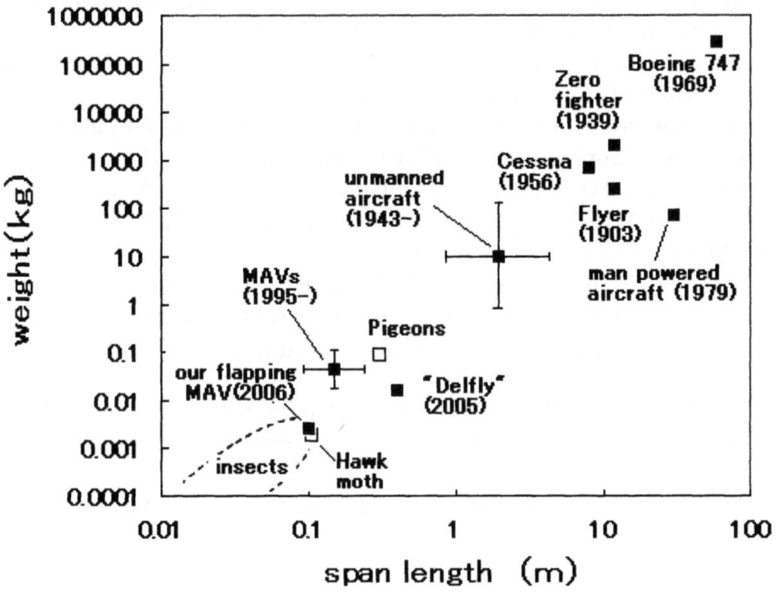

Fig. 11. Spans and weights of various aircraft in aviation history. Pigeons and hawkmoths are also plotted for comparison. Our clapping wing MAV of insect size has just entered into the boundary of the insect flight region

7 Summary

In Fig. 11, the span of various air vehicles is plotted against their weight . Solid squares represent man-made vehicles, while the open squares represent pigeons and hawkmoths for comparison with the clapping wing MAV. It can be seen that our clapping wing MAV has just entered into the boundary of the insect flight region. We think that the successful flight of our MAV showing similar flight characteristics to a real insect should encourage studies on insect-mimicking MAVs at lower Reynolds numbers.

Acknowledgements
This work is partially supported by the grant of Comprehensive Research Organization、Fukuoka Institute of Technology.

References

Weis-Fogh T (1973), Quick Estimates of Flight Fitness in Hovering Animals, Including Novel Mechanisms for Lift Production, J. exp. Biol, **59** 169-230.

Ellington CP, van den Berg C, Willmott AP and Thomas ALR (1996), Leading-edge vortices in insect flight, Nature, vol., **384** 626-630.

Johansson LC and Norberg RA (2003), Delta-wing function of webbed feet gives hydrodynamic lift for swimming propulsion in birds, Nature, **424** 65-68.

Dickinson, M. H., Lehmann, F.O. and Sane, S. P. (1999), Wing Rotation and the Aerodynamic Basis of Insect Flight Science, **284** 1954-1960.

Fry SN, Sayaman R and Dickinson MH (2002), The Aerodynamics of Free-Flight Maneuvers in Drosophila Science, **300** 495-498

Isogai K and Shinmoto Y (1999), Effects of Dynamic Stall on Propulsive Efficiency and Thrust of Flapping Airfoil AIAA Journal, 37, 1145-1151.

Lian Y and Shyy W (2005), Numerical Simulations of Membrane Wing Aerodynamics for Micro Air Vehicle Applications, Journal of Aircraft, **42** 865-873.

Ellington CP (1999), The novel aerodynamics of insect flight: applications to micro-air vehicles, Journal of Experimental Biology, **202**, 3439-3448.

Willmott AP and Ellington CP (1997A), The mechanics of flight in the hawkmoth Manduca sexta. I. Kinematics of hovering and forward flight, J. exp. Biol., **200** 2705-2722.

Willmott AP and. Ellington CP (1997B), The mechanics of flight in the hawkmoth Manduca sexta. II. Aerodynamic consequences of kinematic and morphological variation,J. exp. Biol., **200** 2723-2745.

Marden JH, Maximum lift production during takeoff in flying animal, J. exp. Biol., **130**, 235-258 (1987).

Part IV

Sports Science

Chapter 27

Study on the Application of Lagrangian Numerical Simulation to Fluid Dynamics in Sports Science

Kyoji Kamemoto[1], Akira Ojima[2], Satoshi Ido[3], and Toshiharu Arai[4]

[1] Department of Mechanical Engineering, Yokohama National University, 79-5 Tokiwadai, Hodogaya-ku, Yokohama, Kanagawa 240-8501, Japan
[2] College Master Hands, Inc., 2-1-31 Midorigaoka, Zama, Kanagawa 228-0021, Japan
[3] Toyota Motor Corporation, 1, Toyota-cho, Toyota, Aichi, 471-8571, Japan
[4] NTT Communications Corporation, 1-1-6 Uchisaiwaicho Chiyoda-ku, Tokyo, 100-8019, Japan

Summary. This paper describes a pioneering work of practical application of an advanced vortex method in the field of fluid dynamics in sports science. The vortex method developed by the present authors is one of vortex element methods based on the Biot-Savart law, and it is known that the method provides a Lagrangian simulation of unsteady and vortical flows. In this study, in order to examine the applicability of the vortex method, three-dimensional, complex and unsteady flows around an isolated 100 m runner and a ski-jumper were calculated. Basic equations and mathematical treatment of the method are explained in this paper, and calculation conditions and panel data of deforming configuration of the athletes are described. As results of the present study, vortical and unsteady flow features around a runner and a ski-jumper are understood, and unsteady variation of aerodynamic forces corresponding to deformation of body configuration due to athletic motion are calculated. And, it is confirmed that the advanced vortex element method is useful for the simulation of unsteady and complex flows around dynamic bodies of athletes.

Key words. Fluid dynamics in sports science, Numerical simulation, Vortex method, Runner, Ski-jumper

1 Introduction

In recent years, scientific research on sports technology becomes more important for improvement of training methods which have been usually based on experiences of athletes and their trainers. Especially, it is strongly expected in the sports concerning in aerodynamics and hydrodynamics to investigate unsteady flows around the moving body of an athlete in order to obtain scientific information of fluid forces acting on it.

On the other hand, the vortex methods have been developed and applied for analysis of complex, unsteady and vortical flows in relation to problems in a wide range of industry, because they consist of simple algorithm based on physics of flow. Nowadays, applicability of the vortex element methods has been developed and improved dramatically, and it has become encouragingly clear that the vortex methods have so much interesting features that they provide easy-to-handle and completely grid-free Lagrangian calculation of unsteady and vortical flows without use of any RANS type turbulence models. Leonard (1980) summarized the basic algorithm and examples of its applications. Sarpkaya (1989) presented a comprehensive review of various vortex methods based on Lagrangian or mixed Lagrangian-Eulerian schemes, the Biot-Savart law or the vortex in cell methods. Kamemoto (1995) summarized the mathematical basis of the Biot-Savart law methods. Recently, Kamemoto (2004) reported several attractive applications involving simulation of various kinds of unsteady flows with an advanced vortex method, and Ojima and Kamemoto (2005) reported interesting results of a study on numerical simulation of unsteady flows around a swimming fish by using their vortex method.

As well as many finite difference methods, it is a crucial point in vortex methods that the number of vortex elements should be increased when higher resolution of turbulence structures is required, and then the computational time increases rapidly. In order to reduce the operation count of evaluating the velocity at each particle through a Biot-Savart law, fast N-body solvers, by which the operation count from $O(N^2)$ to O (N log N) is reduced, have been proposed by Greengard et al. (1987). On the other hand, in order to reduce the computational load in calculation of turbulence structures, Fukuda and Kamemoto (2005) proposed an effective redistribution model of vortex elements with consideration of convective motion and viscous diffusion in a three dimensional core-spreading model as a grid-free Lagrangian large eddy simulation .

In the present study, in order to examine the applicability of the vortex method developed to extend as a tool for a grid-free Lagrangian large eddy simulation by the present authors to fluid dynamics in sports science, numerical simulation of three dimensional, complex and unsteady flows around an isolated 100 m runner and a ski-jumper were performed. From the numerical analyses, vortical flow characteristics around the athletes and fluctuating aerodynamic forces of drag and lift have been investigated.

2 Algorithm of Vortex Method

2.1 Mathematical Basis

The governing equations of viscous and incompressible flow are described by the vorticity transport equation and the pressure Poisson equation which can be derived by taking the rotation and the divergence of Navier-Stokes equations, respectively.

$$\frac{\partial \boldsymbol{\omega}}{\partial t} + (\boldsymbol{u} \cdot grad)\,\boldsymbol{\omega} = (\boldsymbol{\omega} \cdot grad)\,\boldsymbol{u} + \nu \nabla^2 \boldsymbol{\omega} \tag{1}$$

$$\nabla^2 p = -\rho\, div(\boldsymbol{u} \cdot grad\ \boldsymbol{u}) \tag{2}$$

Where $\boldsymbol{u}$ is a velocity vector and a vorticity $\boldsymbol{\omega}$ is defined as follows.

$$\boldsymbol{\omega} = rot\ \boldsymbol{u} \tag{3}$$

As explained by Wu and Thompson (1973), the Biot-Savart law can be derived from the definition equation of vorticity as follows.

$$\boldsymbol{u} = \int_V (\boldsymbol{\omega}_0 \times \nabla_0 G)\,dv + \int_S \{(\boldsymbol{n}_0 \cdot \boldsymbol{u}_0) \cdot \nabla_0 G - (\boldsymbol{n}_0 \times \boldsymbol{u}_0) \times \nabla_0 G\}\,ds \tag{4}$$

Here, subscript "0" denotes variable, differentiation and integration at a location $\boldsymbol{r}_0$, and $\boldsymbol{n}_0$ denotes the normal unit vector at a point on a boundary surface S. And G is the fundamental solution of the scalar Laplace equation with the delta function $\delta(\boldsymbol{r} - \boldsymbol{r}_0)$ in the right hand side, which is written for a three-dimensional field as follows.

$$G = \frac{1}{4\pi R} \tag{5}$$

Here, $\boldsymbol{R} = \boldsymbol{r} - \boldsymbol{r}_0$, $R = |\boldsymbol{R}| = |\boldsymbol{r} - \boldsymbol{r}_0|$. In Eq. (4), the inner product $\boldsymbol{n}_0 \cdot \boldsymbol{u}_0$ and the outer product $\boldsymbol{n}_0 \times \boldsymbol{u}_0$ stand for normal velocity component and tangential velocity vector on the boundary surface, and they correspond to the source distribution on the surface and the vortex distribution that has the rotating axis in parallel to the surface, respectively. Therefore, it is mathematically understood that velocity fields of viscous and incompressible flow are obtained from the field integration concerning vorticity distributions in the flow field and the surface integration concerning source and vortex distributions around the boundary surface. In this study, a boundary surface is represented by the panel method. The source panels and vorticity layers corresponding to a couple of terms in the surface integration in the right hand side of Eq.(4) are distributed on the boundary surface. The strengths of source and vorticity are determined by using the following two conditions; zero normal component of relative velocity to the boundary surface and non-slip condition of the relative flow on the solid surface. The pressure in the field is obtained from the integration equation formulated by Uhlman (1992), instead of the finite difference calculation of the Eq.(2) as follows.

$$\beta H + \int_S H \frac{\partial G}{\partial n} ds = -\int_V \nabla G(\boldsymbol{u} \times \boldsymbol{\omega}) dv - \int_S G\boldsymbol{n} \frac{\partial \boldsymbol{u}}{\partial t} ds - \nu \int_S \boldsymbol{n}(\nabla G \times \boldsymbol{\omega}) ds \tag{6}$$

Here, β=1 in the flow field and β=1/2 on the boundary S. G is the fundamental solution given by Eq.(5). H is the Bernoulli function defined as follows.

$$H = \frac{p}{\rho} + \frac{u^2}{2} \tag{7}$$

The values of H on the boundary surface are calculated from Eq. (6) by using the panel method. Once the pressure distribution around the boundary surface is calculated from Eqs. (6) and (7), integration of both pressure distribution and frictional stress distribution around the body surface yields the fluid force acting on the body.

2.2 Introduction of Vortex Elements

One of the most important schemes in the vortex methods is how to represent the distribution of vorticity in the proximity of the body surface, taking account of viscous diffusion and convection of vorticity under the non-slip condition on the surface. In the present method, a thin vorticity layer is considered along the solid surface, and discrete vortex elements are introduced into the surrounding flow field considering the diffusion and convection of vorticity from discrete elements of the thin vorticity layer. The details of treatments have been explained in the paper by Ojima & Kamemoto (2001).

The discrete vortex element is modeled by a vortex blob which has a spherical structure with a radially symmetric vorticity distribution proposed by Winkelmans & Leonard (1988). The motion of the discrete vortex elements is represented by Lagrangian form of a simple differential equation $d\boldsymbol{r} / dt = \boldsymbol{u}$. Then, trajectory of a discrete vortex element over a time step is approximately computed from the Adams-Bashforth method. On the other hand, the evolution of vorticity is calculated by Eq.(1) with the three-dimensional core spreading method proposed by Nakanishi & Kamemoto (1992). It should be noted here that in order to keep higher accuracy in expression of a local vorticity distribution, a couple of additional schemes of re-distribution of vortex blobs are introduced in the present advanced vortex method. When the vortex core of a blob becomes larger than a representative scale of the local flow passage, the vortex blob is divided into a couple of smaller blobs. On the other hand, if the rate of three-dimensional elongation becomes large to some extent, the vortex blob is divided into plural blobs to approximate the elongated vorticity distribution much more properly. The code used in the present study is a modified version of Virtual Fluids System 3D.

3 Flow around a 100 m Runner

A static configuration of a man was represented by 3,020 source panels by modifying a sample data set contained in a three dimensional graphic tool (Shade 7.5 Professional). Figure 1 shows the configuration of his whole body expressed by the set of panels.

As a typical running style, the running motion of a famous Japanese 100 m sprinter, Mr. Hideaki Miyata, was adapted to the body configuration. Figure 2 shows a series of stick pictures of his one cycle motion during sprint running of a 100 m race reported by Sekioka et al (1990). In the present study, the one cycle motion was produced by 320steps of instantaneous body configuration based on the stick pictures. Figure 3 shows the one cycle motion of sprint running represented by characteristic eight steps of instantaneous running style.

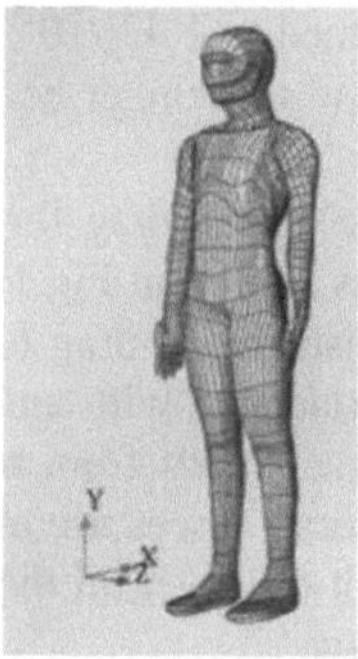

Fig. 1. Panel distribution representing a standing configuration.

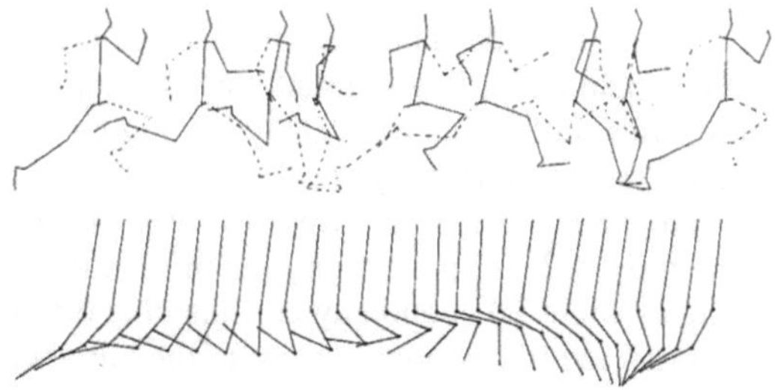

Fig. 2. Stick pictures of one cycle motion of a 100 m runner. (Sekioka et al (1990))

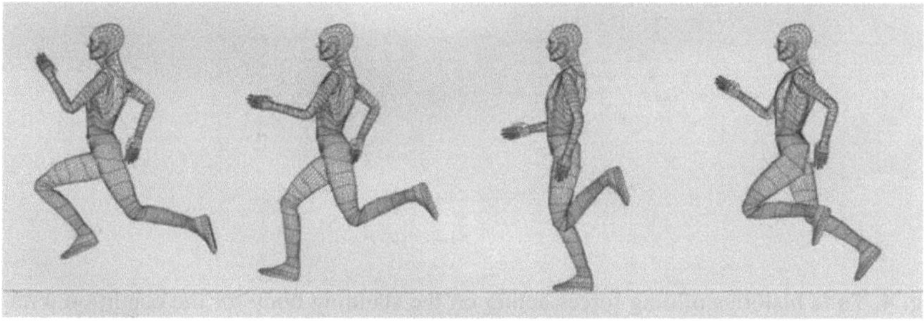

Fig. 3. One cycle of running motion represented by deformation of panel distribution.

Considering the world record of the 100 m race, the running speed of the model runner was assumed to be constant as 10.0 m/s for the calculation of sprint running mode. In the present calculation, the relative flow around the configuration of the runner was dealt with, and therefore, the surface of ground represented by 1,796 panels was moved together with the atmosphere at a constant speed of 10.0 m/s instead of giving the running speed to the runner model itself.

In the redistribution of panels at every moment of calculation steps, it is necessary to consider connection of foot panels to moving ground panels according to alternative stamping of feet with the ground surface. In the present calculation, for the case of stamping the ground, the whole panels of the underside surface of a foot were connected with the ground surface at a time step when the gap between them became less than a small value, and for the case of leaving ground, a part of underside panels were redistributed from the heel at every step of calculation.

In the present study, in order to normalize length scale, the breadth of the runner's shoulders 0.4 m was used as the representative length L for normalization of the length scale. The running speed 10.0 m/s was used as the representative velocity U for normalization of the velocity scale. As one cycle motion was represented 320 steps of instantaneous body configuration, and as it is known that one cycle of sprint running motion of a first class runner takes approximately 0.45 s, the size of time step of the present time marching calculations was taken to be 1.40×10^{-3} s. The kinematic viscosity ν and density ρ of the atmosphere were respectively assumed as 1.43×10^{-5} m^2/s and 1.2 kg/m^3. Therefore, the Reynolds number of the flow around the runner becomes $Re = UL/\nu = 2.8\times10^5$.

In order to understand the influence of the ground surface on the drag force acting on a human body, flows around the standing configuration as shown in Fig.1 in a uniform stream of 10m/s were analyzed. Figure 4 shows time histories of drag forces acting on the standing body for the two cases of calculation condition with and without the ground surface, where the flow was assumed to impulsively start from rest at the time t=0.0 s. It is known that time mean values of the both drag forces are almost the same as about 20 N and any considerable differences are not observed in the features of drag force fluctuations except for the initial developing stage.

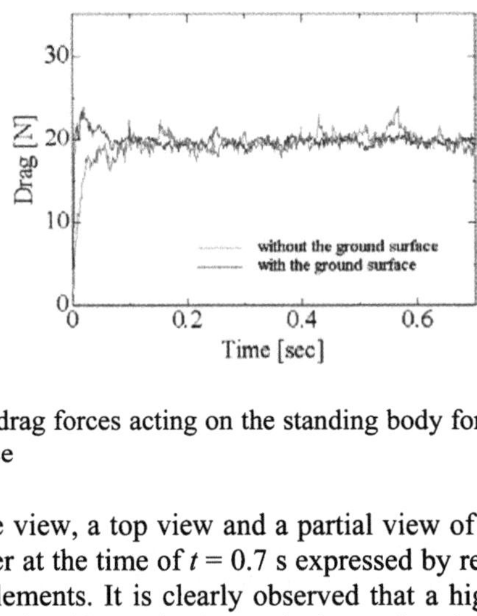

Fig. 4. Time histories of drag forces acting on the standing body for the condition with and without the ground surface

Figure 5 shows a side view, a top view and a partial view of an instantaneous flow pattern around the runner at the time of t = 0.7 s expressed by relative velocity vectors at positions of vortex elements. It is clearly observed that a higher velocity region is formed around the right shoulder and a low velocity region due to three dimensional flow separation is formed behind the back. Figure 6 shows three views of the pressure distribution around the runner's body surface at the same instance shown in Fig.5.

Figure 7 shows the time history of unsteady drag force acting on the runner body compared with the drag force on the body simply standing on the ground shown in Fig. 4. It is clearly known that the time mean value of the drag force on the runner is not so different from that on the standing body, but the fluctuation of the drag force is considerably larger than that of the standing case. It is also interestingly observed in Fig. 7 that the drag force is rather low around the time of t = 0.30 s in comparison with the drag on standing body, (hereafter which is called Case-1), and, after that, it increases and becomes almost equal to the drag on the standing style body or rather large around the time of t = 0.43 (hereafter which is called Case-2).

(b) Top view

(a) Side view

(c) Partial view

Fig. 5. Three views of an instantaneous flow pattern around a runner at $t = 0.7$ s.

(a) Front view (b) Side view (c) Rear view

Fig. 6. Pressure distribution around the body surface of a runner at $t = 0.7$ s.

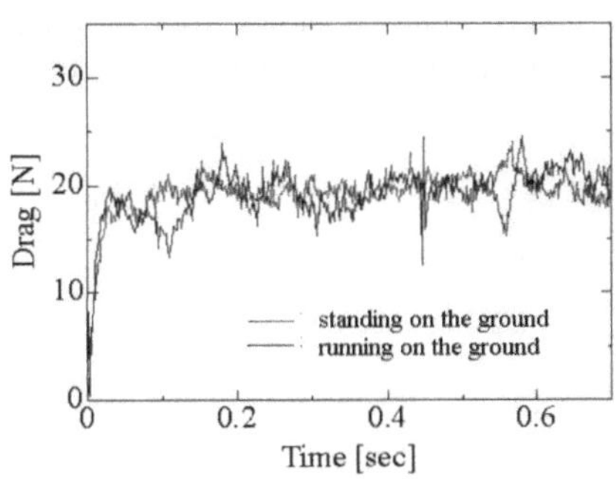

Fig. 7. Time history of drag force acting on a runner compared with drag force on standing configuration.

In Fig. 8, side views of flow patterns and rear views of pressure distributions for both characteristic cases of Case-1 and Case-2 are shown for comparison each others. In Case-1, the left hand is moving forward and the left foot is stamping. At the same time, the right hand is moving backward and the right foot is going to move forward. It is observed that the flow separation causes uniformly low pressure distribution on the back surface. On the other hand, in Case-2, the left hand is approaching to the forward dead point of swinging and the left foot is in the phase just after kicking the ground surface. At the same time, the right hand is approaching to the backward dead point of swinging and the right foot is moving forward just under the runner's body. It is interesting that a rather high velocity region is formed in the narrow space between the left side of body and the swinging left arm, and according to the appearance of the higher velocity region, some spots of very low pressure are distributed on the left hand side surface of the back and the rear surface of the left arm. The appearance of the higher velocity region in the space under the arm seems to result the slightly larger drag force in the running phase of Case-2.

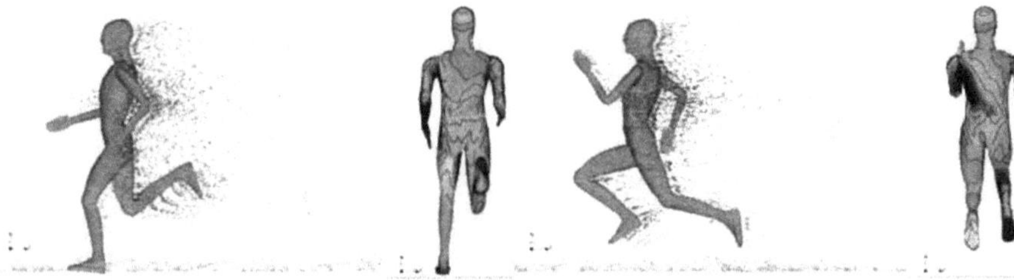

Case-1 at 0.3 s   Case-2 at 0.43 s

Fig. 8. Comparison of flow patterns and back pressure distributions between the lower drag phase (Case-1) and higher drag phase (Case-2).

4 Flow around a Ski-jumper

A configuration of a ski-jumper with a pair of skis was formed by 2,748 source panels which consisted of eight sets of panels representing a head, a body, a pair of arms, a pair of legs and a pair of skis. Figure 9 shows front and side views of a panel distribution representing an instantaneous style at a moment of flight.

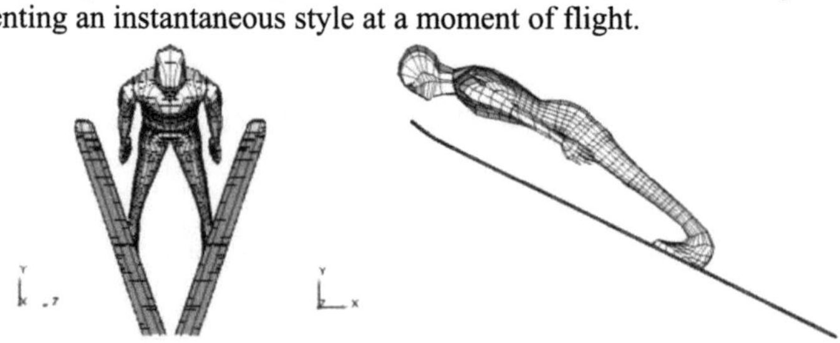

(a) Front view   (b) Side view

Fig. 9. Panel distributions representing a flight style of a ski-jumper.

Referring to the characteristic data obtained by Schmolzer and Muller (2005) for flight styles of the top ten ski jumpers in the Salt Lake City Olympic Games in 2002, flight style at every moment after jumping was formed by using the three dimensional graphic tool (Shade 7.5 Professional). It is known from the characteristic data that it takes about 4 seconds to fly from jumping to landing along the flight course shown in Fig.10, which shows the sectional configuration of the hill (K point: 120 m) and the flight course introduced in this study.

It is known that although both skis are parallel in the period from t=0.0 (jumping) to 0.04 s, they move and take a formation of so called V-shape mainly during the period between 0.04 s and 0.64 s. Therefore, in the present study, paying attention to dynamic characteristics of fluid force acting on the ski jumper during the movement of skis from parallel state to the V-shape, 700 flight styles were introduced to provide a set of panels of instantaneous flight style at every time step of 0.001 s for simulation of the ski flight during the period from t=0.0 s to 0.7 s. Figure 11 shows the oblique view of representative six steps of flight style.

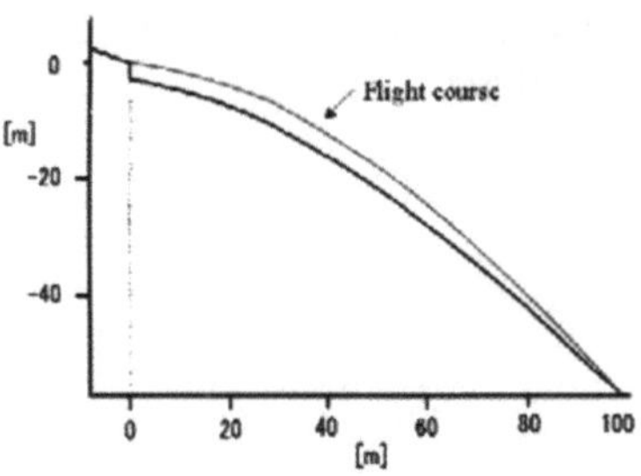

Fig. 10. Sectional configuration of a hill and calculation flight course.

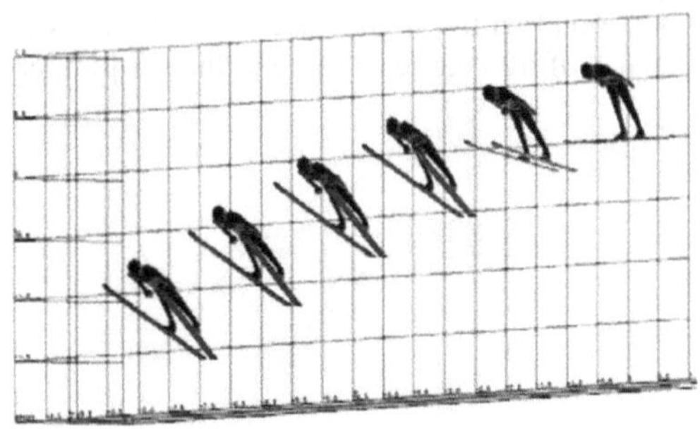

Fig. 11. Representative four steps of flight style

In the present study, a ski jumper, who is 1.73 m tall in height, flies the course shown in Fig. 10, in the stationary atmosphere, where the mean speed of the flight is about 25 m/s. The relative and unsteady flow around the ski jumper was simulated in the period of V-shape formation from t=0.0 s to 0.7 s. Length of a pair of skis is 2.48 m, and the maximum width is 0.102 m and the width at the ski tail is 0.091 m. The size of time step of the present unsteady calculations was taken to be 1.0×10^{-3} s. Kinematic viscosity ν and density ρ of the atmosphere were respectively assumed as 1.45×10^{-5} m^2/s and 1.2 kg/m^3. Therefore, taking the mean speed U of flight and the height L of the ski jumper, the Reynolds number of the relative flow becomes $Re = U L/\nu = 3.0\times10^6$. In this paper, both vertical component F_v and horizontal component F_h of the calculated fluid force acting on the ski jumper including a pair of skis are respectively normalized as $F_v/(0.5\rho U^2)$ m^2 which is called "lift force area" and $F_h/(0.5\rho U^2)$ m^2 which is called "drag force area" in this paper.

Figure 12 shows a couple of side views of instantaneous relative flow patterns around the ski jumper at the time of t = 0.04 s and 0.25 s which are expressed by relative velocity vectors at positions of vortex elements at each moment. It is clearly observed that remarkable separation of the relative flow appears only around the legs at t=0.04 s because both skis are parallel and attack angle of the skis to the relative flow is still very small. On the other hand, at t=0.25 s, the flow separation becomes considerable around the whole configuration including the pair of skis because skis move from parallel phase to the V-shape phase and the attack angle becomes large. Therefore, it is thought that the vortical and complex flow through skis strongly affects the flow separation from the body of the ski jumper.

Figure 13 shows time history of normalized aerodynamic force acting on the ski jumper after jumping, where t_s is the time 0.04 s when movement of skis stars from the initial parallel-phase to the V-shape phase and t_f is the time 0.25 s when the ski jumper finishes the ski-movement. It is very interesting that both values of lift and drag force areas remarkably increases during the transition period of ski-movement.

Figure 14 shows front views and top views of pressure distributions along the surfaces of the ski jumper and both skis in the parallel phase at t=0.04 s and in the V shape phase at t=0.25 s. It is observed in the front view of pressure distribution at t=0.25 s that considerably higher pressure caused by frontal stagnation of the relative flow widely distributes not only along the front surface of the ski jumper's body but also on the back surfaces of both skis, whereas there are not so remarkable differences between top views of pressure distributions at t=0.04 s and 0.25 s. Therefore, it is understood that the flight style of V-shape is certainly effective to obtain considerable increase of lift (vertical) force to the ski jumper during the flight.

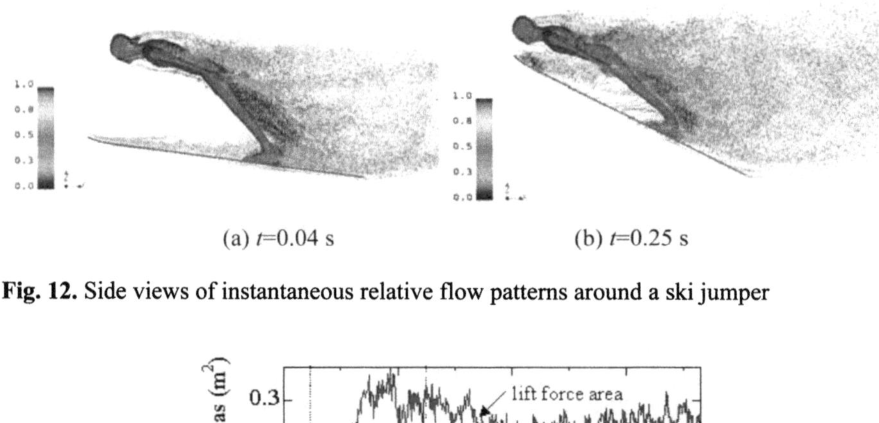

(a) t=0.04 s (b) t=0.25 s

Fig. 12. Side views of instantaneous relative flow patterns around a ski jumper

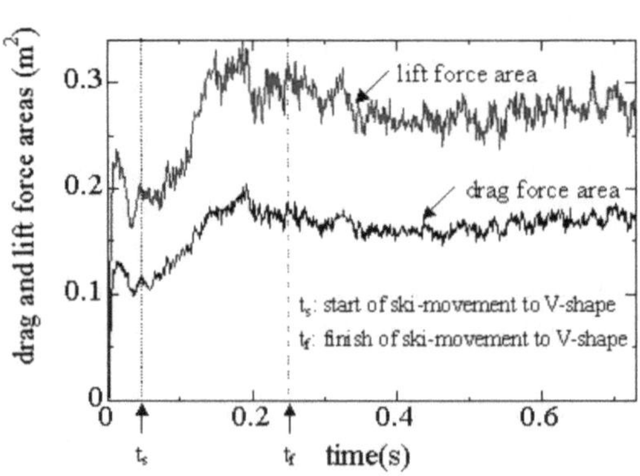

Fig. 13. Time history of lift (vertical) force area and drag (horizontal) force area corresponding to aerodynamic force on a ski-jumper.

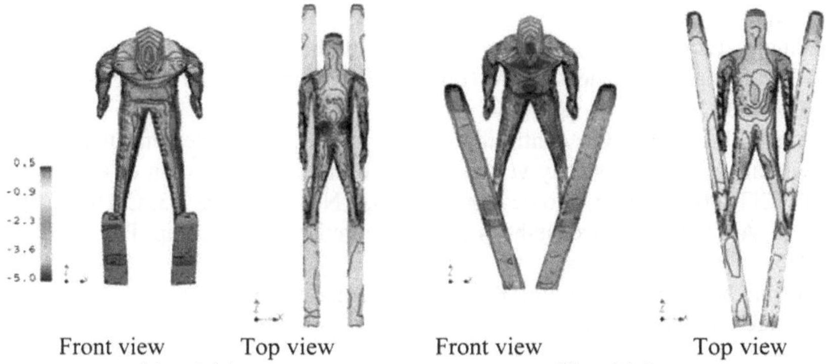

(a) t=0.04 s (b) t=0.25 s

Fig. 14. Comparison of pressure distributions along the surfaces of a ski jumper and skis in the parallel phase and in the V-shape phase.

5 Conclusions

In this study, complex and unsteady flows around an isolated 100 m runner and a ski-jumper were calculated by applying an advanced vortex method based on the Biot-Savart law. Since the method provides completely grid-free Lagrangian calculation of unsteady and vortical flows, dynamic motions of the athletes were easily introduced into the calculation by importing an instantaneous configuration of moving body expressed by a number of source panels at every moment of calculation steps.

From the calculation of flow around a 100 m runner during sprint running, it becomes clear that the aerodynamic drag force acting on the runner is about 20 N and it is always fluctuating according to interaction of flow separation around the body with individual motions of arms and legs.

From the calculation of flow around a ski jumper, it is clarified that the motion of skis to form the V-shape just after jumping brings significant increases in both lift force and drag force on the ski-jumper. This fact seems to suggest that optimization of timing of the ski movement and matching of the style of V-shape with the flight style of the ski jumper's body are very important to elongate the flight distance.

Finally, the present authors would like to conclude the present study by pointing that the advanced vortex method will become one of the most capable methods to contribute to the developments of fluid dynamics in sports science in the very near future.

References

Fukuda K. and Kamemoto, K. (2005) Application of a redistribution model incorporated in a vortex method to turbulent flow analysis. Proc. of ICVFM2005, Yokohama, Japan, Nov.21-23, pp 131-136.

Greengard, L. and Rohklin, V. (1987) A fast algorithm for particle simulations. J. Comp. Phys. 73, 325.

Kamemoto, K. (1995) On attractive features of the vortex methods. Computational Fluid Dynamics Review, ed. M.Hafez and K.Oshima, JOHN WILEY & SONS, pp 334-353.

Kamemoto, K. (2004) On contribution of advanced vortex element methods toward virtual reality of unsteady vertical flows in the new generation of CFD. Proc. of ENCIT2004, ABCM, Rio de Janeiro, Brazil, Nov.29 – Dec. 03, CIT04-IL04.

Leonard, A. (1980) Vortex methods for flow simulations. J. Comp. Phys. 37, pp 289-335.

Nakanishi, Y. and Kamemoto, K. (1992) Numerical simulation of flow around a sphere with vortex blobs. Journal Wind Eng. and Ind. Aero, Vol. 46 & 47, pp 363-369.

Ojima, A. and Kamemoto, K.(2001) Numerical simulation of unsteady flow through a horizontal axis wind turbine by a vortex method. Proc. of ICVFM2001, Istanbul, pp 173-180.

Ojima A. and Kamemoto, K. (2005) Numerical simulation of unsteady flows around a fish. Proc. of ICVFM2005, Yokohama, Japan, Nov.21-23, pp 96-101.

Sarpkaya, T. (1989) Computational methods with vortices - the 1988 Freeman scholar lecture. J. Fluids Engng., 111, pp 5-52.

Schmolzer B. and Muller W. (2005) Individual flight styles in ski jumping: results obtained during Olympic Games competitions. Journal of Biomechanics, Vol. 38, Issue 5, pp 1055-1065.

Sekioka Y., Suganuma F., Nagasawa M., Ogata M., Morita M., Kobayasi T. and Ohgusi K. (1990) Comparison of sprint running motions between representative sprinters; Mr. Hideaki Miyata and Mr. Shinji Aoto. Research report of athletic sports, Science Committee of Japan Association of Athletics Federations, 10, pp 170-174.

Uhlman, J.S. (1992) An integral equation formulation of the equation of motion of an incompressible fluid. Naval Undersea Warfare Center T.R., 10, 086.

Winkelmans, G. and Leonard, A. (1988) Improved vortex methods for three-dimensional flows", Proc. Workshop on Mathematical Aspects of Vortex Dynamics, pp 25-35, Leeburg, Virginia.

Wu, J.C. and Thompson, J.F. (1973) Numerical solutions of time-dependent incompressible Navier-Stokes equations using an integro-differential formulation. Computers & Fluids, Vol. 1, pp 197-215.

Chapter28

Rowing Velocity Prediction Program with Estimating Hydrodynamic Load Acting on an Oar Blade

Takeshi Kinoshita[1], Masaki Miyashita[1], Hiroshi Kobayashi[2], and Takanori Hino[2]

[1] Institute of Industrial Science, University of Tokyo, 4-6-1 Komaba, Meguro-ku, Tokyo, 153-8505, Japan
[2] National Maritime Research Institute, 6-38-1 Shinkawa, Mitaka, Tokyo, 181-0004, Japan

Summary. An oar blade behaves as unsteady hydrofoil and its load strongly depends on a reduced frequency of the rowing motion. The tank test is carried out using a circulating water channel to estimate the load.
Computational Fluid Dynamics (CFD) simulation of a rotating flat plate in uniform flow, which corresponds to the tank test in circulating water channel, reveals transient phenomena of flow field around the plate and explains the reason why the normal force coefficients strongly depend on the reduced frequency.
The program which simulates the rowing motion of a single scull is developed. The force that pulls the oar handle and the motion of the rower's centre of mass are input values for the simulation. The input values represent the rower's performance, that is, the rower's power and body movements. The results of the simulation are in good agreement with the *in-situ* experiments, and the method can be used as a VPP (Velocity Prediction Program).

Key words. rowing, oar blade, reduced frequency, velocity prediction program, CFD

1 Introduction

Although sport rowing has a long history, few engineering studies have been made on the subject. For a realistic simulation of rowing boat motion, it is essential to accurately evaluate hydrodynamic force acting on an oar blade and to fully and reasonably model the rower's body motion.
Fluid motion around an oar blade is transient and the hydrodynamic force acting on an oar blade should be unsteady. Both the relative speed of an oar blade to water and the angle of attack of an oar blade to the water are transiently changed during a stroke phase in which a rower pulls a handle of an oar to propel a boat.
In this study we evaluate the unsteady blade force from a "reduced frequency", as well as the relative speed of the oar blade to the water and the angle of attack of the oar blade to the water by using a circulating water channel. We carry out a simulation to estimate the boat speed and oar motion by solving energy equations for the hull, rower, and oars.

In order to know why the normal force coefficient so strongly depends on the reduced frequency, we carried out CFD simulation of the flow around the flat plate. Numerical results reveal unsteady flow field such as vortex shedding from side edge of the rotating plate.
Hayashi et al. (1990, 1991) modeled the rower's mass as a point mass. Hayashi et al. (1990, 1991) and Doi et al. (1999) input the oar motion to the simulation system. In reality the use of legs, back swing and arms are very important parameters of rowing motion, and they cannot be considered as a point mass. The oar motion is not an input, but an output of rowing. The input to the simulation should be the rower's motion and the inboard torque of the oar caused by pulling the oar handle. The inputs to the simulation are inboard torque (the force that pulls the oar handle) and the rower's motion; leg movement, back swing and arm movement, identified from a skilled rower. They represent the rower's performance, that is, the rower's power and body movements. The simulated boat speed and oar motion are compared with *in-situ* measurements and are in very good agreement, and the method can be used as a VPP (Velocity Prediction Program).

2 Blade Force Characteristics of Rowing

2.1 Blade Motion in the Water

Velocity of an oar blade V_B in the water is written on the space-fixed coordinate system as follows,

$$V_B = \dot{\theta} \times (l_{out} \boldsymbol{u}) + \dot{\boldsymbol{R}}_h \tag{1}$$

where $\dot{\boldsymbol{R}}_h$ is velocity of a hull. As shown in Fig. 1, the oar blade behaves as a hydrofoil in unsteady flow field. Flow velocity V_B and the angle of attack α transiently varies due to the advancing motion of the hull and the rotation of the oar.

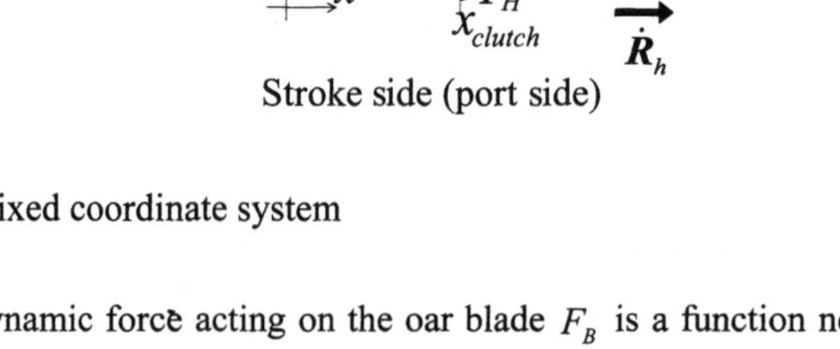

Fig. 1. Hull fixed coordinate system

The hydrodynamic force acting on the oar blade F_B is a function not only of V_B and α, but also a reduced frequency f_r. That is,

$$F_B = F_B(f_r, \dot{V}_B, \alpha) = F_B(f_r, \dot{R}_h, \theta, \dot{\theta})$$
$$\text{where} \quad \alpha = \tan^{-1}\left(\frac{u \times (-V_B)|_z}{u \cdot (-V_B)}\right), \quad f_r = \frac{2\pi c}{TU} \tag{2}$$

where T is representative time scale, which is the duration of the stroke phase in this case. A reduced frequency f_r is a ratio of typical speeds, the relative flow speed and the blade-tip rotational speed. f_r is identical with the Strouhal number, if we take T as the vortex shedding interval, and just a reciprocal of the Swimming number proposed by Tanaka et al.(1996) as a parameter of fishtail motion. The Keulegan-Carpenter (1958) number also has similar meaning.

2.2 Blade Force from *in-situ* Measurements of a Single Scull Motion in the Water

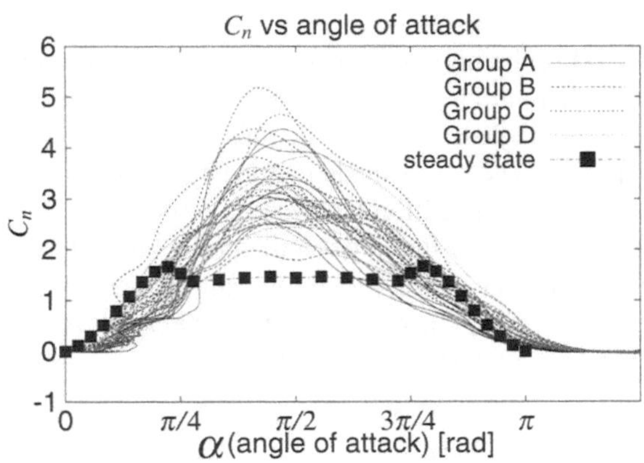

Fig. 2. Normal blade force coefficient vs the angle of attack

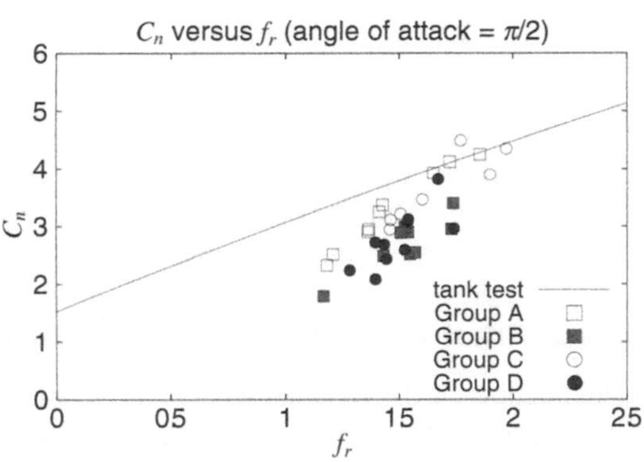

Fig. 3. Normal blade force coefficient $C_n(\pi / 2)$ vs reduced frequency

In-situ measurements of a single scull rowed by a skilled athlete were carried out. These items were measured:

- Two direction (normal and lateral) forces acting on the pivot axis of both sides of the crutch using small load cells specially designed for this study
- Bending moments of the inboard oars for both sides by strain gauges
- Oar swing angles on both sides by potentio meters
- Pushing force on the stretcher using small load cells specially designed for this study
- Acceleration of the hull by an accelerometer
- Mean hull speed by a propeller type velocity meter and integration of the acceleration

The normal component of F_B is normalized as follows,

$$C_n(\alpha) = \frac{F_{Bn}}{\frac{1}{2}\rho|V_B|^2 S} \tag{3}$$

where F_{Bn}, ρ, V_B, S are the normal component of F_B, the density of water, the velocity of the oar blade in the water, and the area of the oar blade respectively. Fig. 2 shows C_n versus the angle of attack. The measurements were carried about four different target speeds. The rower made every effort to keep the averaged hull velocity to designated target speed for each cases. Groups A ~ D show the different measurements of group where the target hull speeds are different as shown in Table 1. The normal force has a peak around the angle of attack $\alpha = \pi/2$, which is the middle of the stroke. In Fig. 2, data for steady state measurements with a fixed angle of attack are also sown as steady state.

Table 1. Mean hull speed of each Group

Group	A	B	C	D
Target speed	3.1[m/s]	3.3[m/s]	3.5[m/s]	3.7[m/s]
Measured actual speed	3.097[m/s]	3.394[m/s]	3.571[m/s]	3.698[m/s]

Fig. 3 shows $C_n(\alpha = \pi/2)$ versus the reduced frequency, and it clearly shows the dependency of C_n on the reduced frequency. The blade force is totally different from the one in a steady state. The "tank test" denotes the result of tank tests using a circulating water channel as explained in the following sections.

2.3 Blade Tests Using a Circulating Water Channel

To confirm the dependency on the reduced frequency of the blade force, we carried out forced rotating tests in a circulating water channel using a flat plate approximating a real oar blade. The water channel was 1.8[m] in width and 1.0[m] in depth. The Planar Motion Mechanism (PMM) was set on the channel with a flat plate of the same area and the same aspect ratio as the actual oar blade, and scale models, connected to the PMM through the force gauge, as shown in Fig. 4.

attached to PMM (Planar Motion Mechanism)

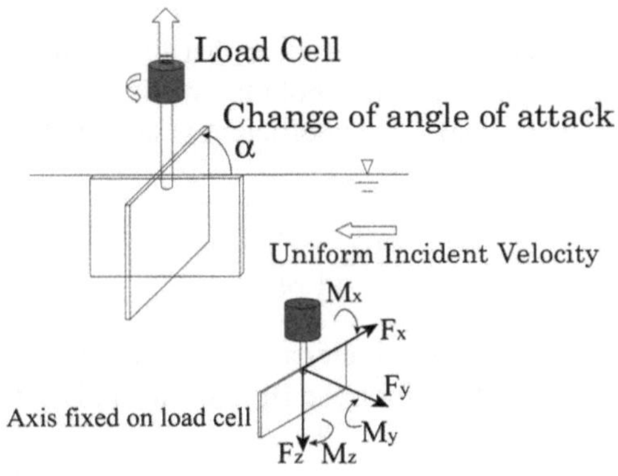

Fig. 4. Schematics of the set up

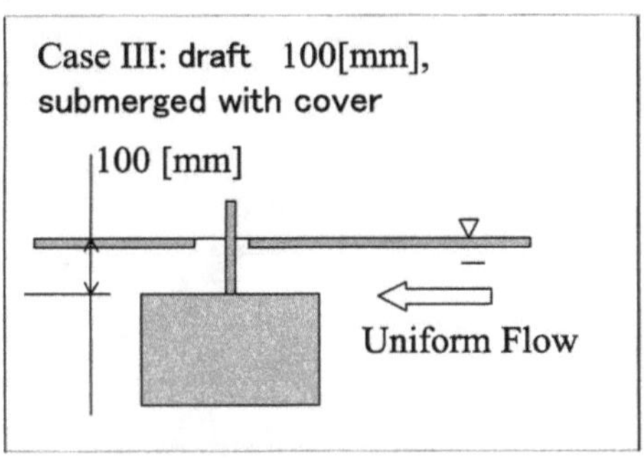

Fig. 5. Setu-up of the flat plate

Table .2 Parameters of the water channel tests

Parameter	Value, type of plate
Flow velocity[m/s]	0.8, 1.0, 1.2
Frequency of the input cosine signal to PMM[Hz]	0.10, 0.15, 0.20, 0.25, 0.30, 0.40, 0.50, 0.60
Type of flat plate (chord × span)[mm]	Type N: (200 × 350)
	Type S: (100 × 175)
	Type L: (300 × 525)

The parameters of this experiment are the flow velocity, the angular velocity of the rotation, and the scale ratio of the flat plate as shown in Table 2. The scale models (Type N, Type S and Type L) were used because of the limited ability of the flow velocity of the circulating water channel. Type N is of the same area as the oar blade.

The flat plate rotates from zero angle of attack until π using a part of cosine signal and the upper edge was 100[mm] under the water level and a flat plate which covers the water surface was placed, as shown in Fig. 5. Considerable ventilation could be observed in Case I, and even in Case II. We used the data of Case III in the following analysis.

From the measurement, it becomes clear that the normal force F_y of the plate is dominant among all forces and that both the vertical force F_z and tangential one F_x are negligible. The normal force coefficient is obtained by normalizing F_y as follows,

$$C_n = \frac{F_y}{\frac{1}{2}\rho U^2 S} \tag{4}$$

where U, S are uniform flow velocity and the area of the flat plate respectively. C_n versus the angle of attack for Case III is shown in Fig. 6.

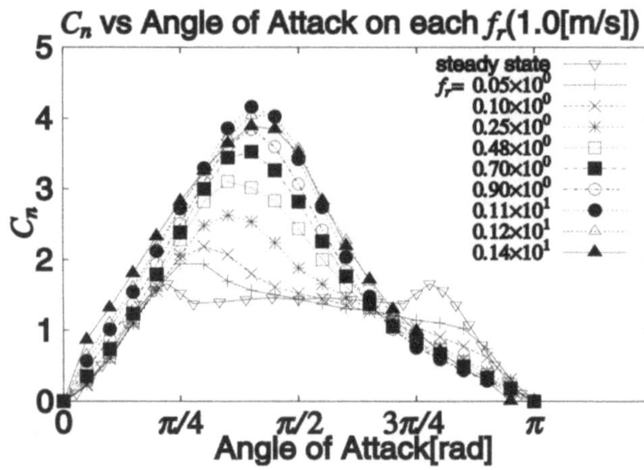

Fig. 6. Normal force coefficient vs. angle of attack (Case III)

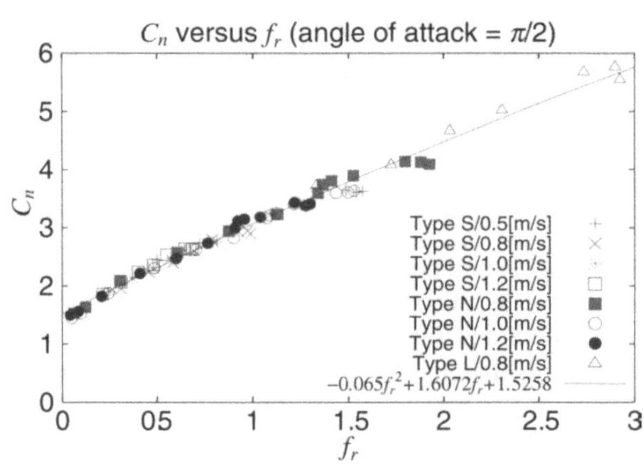

Fig. 7. Normal force coefficient vs. reduced frequency (Case III)

In Fig. 6, the normal force coefficients clearly show the difference from the steady state case. They are asymmetric with respect to the angle of attack and maximum values increase as the reduced frequency comes higher, although they are symmetric about $\pi/2$ in the case of the steady state. Fig. 7 shows $C_n(\alpha = \pi/2)$ versus the reduced frequency, and the dependence on the reduced frequency even in different scale models and at different flow speeds is clearly observed. The ventilation impedes the growth of the normal force. Here we can confirm the dependency on the reduced frequency of the blade force. The tank test data are compared with the *in-situ* data in Fig. 3 and are in qualitatively good agreement with each other. The tank test data can be utilized as an input of the rowing simulation in section 4 with some correction.

3 CFD Simulation of the Flow Around The Blade

3.1 Computational Conditions

In order to know why the normal force coefficient so strongly depends on the reduced frequency, and to obtain a tool to design better shape and camber of the blade, we carried out Computational Fluid Dynamics (CFD) simulation of the flow around the flat plate which rotates from zero angle of attack until π angle of attack in a uniform flow. This CFD simulation corresponds to the hydrodynamic force measurement in section 2. We use the unstructured grid based Navier-Stokes solver for computing turbulent viscous flows, called **SURF**(Hino, 1997 and Sato et al., 2003).

Table 3. Calculation conditions

Parameters		Value
Reynolds number		3.5×10^{5}
Froude Number		0
Reduced Frequency		0, 0.5, 1.0, 2.0
Number of cells (i*j*k)		69654(39*38*47)
Minimum spacing	i	5.0×10^{-3}
	j	1.4285×10^{-2}
	k	7.5×10^{-6}

i: chordwise direction , j: spanwise direction , k: normal direction from plate surface to outer boundary

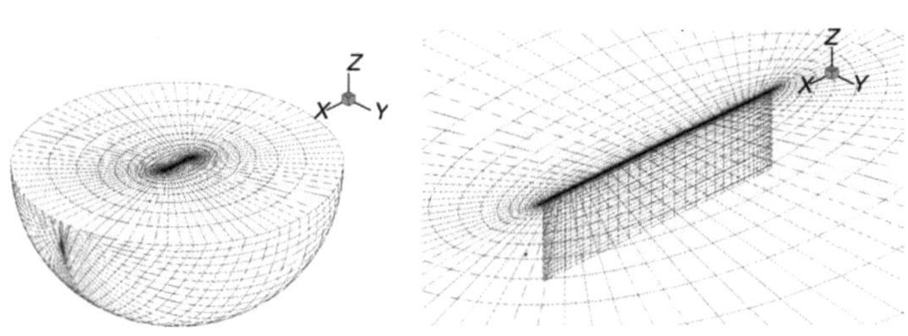

Fig. 8. Computational grid

SURF is developed at National Maritime Research Institute (NMRI).. The governing equations are the three dimensional Reynolds averaged Navier-Stokes equations for incompressible flows. Coupling between pressure and velocity is made by artificial compressibility approach. Spatial discretization is based on a finite volume method for an unstructured grid. In the present approach, a cell centered layout is adopted in which flow variables are defined at the centroid of each cell and a control volume is a cell itself. Also it has the capability to compute unsteady flows.

Computational conditions and grid sizes are shown in Table 3 and the computational grid is shown in Fig. 8. In the computation, no turbulence model is used. An O-O type grid is adopted for this simulation. A Solution domain is a hemispheric shape which covers a half side of whole domain for spanwise direction since a flow filed can be assumed to be symmetric due to a rectangular planform of the flat plate. An aspect ratio of the flat plate is 0.57 which is the same as the flat plates shown in Table 2 tested in section 2.3. A radius of the hemispheric domain is 5 times of chord length of the flat plate.

The whole grid rotates around the z -axis from angle 0 to π [rad] to the uniform flow as the time step advances. The number of time steps for the rotation is 8000 in each computational case. Therefore a time step is defined according to the reduced frequency as follows:

$$\Delta t = \frac{T}{8000} = \frac{\pi}{4000 f_r} \tag{5}$$

where $T(=2\pi / f_r)$ is period in which the angle of attack varies from 0 to π [rad]. T is normalized by uniform flow speed and the chord length of the flat plate.

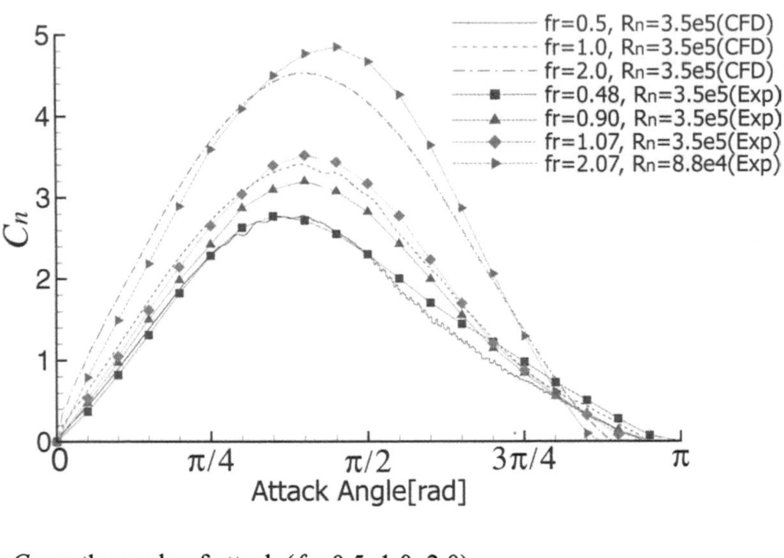

Fig. 9. C_n vs the angle of attack (f_r =0.5, 1.0, 2.0)

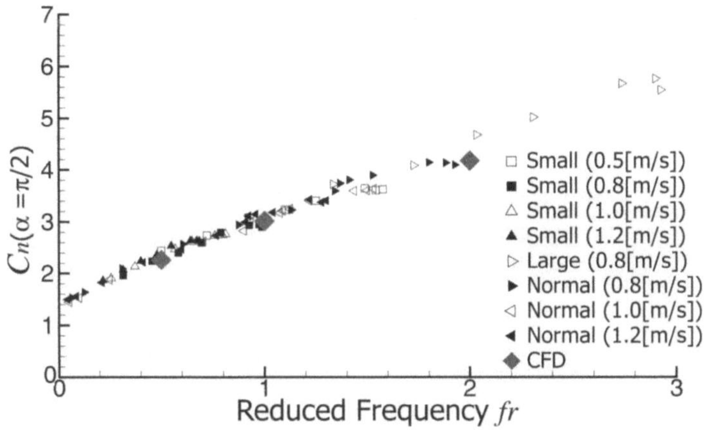

Fig. 10. C_n vs reduced frequency

3.2 Computational Results and discussion

The numerical results of the normal force coefficients C_n versus the angle of attack are shown in Fig. 9 in comparison with experimental data of tank test in the circulating water channel. Fig. 10 shows $C_n(\alpha = \pi/2)$ versus the reduced frequency also compared with measured data. The computations and measurements show very good agreement. It means that numerical simulation can solve unsteady flow field around a rotating plate and can estimate the load that acts on the plate reasonably.

In order to know the effect of the reduced frequency to unsteady hydrodynamic force in detail, development of vortex and pressure distributions are examined. For example, in case of f_r =1.0, which approximately corresponds to the reduced frequency in the real rowing condition, pressure distributions on an equi-vorticity ($|\omega|$=30) surface and on a surface of the plate at a suction side are shown in Fig. 11. Pressure and vorticity are normalized by uniform flow speed and the chord length of the flat plate. From Fig. 11, we can see that

1. long "side vortex" is created along the bottom edge of the blade (see (a))
2. the "side vortex" is growing along the chord of the blade, streaming to the tip, at the same time "tip vortex" created along the tip (leading edge) (see (b,c))
3. "side vortex" is streaming toward the tip and combined with the "tip vortex" (see (d))
4. the combined vortex is shed from the plate and streaming out. (see (e,f))

It means that the "side vortex" created along the bottom edge and the "tip vortex" created at the tip produce negative pressure on the surface of the blade at the suction side. The former contribution is significant to the normal force of the blade.

The size and strength of the "side vortex" strongly depend on the reduced frequency.

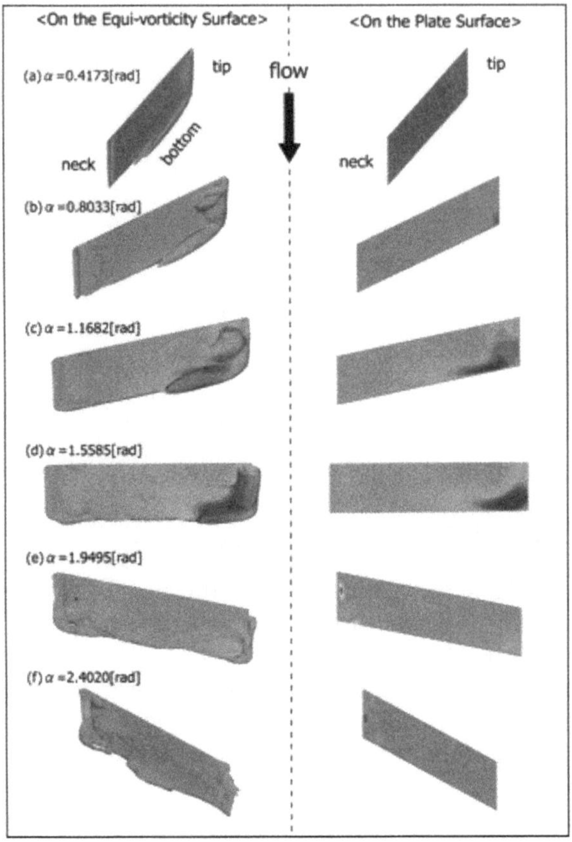

Fig. 11. Pressure distribution on equi-vorticity surface ($|\omega|$=30) and on the plate surface at suction side

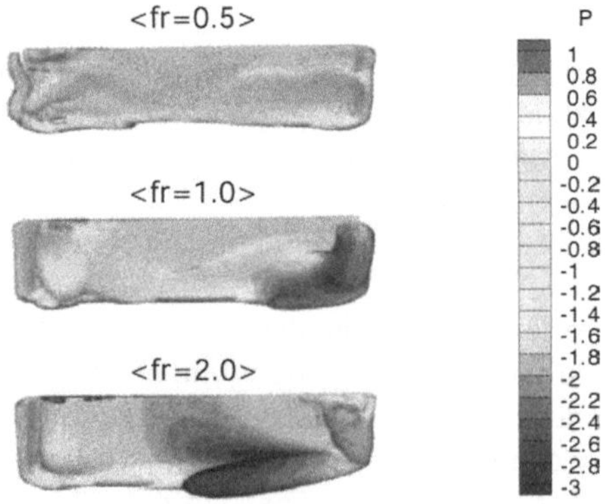

Fig. 12. Pressure distribution on equi-vorticity surface($|\omega|$=30), ($\alpha = 1.5585[rad]$)

The pressure distributions on an equi-vorticity ($|\omega|$=30) surface at $\alpha = \pi/2$ in the different reduced frequencies are shown in Fig. 12. The larger reduced frequency yields the longer and stronger "side vortex". It is revealed by the numerical simulation that the development and shedding of three dimensional vortices play a dominant role in this phenomenon.

4 Hull Speed Prediction for Single Scull Rowing

The equations of motion of a hull, a rower and an oar on the space fixed coordinate system are as follows (Kinoshita et al., 2004),

$$\begin{aligned} M_h\ddot{\boldsymbol{R}}_h &= 2\boldsymbol{F}_C + \boldsymbol{F}_S + \boldsymbol{D} \\ M_r\ddot{\boldsymbol{R}}_r &= -\boldsymbol{F}_S - 2\boldsymbol{F}_H \\ 2M_o\ddot{\boldsymbol{R}}_o &= 2\boldsymbol{F}_B - 2\boldsymbol{F}_C + 2\boldsymbol{F}_H \\ 2I\ddot{\boldsymbol{\theta}} &= 2(l_{out}\boldsymbol{u})\times\boldsymbol{F}_B + 2(-l_{in}\boldsymbol{u})\times\boldsymbol{F}_H \end{aligned} \tag{6}$$

Subscript h, r, o, B, H, C and S denote hull, rower, oar, oar blade, oar handle, crutch, and stretcher fixed to the hull, respectively. (see Fig. 13) As shown in Fig. 1, θ is the swing angle of the oar.Only longitudinal motion of the hull is considered here, and vertical motion is neglected. It means this simulation includes surging, but not pitching and heaving. This is because heaving and pitching motions are very slight in comparison to longitudinal motion. Math of the hull M_h includes added mass due to surging.
The increase of the kinetic energy can be obtained integrating Equations (6) as follows,

$$\begin{aligned} dK &= d\left(\frac{1}{2}M_h|\dot{\boldsymbol{R}}_h|^2\right) + 2d\left(\frac{1}{2}M_o|\dot{\boldsymbol{R}}_o|^2\right) + d\left(\frac{1}{2}M_r|\dot{\boldsymbol{R}}_r|^2\right) + 2d\left(\frac{1}{2}I|\dot{\boldsymbol{\theta}}|^2\right) \\ &= 2(\boldsymbol{F}_B - \boldsymbol{F}_C + \boldsymbol{F}_H)\cdot d\boldsymbol{r}_o + (-\boldsymbol{F}_S - 2\boldsymbol{F}_H)\cdot d\boldsymbol{r}_r + \boldsymbol{D}\cdot d\boldsymbol{R}_h \\ &\quad + 2\boldsymbol{F}_B\cdot d\boldsymbol{R}_o + 2d\boldsymbol{\theta}\cdot\{(l_{out}\boldsymbol{u})\times\boldsymbol{F}_B\} + 2d\boldsymbol{\theta}\cdot\{(-l_{in}\boldsymbol{u})\times\boldsymbol{F}_H\} \\ &= 2(\boldsymbol{F}_B - \boldsymbol{F}_C + \boldsymbol{F}_H)\cdot d\boldsymbol{r}_o + (-\boldsymbol{F}_S - 2\boldsymbol{F}_H)\cdot d\boldsymbol{r}_r + \boldsymbol{D}\cdot d\boldsymbol{R}_h \\ &\quad + 2\boldsymbol{F}_B\cdot d\boldsymbol{R}_B + 2d\boldsymbol{\theta}\cdot\{(-l_{in}\boldsymbol{u})\times\boldsymbol{F}_H\} \end{aligned} \tag{7}$$

$$\text{where}\quad \begin{aligned} d\boldsymbol{r}_r &= d\boldsymbol{R}_r - d\boldsymbol{R}_h \\ d\boldsymbol{r}_o &= d\boldsymbol{R}_o - d\boldsymbol{R}_h \\ d\boldsymbol{R}_B &= d\boldsymbol{R}_o + d\boldsymbol{\theta}\times(l_{out}\boldsymbol{u}) \end{aligned}$$

where $d\boldsymbol{r}_r$, $d\boldsymbol{r}_o$ and $d\boldsymbol{R}_B$ are movement of the rower, movement of the oar center of gravity on the hull-fixed coordinate system, and the blade movement in the water, respectively. The Equation (7) means that the total kinetic energy increase consists of following terms:

- $(\boldsymbol{F}_B - \boldsymbol{F}_C + \boldsymbol{F}_H)\cdot d\boldsymbol{r}_o$: work caused by the motion of the center of gravity of the oar
- $(\boldsymbol{F}_S - 2\boldsymbol{F}_H)\cdot d\boldsymbol{r}_r$: work caused by the motion of the center of gravity of the rower
- $\boldsymbol{D}\cdot d\boldsymbol{R}_h$: work spent by hull resistance
- $\boldsymbol{F}_B\cdot d\boldsymbol{R}_B$: slip loss of the oar blade
- $d\boldsymbol{\theta}\cdot\{(-l_{in}\boldsymbol{u})\times\boldsymbol{F}_H\}$: input of the inboard torque by pulling the oar handle

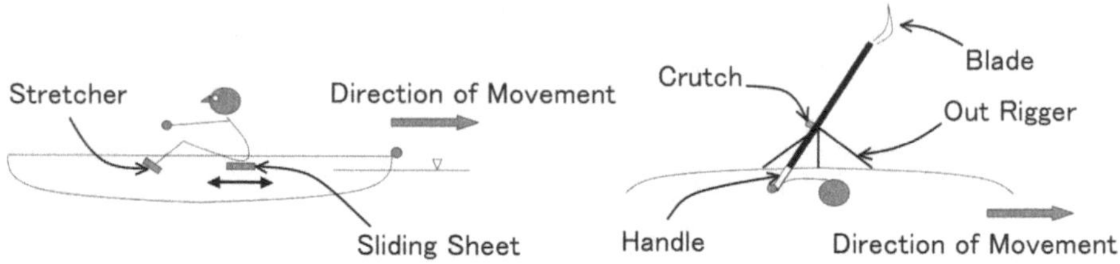

Fig. 13. Constitution of rowing boat

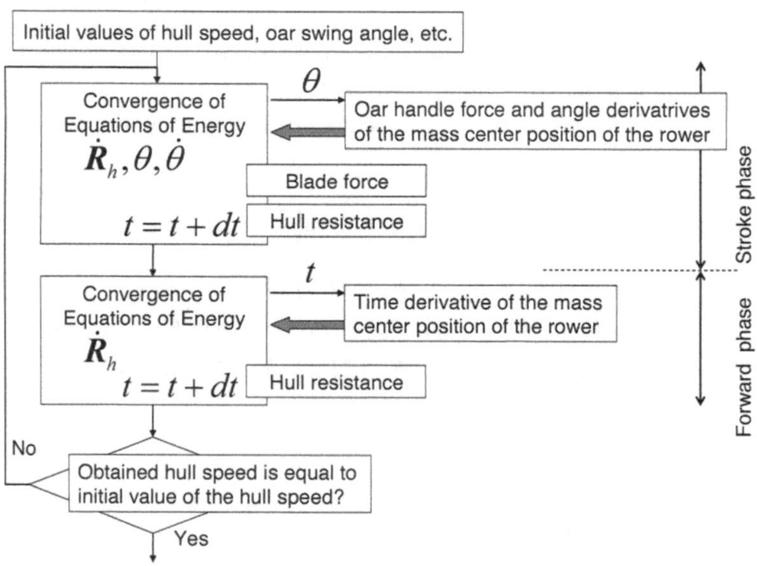

Fig. 14. Simulation flowchart

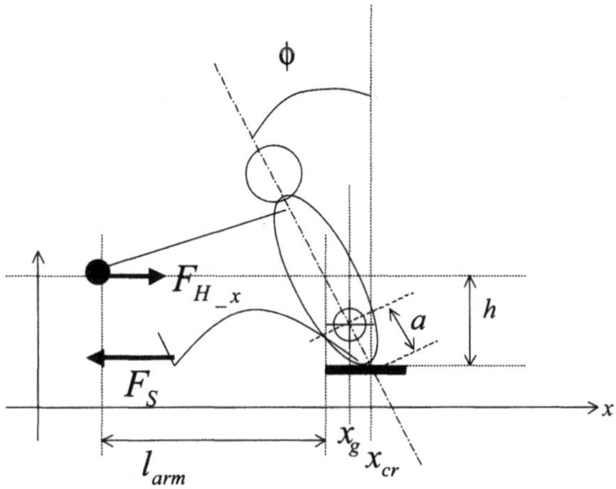

Fig. 15. Definition of variables of the rower's position

Stiffness of the oar shaft is taken into account by the beam theory. The rowing rate and the range of the oar swing are also constrained. The load acting on the oar blade can be estimated by using the results of tank test in section 2.3. The flow chart of the simulation is shown in Fig. 14.

The simulation starts from the catch point with the initial condition of the hull speed, the oar swing angle, and the swing angular velocity. The time interval *dt* was set as 0.01 sec. The stroke finishes when the oar swing angle reaches the constrained angle of the rowing range. For the "forward" phase, which means after the finish of the stroke until the next catch, no hydrodynamic force acts on the oar blade. Then we drop the terms relating θ, and obtain the hull speed. The forward phase period is determined from the constrained rowing rate and the simulated stroke phase period. The final hull speed of the forward phase is compared with the initial value of the hull speed at the catch point. If they do not coincide with each other, we continue the simulation until they coincide. In the simulation we need the following input data. For the stroke phase, the oar handle force F_H ($F_H \cdot l_{in}$ is the inboard torque) and the angle derivative of the mass center of the rower $\frac{dx_g(\theta)}{d\theta}$ are given as functions of θ for the input data as well as the hull resistance. For the forward phase, the time derivative of the mass center of the rower and the hull resistance are given.

The seat position x_{cr}, the stooping angle φ, and the shortened arm length l_{arm} should hold the following equation on the hull fixed coordinate system (See Fig. 15).

$$x_{crutch} + l_{in} \sin\theta + l_{arm}(\theta) = x_{cr}(\theta) + h\tan(\varphi(\theta)) \tag{9}$$

where x_{crutch} is position of the crutch and h is the height of the handle. The angle derivative of the mass center of the rower x_g is

$$\frac{dx_g(\theta)}{d\theta} = \frac{dx_{cr}(\theta)}{d\theta} + a\cos\frac{d\varphi(\theta)}{d\theta} \tag{10}$$

where *a* is the height of the rower's center of gravity from the seat.
The simulated and measured hull speeds and oar swing angles are compared. Very good agreement are obtained for all cases, Groups A ~ D. Results are shown in Fig. 16 and Fig. 17 for Group A.

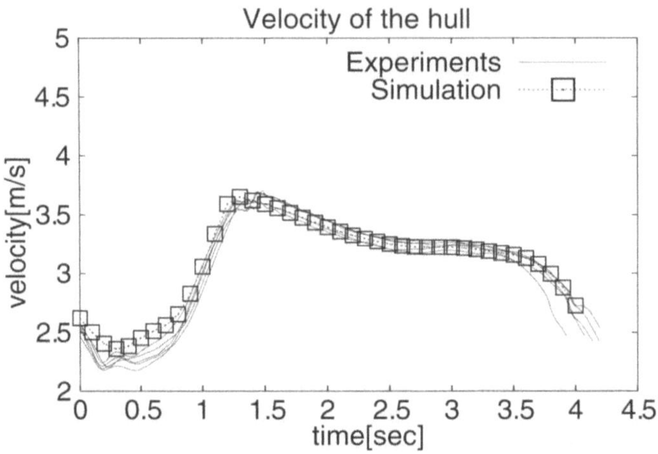

Fig. 16. Comparison of hull speed: Group A

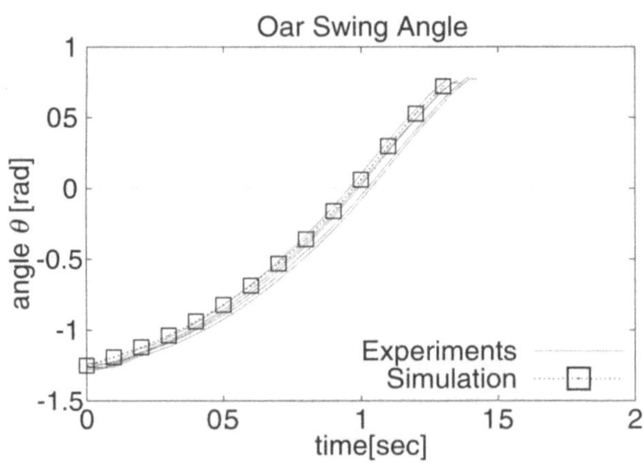

Fig. 17. Comparison of oar swing angle: Group A

5 Conclusions

The blade force characteristics of rowing are clarified by both experiments in the circulating water channel and CFD simulation. The normal force coefficient of the blade strongly depends on the reduced frequency. It is because that the longer and stronger "side vortex" of the blade is formed at the higher reduced frequency.

A VPP (Velocity Prediction Program) for a rowing boat is developed with input values of the force that pulls the oar handles and the motion of the rower's centre of mass.
It is proved that CFD simulation has capability to solve fluid motion around the oar blade and VPP can predict the motion of the hull and the oar. It will be possible to obtain optimal shape of the oar blade by combining CFD simulation and VPP. To achieve this object, an effect of free surface should be taken into account in the CFD simulation, which is not considered in the present simulation. Reynolds number also affects the hydrodynamic load significantly and it should be studied as well in the next step.

6 Acknowledgements

The authors wish to express their gratitude to Mr. Koutarou Horiuchi, Human and Solar Power Boat Association and Mr. Tadashi Abe, Japan Rowing Boat Association for their cooperation of it *in-situ* measurement. They are also grateful to Dr. Nobuyuki Hirata, National Maritime Research Institute for helpful cooperation in CFD simulation.

References

DOI, Y. and UEDA, T. and MORI, K. and NINOMIYA, S., "Study on Rowing Simulation and Its Application to Evaluate Oar Size and Rowing Pattern", "Journal of The Society of Naval Architects of Japan, Vol.186, pp.89-96", 1999.

HAYASHI, I. and FUKUDA, J. and YAMADA, H., "Effects of oar shape on thrust of a shell four in the numerical model, 1st report", "Human Science, Vol.3, No.2, Synthetic Human Research Center, Waseda University", (1990).

HAYASHI, I. and FUKUDA, J. and YAMADA, H., "Effects of oar shape on thrust of a shell four in the numerical model, 2nd report", "Human Science, Vol.4, No.2, Synthetic Human Research Center, Waseda University", 1991.

Hino T., A 3D Unstructured Grid Method for Incompressible Viscous Flows, Journal of the Society of Naval Architects of Japan, Vol.182, pp.9 --15, 1997.

Keulegan, G. H. and Carpenter, L. H., Forces on Cylinders and Plates in an Oscillating Fluid, J. res. National Bureau of Standards, 1958.

Kinoshita. T. and Kobayashi, H. "Improving rower motion and rowing equipment by using rowing velocity prediction program with estimating hydrodynamic load acting on an oar blade," International Journal of Small Craft Technology, vol. 146, RINA, pp16-26, 2004.

Sato, Y. and Hino, T. "Estimation of Thrust of Swimmer's Hand Using CFD," Proceedings of Second International Symposium on Aqua Bio-Mechanisms, 2003

TANAKA, I. and NAGAI, M., "Hydrodynamics of Resistance and Propulsion", Ship and Ocean Foundation, 1996.

Chapter 29

Analysis of Breast, Back and Butterfly Strokes by the Swimming Human Simulation Model SWUM

Motomu Nakashima[1]

[1] Department of Mechanical and Environmental Informatics, Graduate School of Information Science and Engineering, Tokyo Institute of Technology, 2-12-1 Ookayama, Meguro-ku, Tokyo 152-8552, Japan

Summary. In the preceding study, the swimming human simulation model "SWUM" has been developed by the authors, and the analysis of standard six beat crawl stroke has been conducted. In this paper, the analyses of three strokes, that is, breast, back and butterfly strokes were conducted. The swimming motion, velocity in the propulsive direction of the human body, and the fluid force in the propulsive direction acting on the upper and lower limbs were respectively shown for those three strokes. In addition, the stroke length and the propulsive efficiency of the four strokes were investigated by the simulation. It was found that the stroke lengths of the crawl, back and butterfly strokes in the simulation agree significantly with the actual values, and that of the breaststroke becomes relatively smaller. It was also found that the propulsive efficiency of the crawl, back and butterfly strokes becomes around 0.2, although that of the breaststroke becomes only 0.036.

Key words. Swimming, Simulation, Fluid Force, Propulsive Efficiency, SWUM

1 Introduction

The authors have recently developed a simulation model for a self-propelled swimmer "SWUM" (SWimming hUman Model), which can represent the dynamics of the whole body and which has the potential to be a widely used tool for the analysis of various mechanical problems in human swimming (Nakashima et al. 2007). In this model, the fluid force is considered as simplified modeled force, and the equations of motion of the self-propelled swimmer's body in the translational and rotational directions are solved by the time integration. As inputs, the body geometry and relative body motion (joint motion) are given. As outputs, the swimming speed, rolling and yawing motions, joint torque and so on are obtained. An analysis of the six beat crawl stroke has been already conducted, in which the joint motion data was input based on a video taken of an actual athlete swimming. It was found that the swimming speed obtained in the simulation agrees significantly with that of the video, indicating the validity of the simulation model. The authors have also investigated detailed characteristics of the six beat crawl, that is, the contribution of each fluid force component and each body part to the thrust, and computed the propulsive efficiency (Nakashima 2007). In this study, investigation similar to that conducted on the crawl stroke was performed on the other three strokes; those being: breast, back and butterfly strokes. In addition, comparison among the four strokes was made from the viewpoint of mechanics.

2 Analysis Method

2.1 Overview of SWUM

In SWUM, the relative body motion as the joint angles is given for the human body, which is modeled as rigid link segments. The absolute motion of the whole body is computed considering the unsteady fluid force acting on the human body. The human body is represented by a series of 21 truncated elliptical cone segments. With respect to the fluid force, the inertial force due to the added mass of the fluid, the drag force in the normal and tangential directions, and buoyancy are taken into account. In order to compute the inertial force due to the added mass and the drag forces in the normal and tangential directions, fluid force coefficients are invoked. These coefficients have been determined by an experiment in which a model of a human limb was flapped in the water, measuring its motion and fluid force. The details are described in the previous paper (Nakashima et al. 2007).

2.2 Analysis Condition

The joint motions as input data for the simulation model were determined based on videos of 'model swimming' (IPA 2004), which were produced by the Japan Swimming Federation and open to the public for educational purposes. Note that the swimming motions in the videos are not full strength competitive ones because of their educational purposes. One cycle of the swimming motion was divided into 18 frames. The joint angles at each frame were manually determined so that the body position in the video at the corresponding frame becomes as close to that in the simulation as possible. The given number of degrees-of-freedom of the joint rotations for breast stroke was 34 ; for back, 28 ; and for butterfly, 35.

All the quantities in the simulation were computed in a system normalized by time, length, and density. Their reference values are, respectively, stroke cycle, swimmer's stature, and the water density. The body geometry was set as that of an average 20-29 year-old Japanese male. The body height and weight are 1.705 m and 64.9 kg, respectively. The hydrostatic lift, which is defined as the difference between the net buoyancy and body weight, is 22.5 N (2.3 kgf). According to Chatard et al. (1990), the hydrostatic lift for swimmers are 17~29 N (1.7~3.0 kgf). The value of 22.5 N in the present model is almost at the center of that range. The number of divisions of each truncated elliptical cones for longitudinal and circumferential directions were, respectively, 10 and 36. The time step was set as 0.002, that is, 500 steps for one stroke cycle.

3 Results and Discussion

3.1 Breaststroke

Figure 1 shows the simulation results of the swimming motion for one cycle after nine stroke cycles, that is, after the effect of initial condition disappears. The water surface is colored with dark and pale colors alternatively. The area width of one color represents half of the stroke length. Therefore, the total width of one set of dark and pale areas corresponds to the stroke length. The dark sticks from each of the swimmer's body parts represent the direction and magnitude of the fluid force, except buoyancy. Figure 2 shows the velocity of the center of mass for the swimmer's body in the propulsive direction. Figure 3 shows the fluid force acting on the right upper and lower limbs in the same direction. Note that the negative value in these graphs corresponds to the positive propulsive speed and thrust. From these figures, it is found that the swimmer's body first accelerates due to the thrust generated by the hand at t = 9.3, and accelerates again significantly due to the large thrust generated by the kick of the legs. For the breaststroke, it is generally known that the propulsive velocity has two positive major peaks due to the upper and lower limbs. The peak value due to the lower limb (kick) is known to be almost equal to or slightly greater than that due to the upper limb (hand stroke) (Maglischo 2003). However, the peak due to the lower limb in the simulation at t = 9.7 is smaller than that due to the upper limb at t = 9.4. One possible reason for this inconsistency may be that the thrust generated by the kick is underestimated in the simulation due to the modeling error for the fluid force.

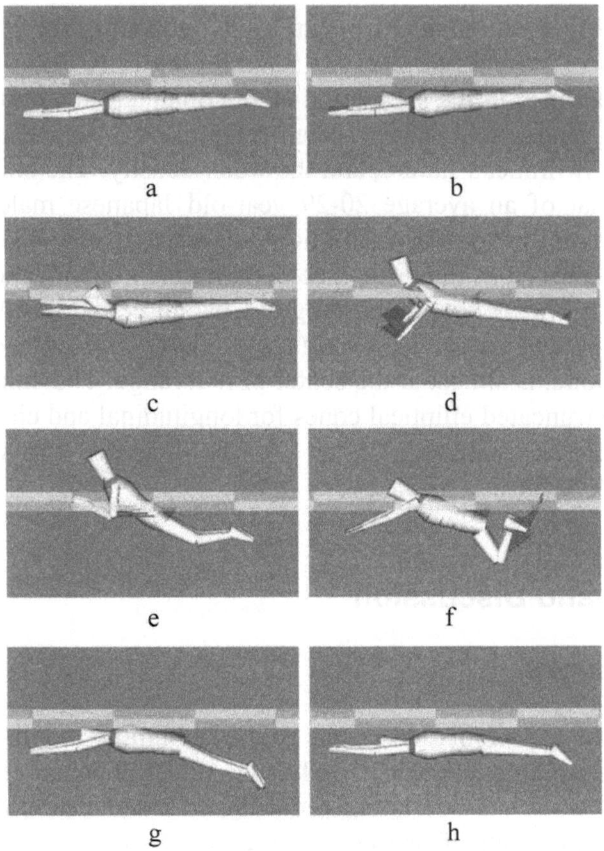

Fig. 1. Simulation results of swimming motion for one cycle from the side view (Breaststroke). a: $t = 9.0$. b: $t = 9.125$. c: $t = 9.25$. d: $t = 9.375$. e: $t = 9.5$. f: $t = 9.625$. g: $t = 9.75$. h: $t = 9.875$.

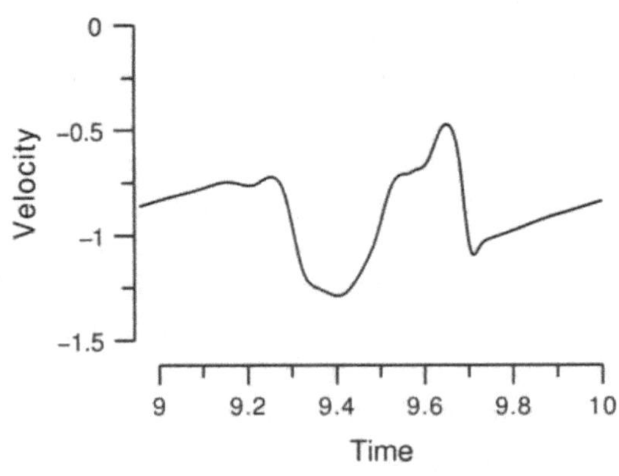

Fig. 2. Velocity of center of mass in propulsive direction (Breaststroke).

With respect to the fluid force components to the thrust, it is found that the thrust of the upper limb is mainly generated by the normal drag force, and the thrust of the lower limb is generated by the inertial force due to the added mass of the fluid as well as the normal drag force. It is also found that both the upper and lower limbs produce negative thrust at $t = 9.5$ and 9.6, respectively, that is, at the recovery motions for them. In particular, the negative thrust by the upper limbs around $t = 9.5$ seems to be unnaturally large. This causes deceleration during $t = 9.4$~9.6, resulting the reduction of the averaged swimming speed.

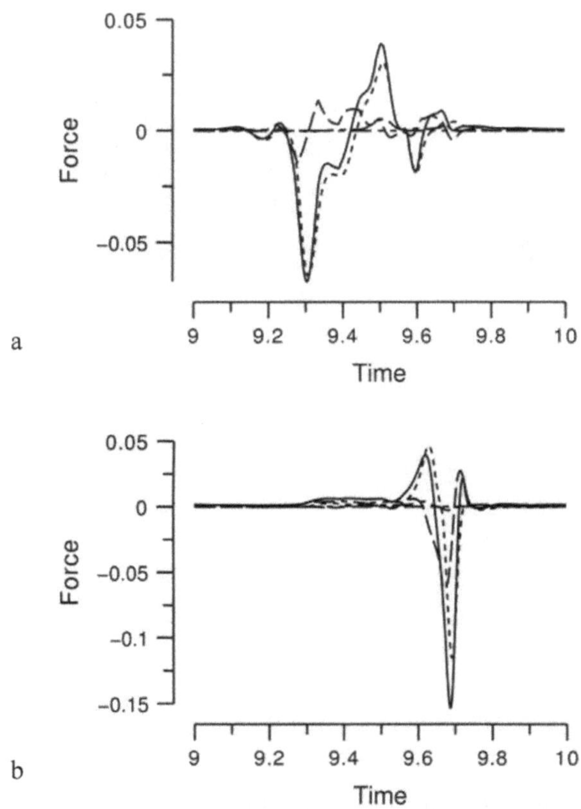

Fig. 3. Fluid force acting on right limbs in propulsive direction (Breaststroke). **a:** upper limb. **b:** lower limb. Solid line: total of all components. Long-dashed line: inertial force due to added mass. Short-dashed line: normal drag. Dot-dashed line: tangential drag. Two-dot-dashed line: buoyancy.

3.2 Backstroke

Figure 4 shows the simulation results of the swimming motion for a half cycle. Figure 5 shows the velocity of the center of mass for the swimmer's body in the propulsive direction. Figure 6 shows the fluid force acting on the right upper and lower limbs in the same direction. It is found that the thrust by the hand is large, the same as in the crawl stroke (Nakashima 2007), and it reaches its maximum, due to the normal drag force, at $t = 9.3$, that is, at the hand's 'push' motion. Immediately following, the velocity reaches its maximum. With respect to thrust of the lower limb, three major negative peaks appear in the flutter kick, and the primary components of them are found to be the inertial force due to the added mass and the normal drag.

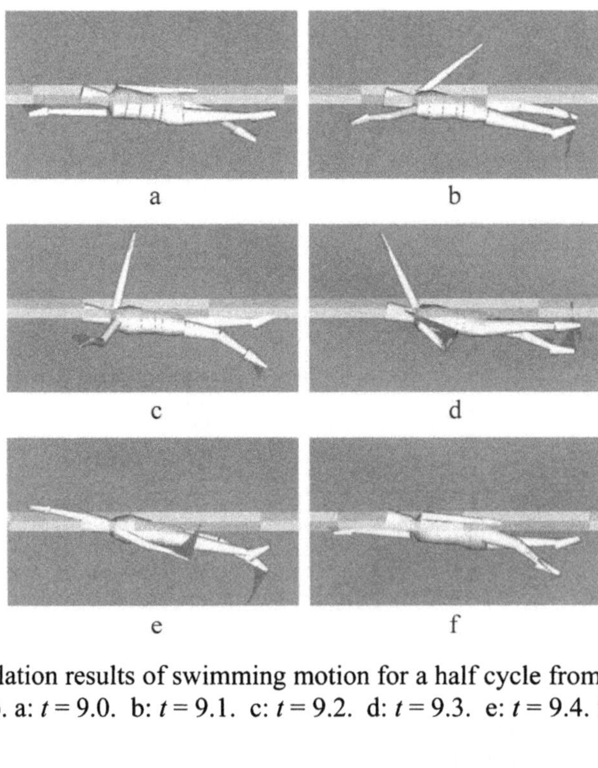

Fig. 4. Simulation results of swimming motion for a half cycle from the side view (Backstroke). a: $t = 9.0$. b: $t = 9.1$. c: $t = 9.2$. d: $t = 9.3$. e: $t = 9.4$. f: $t = 9.5$.

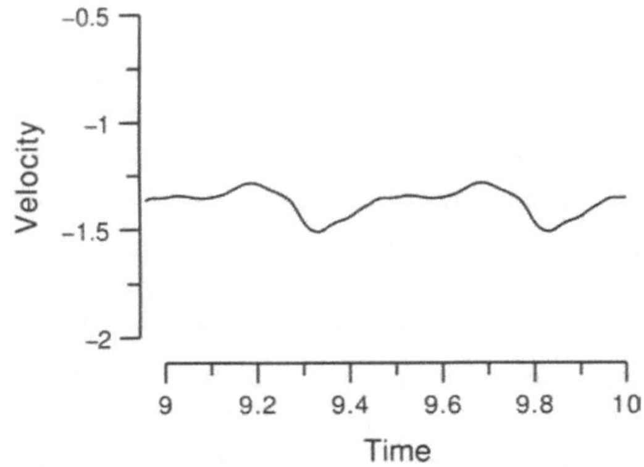

Fig. 5. Velocity of center of mass in propulsive direction (Backstroke).

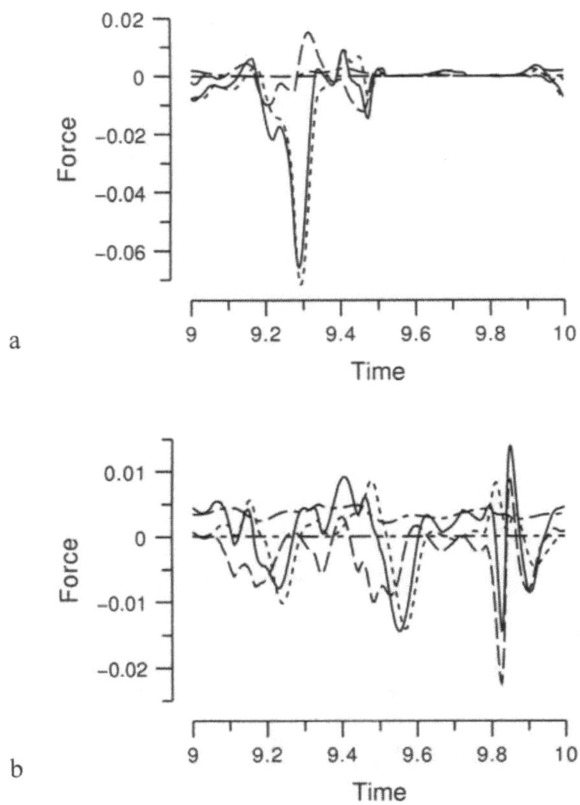

Fig. 6. Fluid force acting on right limbs in propulsive direction (Backstroke). **a:** upper limb. **b:** lower limb. *Solid line*: total of all components. Long-dashed line: inertial force due to added mass. Short-dashed line: normal drag. Dot-dashed line: tangential drag. Two-dot-dashed line: buoyancy.

3.3 Butterfly Stroke

Figure 7 shows the simulation results of the swimming motion for one cycle. Figure 8 shows the velocity of the center of mass for the swimmer's body in the propulsive direction. Figure 9 shows the fluid force acting on the right lower and upper limbs in the same direction. With respect to the upper limb, it is found that the large thrust is generated by the hand's motion, which starts from $t = 9.2$, reaches its peak at $t = 9.55$; with the velocity reaching its peak concomitantly. With respect to the lower limb, it is found that the thrust is generated by the weaker kick at $t = 9.45$, and by the stronger one at $t = 9.9$. It is also found that almost all the thrust by the upper limb is due to the normal drag force, and the thrust by the lower limb is generated by the inertial force due to added mass and the normal drag force.

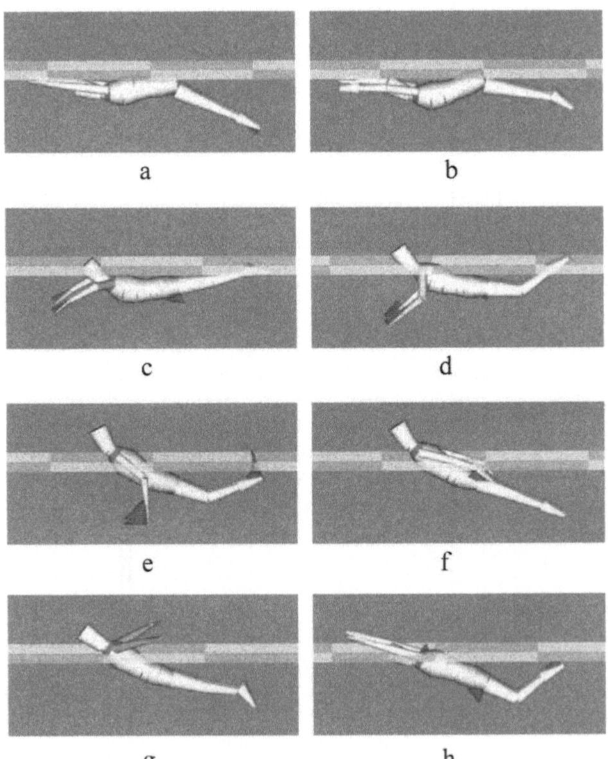

Fig. 7. Simulation results of swimming motion for one cycle from the side view (Butterfly stroke). **a:** $t = 9.0$. **b:** $t = 9.125$. **c:** $t = 9.25$. **d:** $t = 9.375$. **e:** $t = 9.5$. **f:** $t = 9.625$. **g:** $t = 9.75$. **h:** $t = 9.875$.

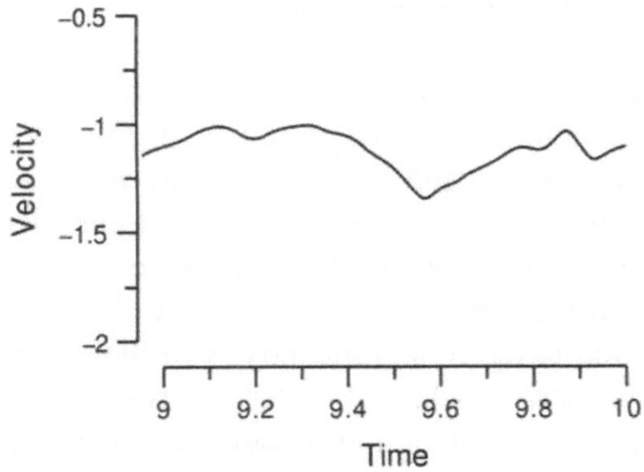

Fig. 8. Velocity in propulsive direction (Butterfly stroke).

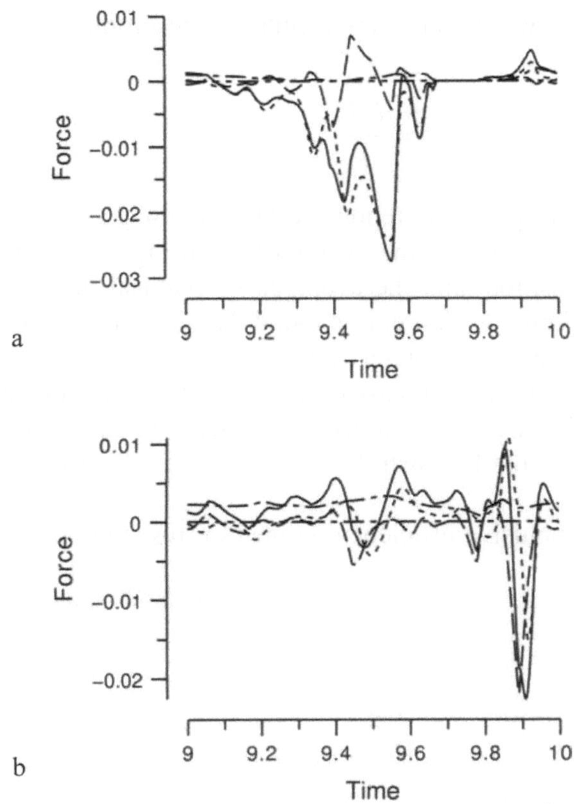

Fig. 9. Fluid force acting on right limbs in propulsive direction (Butterfly stroke). **a:** upper limb. **b:** lower limb. Solid line: total of all components. Long-dashed line: inertial force due to added mass. Short-dashed line: normal drag. Dot-dashed line: tangential drag. Two-dot-dashed line: buoyancy.

3.4 Comparison Among Four Strokes

Based on the analyses in the previous sections, the comparison between the four strokes including the crawl is made in this section. Figure 10 shows the results of comparison for the normalized stroke length, which represents the distance traveled in one stroke cycle divided by the swimmer's stature. The actual values of the normalized stroke length, which were estimated from the video, are also shown in the figure. The simulation results are found to agree significantly with the actual values for the three strokes of the crawl, back, and butterfly strokes. This agreement indicates the validity of the simulation model. However, the simulation result for the breaststroke is considerably smaller (about 35%) than the actual value. Two possible reasons for this discrepancy are that the thrust due to the kick is underestimated, and that the drag of the upper limbs in the recovery motion reduces the stroke length, as stated in section 3.1.

Next, Fig. 11 shows the analysis results of the propulsive efficiency; which is defined by:

$$\eta = \frac{UT}{P} \tag{1}$$

where U, T, and P respectively represent the swimming speed averaged in one cycle, the drag acting on the swimmer's body towed at the speed U with the gliding position, and the averaged mechanical power consumed by the swimmer. From the figure, it is found that the values of η are about 0.2 for the crawl, back, and butterfly strokes, whereas it is considerably smaller (about 0.036) for the breaststroke. Even if it is acknowledged that the stroke length in the simulation becomes smaller as shown in Fig. 10, it can still be said that the breaststroke is less efficient than the other three strokes.

Note that the propulsive efficiency defined in this study corresponds to

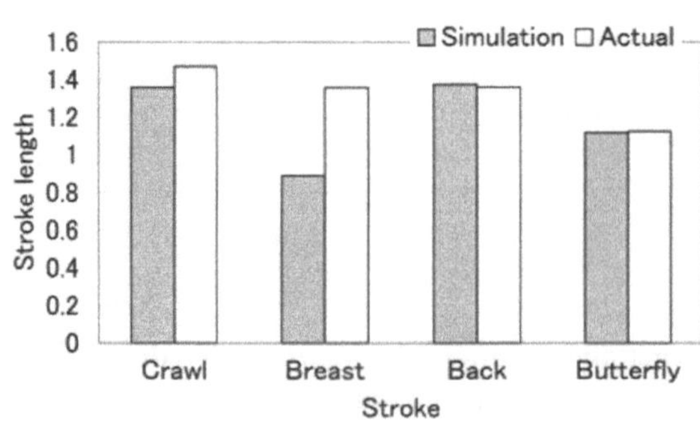

Fig. 10. Results of normalized stroke length for four strokes.

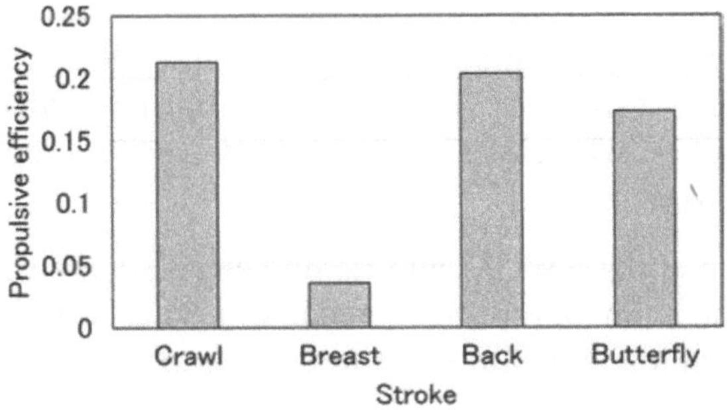

Fig. 11. Results of propulsive efficiency for four strokes.

the 'propelling efficiency' in Toussaint et al. (1988), and they reported its value to be 0.46~0.77 for the crawl due to their subjective experiment. The simulation results in this study are considerably smaller than those. It is possible that the present simulation underestimates the propulsive efficiency by some reasons. However, it may be also possible that the previous results of 0.46~0.77 overestimate it. For example, the propulsive efficiency of a high-performance-type screw propeller is generally known to be about 0.7. The propulsive efficiency of a dolphin robot, which has been developed by the authors, and has a streamlined body and a wing-shaped caudal fin, was also found to be 0.7 in the best condition (Nakashima and Ono 2002). It is hard to believe that the propulsive efficiency of the human swimming may exceed those of the screw propeller and the dolphin-type swimming since human has neither the screw propeller, nor the streamlined body, nor the wing-shaped fin. Further investigation is undoubtedly necessary in order to conclude this discussion.

4 Conclusions

In this paper, the breast, back, and butterfly strokes were analyzed using the swimming human simulation model SWUM. In addition, a comparison was made among the four strokes including the crawl from the viewpoint of mechanics. Findings are summarized as follows:

1. In the simulation of the breaststroke, the thrust due to the kick tends to be underestimated and the negative thrust by the upper lilmbs tends to be overestimated. Therefore, the swimming speed in the simulation becomes 35% less than the actual value.

2. The swimming speeds of the crawl, back, and butterfly strokes agree significantly with the actual values, indicating the validity of the

simulation model.

3. As a result of the comparison among the four strokes, it is found that the propulsive efficiency is about 0.2 for the crawl, back, and butterfly strokes, whereas it is considerably smaller (about 0.036) for the breaststroke.

All information about SWUM and its implementation software Swumsuit, as well as the animation of the simulation results which are shown in this paper, are open to the public at the website (http://www.swum.org/).

References

Chatard JC, Lavoie JM, Lacour JR (1990) Analysis of determinants of swimming economy in front crawl. Eur. J. of Appl. Physiology, 61:88-92

IPA (Information-technology Promotion Agency, Japan) (2004) http://www2.edu.ipa.go.jp/gz/h1swim/ (in Japanese)
(Videos of each stroke are located as below:
Crawl: http://www2.edu.ipa.go.jp/gz/h1swim/h1kn20/h1cr10/h1cr11.mpg,
Breast: http://www2.edu.ipa.go.jp/gz/h1swim/h1br00/h1br10/h1br11.mpg,
Back: http://www2.edu.ipa.go.jp/gz/h1swim/h1ba00/h1ba10/h1ba11.mpg,
Butterfly: http://www2.edu.ipa.go.jp/gz/h1swim/h1bu00/h1bu10/h1bu11.mpg)

Maglischo EW (2003) Swimming fastest. Human Kinetics, pp 228-231

Nakashima M (2007) Mechanical study of standard six beat front crawl swimming by using swimming human simulation model. J. Fluid Science and Technology, (in press)

Nakashima M, Ono K (2002) Development of a two-Joint dolphin robot. In: Ayers J, Davis JL, Rudolph A (Eds) Neurotechnology for biomimetic robots, MIT Press, Cambridge Massachusetts, pp 309-324

Nakashima M, Miura Y, Satou K (2007) Development of swimming human simulation model considering rigid body dynamics and unsteady fluid force for whole body. J. Fluid Science and Technology, 2(1):56-67

Toussaint HM, Beelen A, Rodenburg A et al. (1988) Propelling efficiency of front-crawl swimming. J. Appl. Physiol. 65:2506-2512

Chapter 30

Research in Fluid Dynamical Specification of Hand Palms in Freestyle Swimming

Shinichiro Ito[1]

[1] Department of Mechanical Engineering, The National Defense Academy, 1-10-20 Hashirimizu, Yokosuka, 239-8686, Japan

Summary. The characteristics of lift and drag forces are influenced by the configuration of a hand palm. The fluid dynamical characteristics of hand palms and that of a delta wing are compared experimentally. Delta wing is a simplified model of a hand palm. Three-dimensional velocity distribution of the wake flow was measured by X shaped hot wire probe. The results of these experiments showed clearly that the characteristics of hand palms in swimming were similar to that of delta wings.

Key words. Swimming, Hand, Configuration, Delta wing, Fluid Dynamic Characteristics

1 Introduction

James Counsilman (1968) proposed S-shaped pull swimming style is suitable for the front crawl. This S-shaped pull shown in Fig. 1 uses the resultant force of the lift and drag as an impellent. Lately, Sanders (1997) stated that competitive swimmers seem to use a drag type stroke rather than S-shaped pull stroke in his experimental research. Moreover, the author (Ito & Okuno, 2003) introduced I-shaped pull stroke, also indicated in Fig. 1, obtained larger thrust force than S-shaped. Hand palms generate the main propelling force in swimming. The characteristics of lift and drag forces vary according to the configuration of hand palm as the author investigated (2004). The difference of the characteristics is remarkable when fingers are opened or closed.

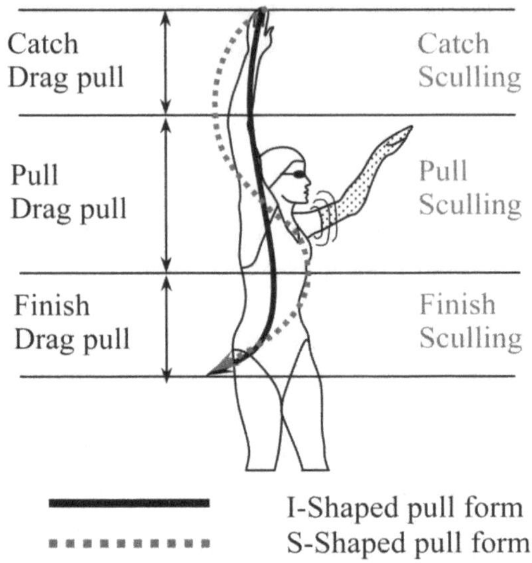

The left hand in this figure is on the water surface. This is modified figure of Maglischo(1993).

Fig. 1. Stroke pattern of I-Shaped and S-Shaped pull.in freestyle swimming

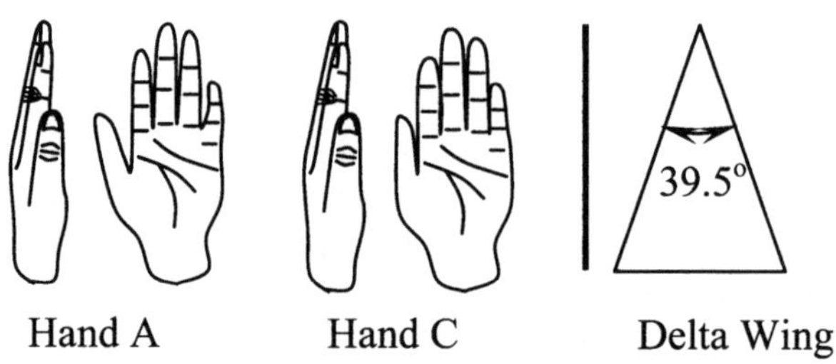

Fig. 2. Test replicas used in the experiment

2 Methods

2.1 Aerodynamic Characteristics

Two kinds of hand replicas and a delta wing of a flat plate with its vertical angle of 39.5 degrees are used as measuring objects shown in Figs.2. Hand

A is a flat-shaped palm with finger spread about 3mm apart. This shape is a popular configuration among competitive swimmers. Hand C is a flat palm with fingers closed. Delta wing is a simplified model of Hand C. The aerodynamic characteristics were measured on the angle of attack α of the above objects from 0° to 90° in every 1° by a 3-component load cells. For the hand replicas, three types of sweepback angle, ψ=45°, 90° and 135°, were adopted to simulate S shaped pull stroke. The sweepback angle ψ was defined by Schleihauf (1979), as middle finger faces to a leading direction as 90° in Fig.3. These sweepback angles, ψ=45°, 90° and 135°, correspond to the movement of arms. With the angle ψ=45°, the arm moves outward from the body axis. With the angle ψ of 90°, it moves parallel to the body axis. With the angle of 135°, it moves inward to the body axis respectively. Concerning the delta wing, the aero-dynamic characteristics were measured on ψ=90°.

2.2 Velocity Distribution in the Wake Flow

Furthermore, 3D-velocity distribution in the wake flow was measured with a X shaped hot wire probe at four sections of 1/4*L* interval of the model length *L* behind the center of the model in each wake flow shown in Fig.4. Velocity was measured at every 1cm of y and z directions in each cross section. Each experiment was performed at the wind velocity of 20m/s

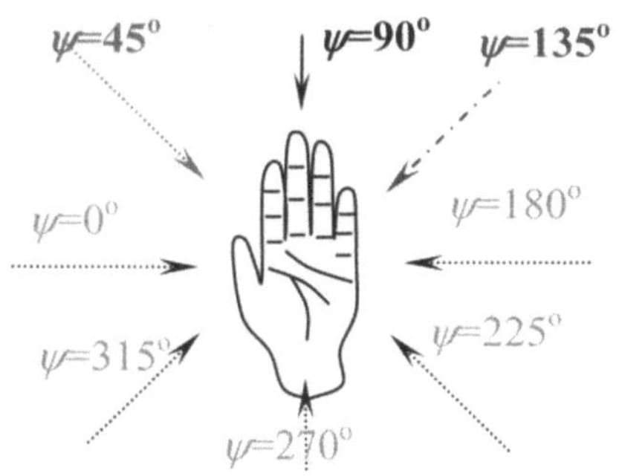

Fig. 3. Sweepback angle ψ defined by Schleihauf(1979)

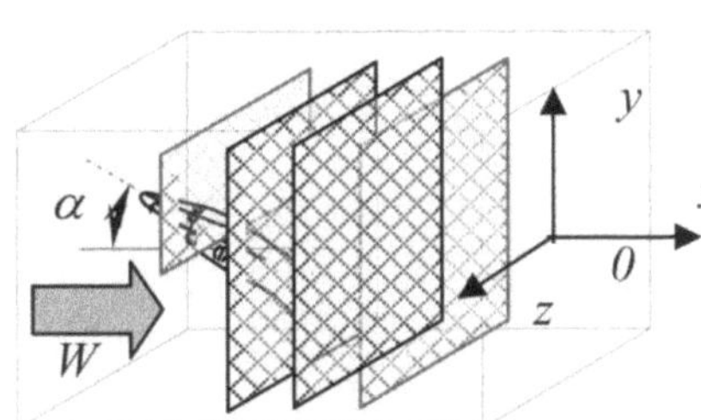

Fig. 4. Velocity measurement area in the wake flow behind the hand replica.

(R_e=3.4×10^5), of which the Reynolds number is almost as same as competitive free style swimming speed underwater.

3 Results and Discussions

3.1 Aerodynamic Characteristics

Each of Figure 5 shows drag-lift curves or polar curves of (a) Hand A with fingers spread, (b) Hand C with fingers closed and (c) Delta wing respectively. Markers are shown every 5 degrees in the angle of attack α in respective curves. The polar curve is utilized to characterize the aerodynamic specification of wings. A radius of C_{R0} is a length from the origin to the point at ψ=90° and the angle of attack α=90° as an index of maximal drag force. Comparing with the configurations of the polar curves of each objects, the curves of Fig. 5(b) and Fig. 5(c) have "lumps" near the point α=45° with discontinuous points in about α=50°. However, the curves of Fig. 5(a) do not have discontinuous points as others have. The difference of these curves may be caused by the leakage of flow from the pressure side (palm) to the suction side (the back of the hand) due to the finger gaps.

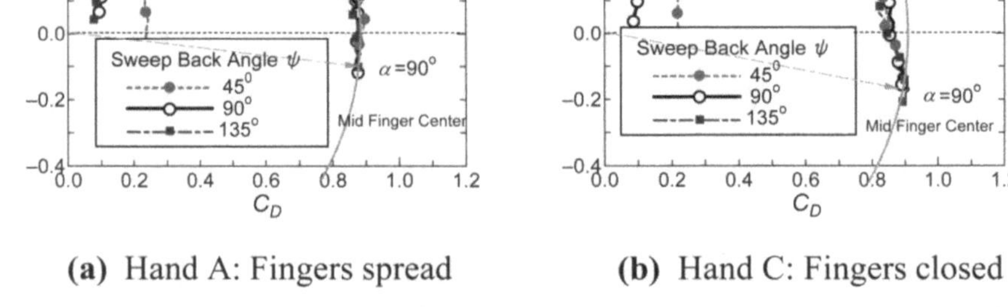

(a) Hand A: Fingers spread **(b)** Hand C: Fingers closed

Fig. 5. Difference of polar curves among the hand shapes and wing shape.

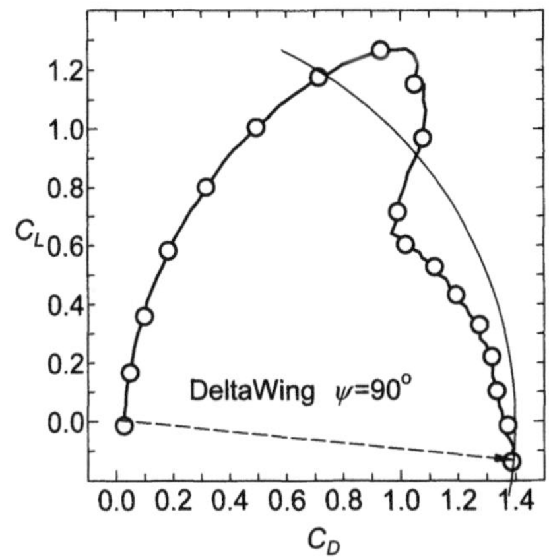

(c) Delta wing

Fig. 5. Difference of polar curves among the hand shapes and wing shape.

3.2 Velocity Distribution and Vorticity Contour in the Wake Flow

Velocity distribution maps and vorticity contour maps of each model are shown in Figs. 6 and Figs. 7 Each of them are the aspects of wake flow located in the center of the objects and in the 3/4 objects length *L* behind from the center of the objects respectively. Although each condition was in the maximum lift, a big pair of vortices appeared in the left and right side of the center of the object in each graph. Especially, these vortices in Figs. 6 showed a big separation of flow occurred in the suction side, the upper side, of the object although the maximum lift occurred. According to the lift occurrence mechanism in two-dimensional wing, separations never occur in the suction side or upper side of the wing. However, three dimensional wings like delta wings, the mechanism of occurrence of the lift force is different from that of two dimensional wings. Flow separation is a big factor to obtain a large lift force. The separation of the leading edge develops to a pair of large vertical vortices. These vertical vortices play a big roll to obtain high lift force.

It is natural that the pair of the vortex of the delta wing is symmetrical because of the symmetric plane form of the wing in Fig 6(c). However, concerning the strength of the vortices of hand palms, the strength on the little finger side is stronger than that on the thumb side. This is because a vertical vortex was dispersed by a flow which passed through a finger gap between a thumb and a forefinger from the palm side to the back side.

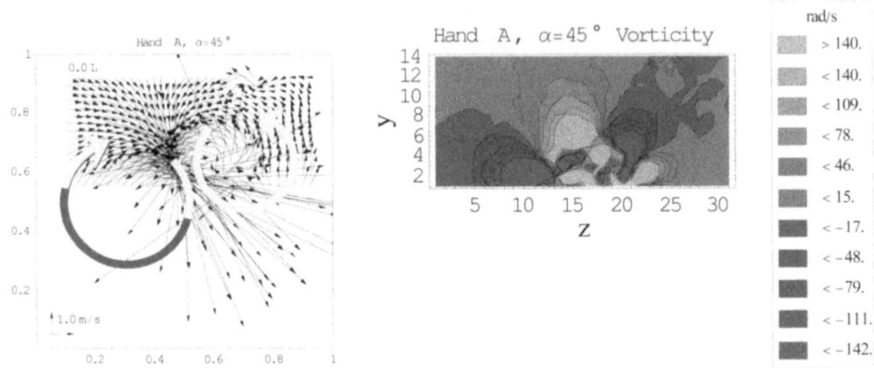

Velocity distribution map Vorticity contour map

(a) Hand A: Fingers spread, α=45°

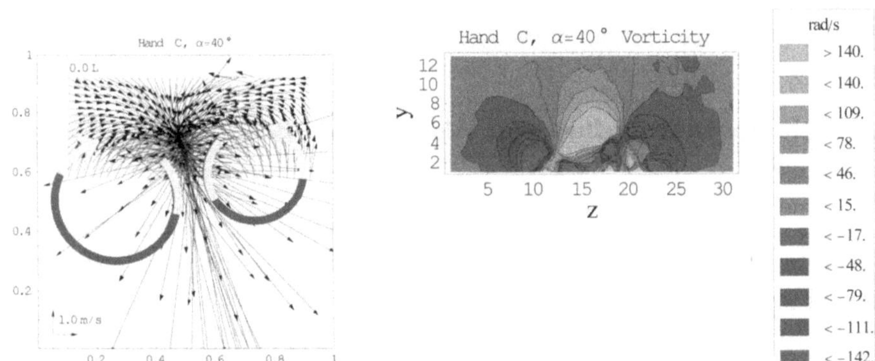

Velocity distribution map Vorticity contour map

(b) Hand C: Fingers closed, α=40°

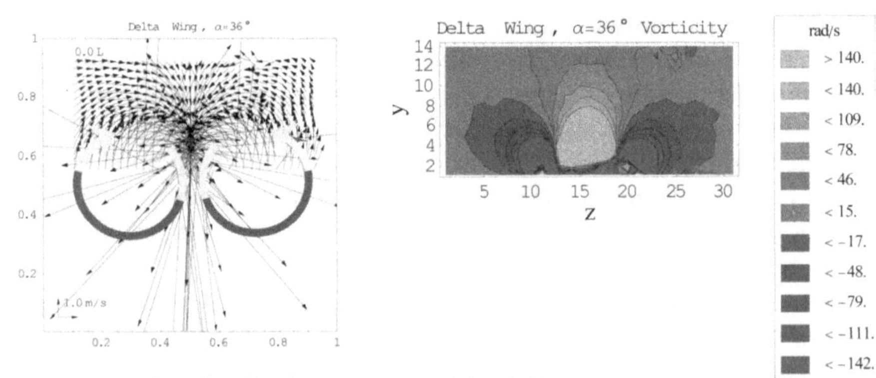

Velocity distribution map Vorticity contour map

(c) Delta wing, α=36°

Fig.6. Velocity vector and vorticity of the wake flowat the center of the hand

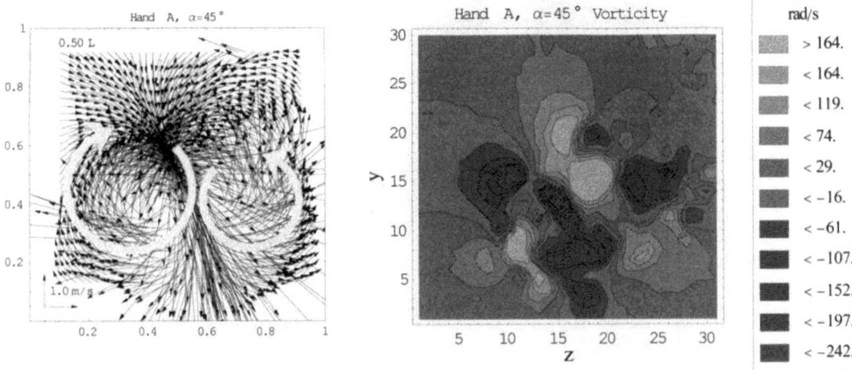

Velocity distribution map Vorticity contour map

(a) Hand A: Fingers spread, α=45°

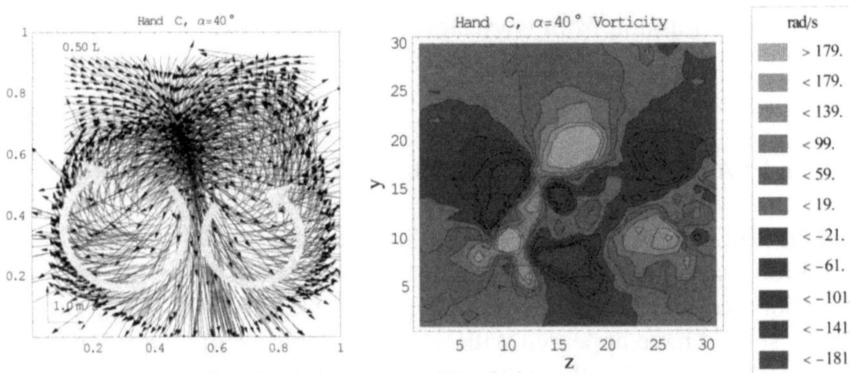

Velocity distribution map Vorticity contour map

(b) Hand C: Fingers closed, α=40°

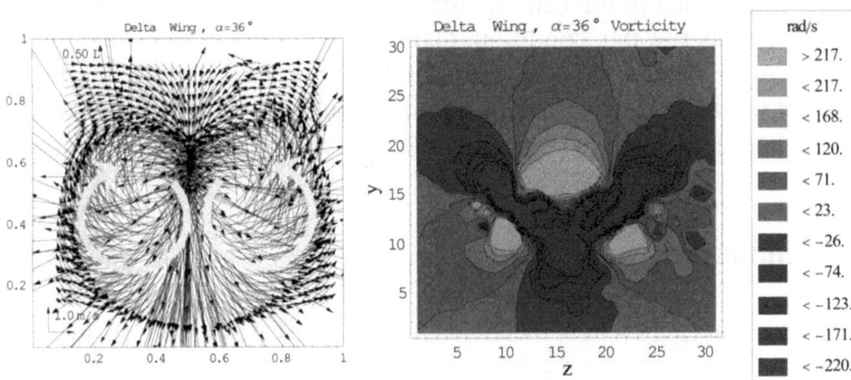

Velocity distribution map Vorticity contour map

(c) Delta wing, α=36°

Fig. 7. Velocity vector and vorticity of the wake at 3/4L behind the center of the hand

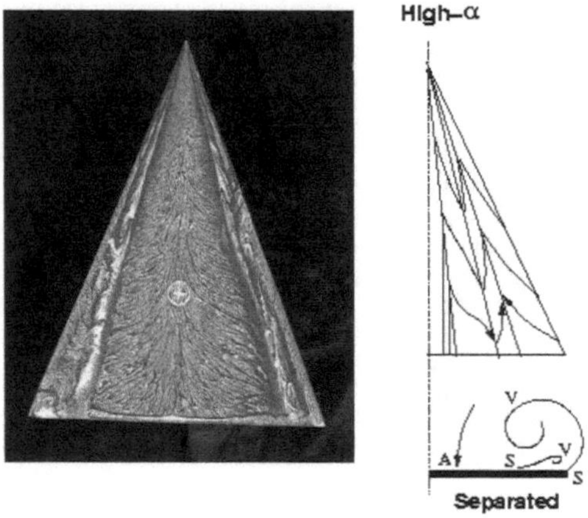

Fig. 8. Oil flow pattern and its explanation.
Used with Permission from **www.aerodyn.org**

3.3 Oil Film Experiment in the Suction Side of Delta Wing

The oil film pattern of the flow in $\alpha=36^{\circ}$ were shown in Fig.8 of which the condition was in the neighborhood of the maximum lift point. As for the neighborhood of the maximum lift point, straight lines along the main stream should have appeared with no separation case of a two-dimensional wing. However, as for the delta wing, separation occurred from the whole leading edge as mentioned in the section of velocity distribution. The separation re-attached in the central side of the wing plane. Figure 8 describes the flow aspect in this condition. Gain of the lift force on a high angle of attack in the delta wing was caused by the pressure decline due to these vertical vortices. In other words, as for the delta wing, lift force is obtained by the big separation-re-attachment mechanism.

4 Conclusions

Lift and drag generation mechanism of a hand palm can be analogized with that of the delta wing because of the similar configurations of polar curves and velocity distribution around the objects.

References

Counsilman, J.E. (1968), *Science of Swimming*. Prentice-Hall, Englewood Cliffs, N.J.

Ito, S., Okuno, K. (2003), *A Fluid dynamical consideration for arm stroke in swimming*, Biomechanics and Medicine in Swimming IX, pp.39-44, Pub. de l'univ. de Saint-Etienne.

Ito, S. (2004). *Optimal Hand Shape for Competitive Free Style Swimming*, The Engineering of Sport 5 Vol.1, pp.580-586, International Sport Engineering Association.

Maglischo, E. W. (1993), *Swimming Even Faster*, Mayfield Publishing Company, Mountain View, CA, p315.

Schleihauf, R.E. (1979), *A hydrodynamic analysis of swimming propulsion,* Swimming III, pp70-109. University Park Press, Baltimore.

Antonio Filippone, http://www.aerodyn.org/Wings/delta.html#separation

Chapter 31

Flexural Vibration of a Jump Ski in Flight

Akira Shionoya[1], Toshio Kobayashi[1], Toshinori Saijo[1], Kenichiro Ogata[1], and Toshiaki Matsuhashi[1]

[1] Department of Sport Engineering and Physical Education, Health Care Center, Graduate School of Engineering, Nagaoka University of Technology, 1603-1, Kamitomioka, Nagaoka, Niigata, 940-2188, Japan

Summary. The purposes of this study were to develop a system for measuring the flexural vibration of a jump ski in flight and to measure the vibration of a ski in flight using this system. To measure the flexural vibration of a jump ski in flight, accelerometers were installed on a ski plate. By means of accelerometers, vibrations of a ski were detected in flight and were recorded to a data recorder. After flight this data was transferred to a computer, after which it was transformed into a power spectrum by means of FFT. In actual flight, an amplifier and a data recorder were inserted into a backpack attached to a jumper's jump suit. The vibrations were analyzed in the following phases; approach, take off, flight and landing. From approach to landing, few natural vibration frequencies between 10 and 200[Hz], as estimated by the impulse response, were detected. A low frequency of about 10 [Hz] was detected in nearly every phases. Furthermore, at landing, a low frequency components band between 40 and 70 [Hz] was detected at the top of the ski, because the jumper landed on the tail of the ski. In addition an acceleration amplitude of over 9800[m/sec^2] was detected at the top of a ski in landing.

Key words. Flexural vibration, FFT, Frequency, Jump Ski, Flight

1 Introduction

Computer simulation is a very useful procedure to understand phenomena in sports. It has been used on various sports subjects as follows; a trajectory analysis of a kicked ball in soccer, a vibration analysis of a tennis racket in impact, a fluid dynamics analysis in swimming, a flight distance optimization in ski jumping and so on. However, the objects in these simulations have been calculated as rigid bodies because it has been easy to make calculations with respect to rigid bodies in simulation but difficult to measure an object's physical phenomenon. For example, in a flight simulation of ski jumping, the body of a human jumper or a ski plate has been calculated as a rigid body.

In actual ski flight studies, human muscular activity has frequently been studied using an EMG (Electromyogram). The EMG measurement in flight is easy because a telemeter system or a small-size and lightweight data recorder for the EMG measurement has already been developed and is on the market. A system for measuring physical phenomena such as the vibration of a ski plate in flight, however, has not been developed because such a system is considerably bigger than the EMG measurement system, and the present system is too big and heavy to use in actual ski flight.

The purposes of this study are to develop a system for measuring the flexural vibration of a jump ski in flight and to measure the vibration of a jump ski in flight using this system.

2 Development the System and Measurement

2.1 Impulse Response of Jump Ski

Fig.1 shows the system for measuring the impulse response of a jump ski. To measure the impulse response of a ski, a ski was suspended and accelerometers (PCB PIEZOTRONICS corporation made; 352B10 type) were installed on the top, center and tail of the ski. An impulse was applied to the ski using an impulse hummer (PCB PIEZOTRONICS corporation) and then, by means of accelerometers, the vibrations of the ski were detected. Through an amplifier (PCB PIEZOTRONICS corporation; Signal conditioner 480B10) and a units (ELMEC corporation), the vibrations were transformed by an A/D converter and were transferred to a personal computer (DELL corporation; INSPIRON 5100) and then transformed into a power spectrum by means of the fast fourier transform (FFT) programming (ELMEC corporation; WAAP-WIN) installed on a personal computer.

Fig.2 shows the jump ski used in this impulse response experiment.

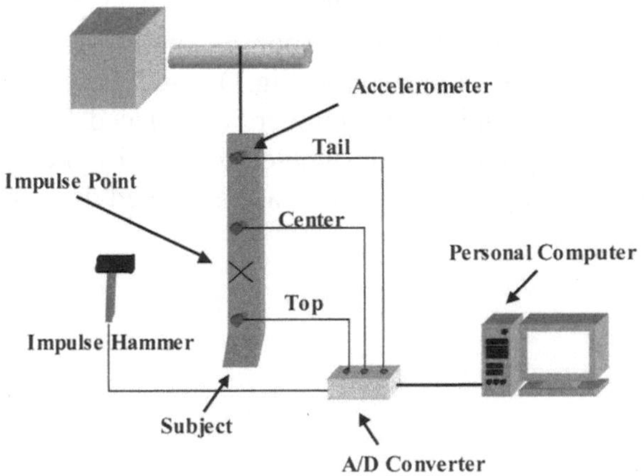

Fig.1. An outline of the system for measuring the impulse response of a jump ski

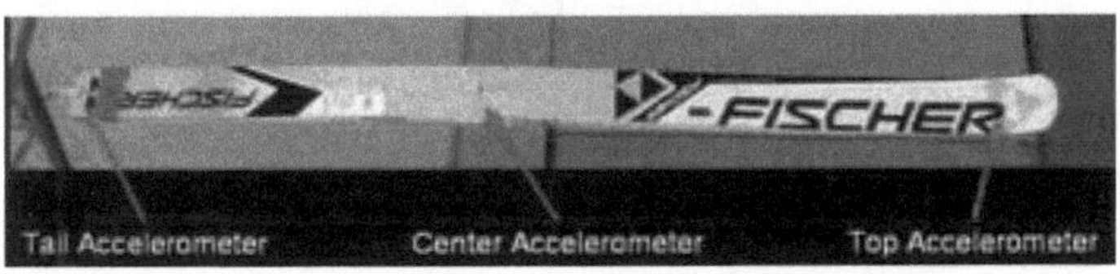

Fig.2. The jump ski used in the impulse response experiment in this study

2.2 Vibration Measurement in Ski Flight

Fig.3 shows the outline of the system developed for measuring the vibration of a jump ski in ski flight. To measure the vibration in flight, two of the above-mentioned accelerometers (PCB PIEZOTRONICS corporation; 352B10 type) were installed to a jump ski. One accelerometer was installed on the top of the ski and another one was on the tail. By means of accelerometers, the vibrations of a jump ski were detected in flight. Through the above-mentioned amplifier (PCB PIEZOTRONICS corporation; Signal conditioner 480B10), vibrations were recorded to a small-size and lightweight data recorder (NF circuit; model EZ7510). In flight, an amplifier and a data recorder were inserted into a backpack and then the subject jumped with this backpack attached to the back on his jump suit. The total weight of this backpack was within 2.0kg. After flight, the vibration data were transferred from a data recorder to a personal computer (DELL corporation ; INSPIRON 5100), after being processed by A/D converter they were transformed into a power spectrum by means of the fast fourier transform (FFT) program (ELMEC corporation; WAAP-WIN) installed into a personal computer as the impulse response experiment.

The length of the ski used in this study (Fig.1) was 2.5m and the weight was 2.5kg. Subject was a junior elite jumper who placed 5th in an inter-high school championship and was designated to train by the Niigata Ski Association. The actual flight experiment was performed using a medium-hill schanze.

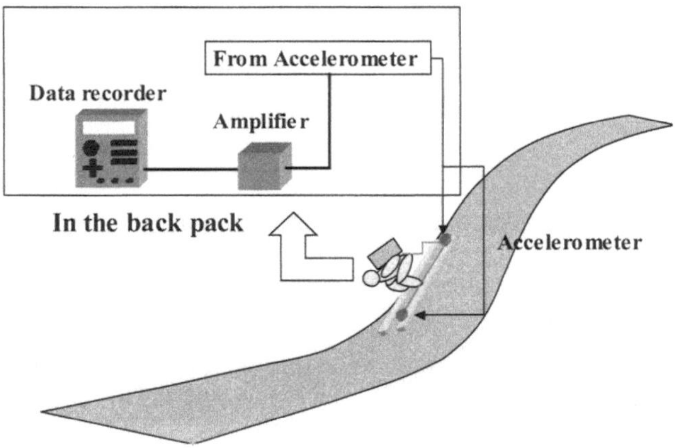

Fig.3. An outline of the system developed for measuring the vibration of a jump ski in actual flight and the experiment using this system.

3 Results

3.1 Impulse Response of Jump Ski

Fig.4 shows the acceleration wave in the top of a jump ski in the impulse response measurement. Damping ratio(δ) was calculated from this acceleration wave as follows;

$$\delta = \frac{1}{m} \cdot \ln \frac{a_n}{a_{n+m}}$$

where a represents amplitude, n represents number of wave, m=n/T(T represents time).

Tabel.1 shows the damping ratio in each position of the ski. The ratio at the top was 0.174; the highest ratio.

Fig.5 shows the vibration frequencies of the jump ski in the impulse response. The upper figure is the frequency in the top of a ski, the middle is figure is that in the center and the bottom figure is that in the tail. From these results, the natural frequencies in this ski were estimated between about 10 (12[Hz]) and 200[Hz] (Fig.6). The frequency of the maximum density in both the top and the center was 153 [Hz]. In the tail, this frequency was 2nd density.

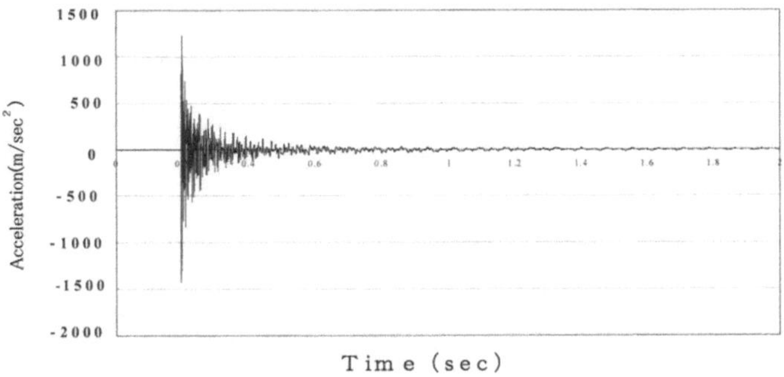

Fig.4. Acceleration wave in top of a jump ski in impulse response

Table.1. Damping ratio in each position of a jump ski.

	First amplitude (m/s^2)	11th amplitude (m/s^2)	Damping ratio
Top	2662.09	469.06	0.174
center	721.87	265.18	0.100
Tail	1626.44	409.33	0.138

3.2 Vibration Measurement in Actual Ski Flight

Fig.7 shows the acceleration amplitude of a jump ski in approach, flight and landing. The acceleration amplitude on approach was higher than that of the flight phase. From immediately after take off, it became lower. In flight, the jumper lifted the top of the ski immediately after take off and then took flight, falling along the landing hill. The base line of the acceleration curve fell in proportion to the flight time (middle figure in Fig.7). Furthermore, on landing, an acceleration amplitude of over 9800[m/sec^2] was detected in the top of a ski.

Fig.8 shows the vibration frequency of a jump ski on approach, flight and landing. On approach, few natural vibration frequencies between 10 and 200[Hz], as estimated by the impulse response, were detected. In flight, a low frequency of about 10 [Hz] was mainly detected (Fig.9). The above-mentioned natural vibration frequencies were detected a moment immediately after take off. However, few such frequencies were

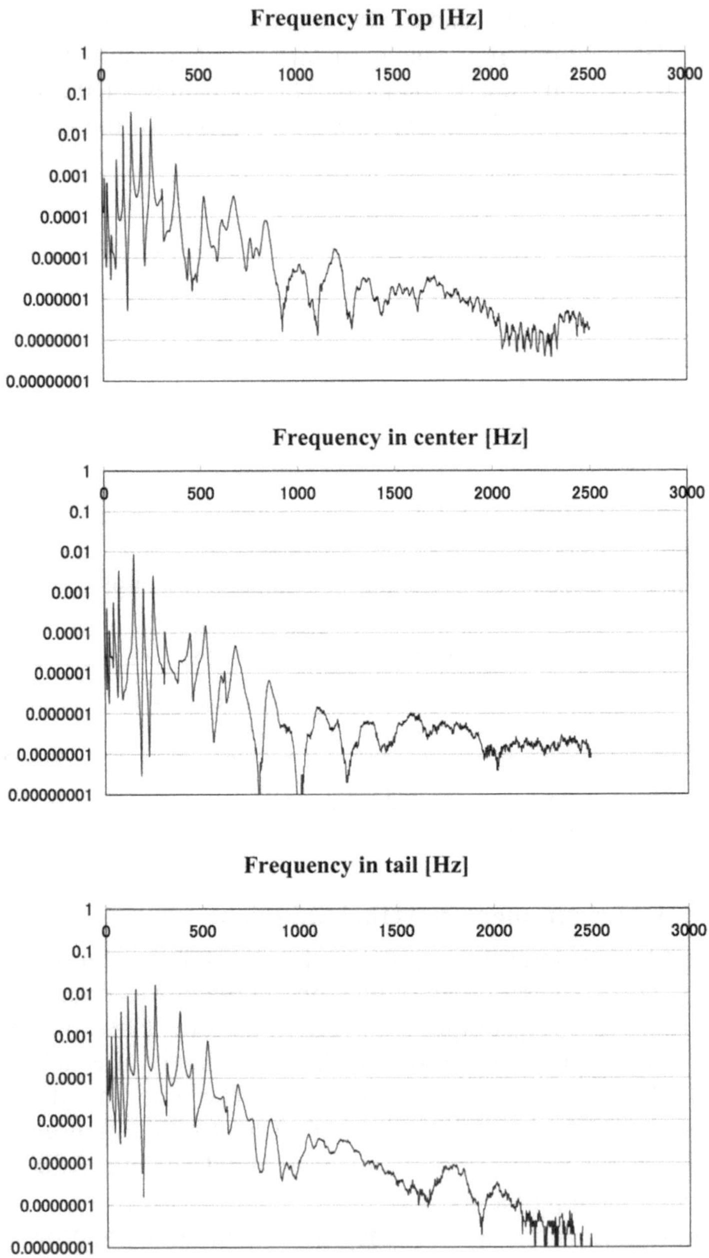

Fig.5. Vibration frequency in the impulse response of jump ski
(Logarithmic axis)

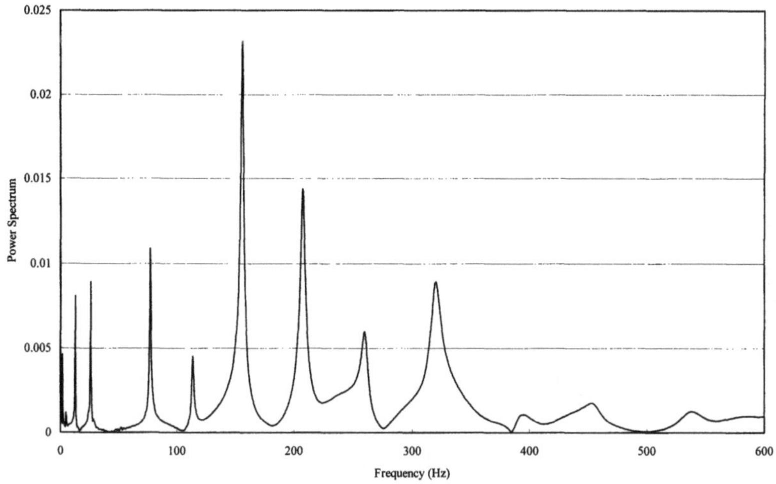

Fig.6. Vibration frequency in the top of a jump ski in impulse response

detected in flight. Furthermore, on landing, low frequency components between 40 and 70 [Hz] were detected in the top of the ski, because the jumper landed from the tail of ski in landing.

In order to clarify the details of the frequency components, the amplitude of the frequency in Fig.6 and 9 is indicated by the logarithmic axis.

Discussion

As mentioned above, in flight simulation of ski jumping, the body of a human jumper or a ski has been calculated as a rigid body, because it has been easy to apply calculations to a rigid body in simulation and difficult to measure a ski's physical phenomena in flight.

In actual ski flight studies, human muscular activity has frequently been studied using the EMG (Electromyogram) measurement. The EMG measurement has been easy to apply in flight, because a telemeter system or a small-size and lightweight data recorder for the EMG measurement has already been developed and is already on the market. A system for measuring physical phenomena such as the vibration of a ski plate in flight, however, has been not developed because such a system is considerably bigger than the EMG measurement system, and subjects have not been able to jump using the present system.

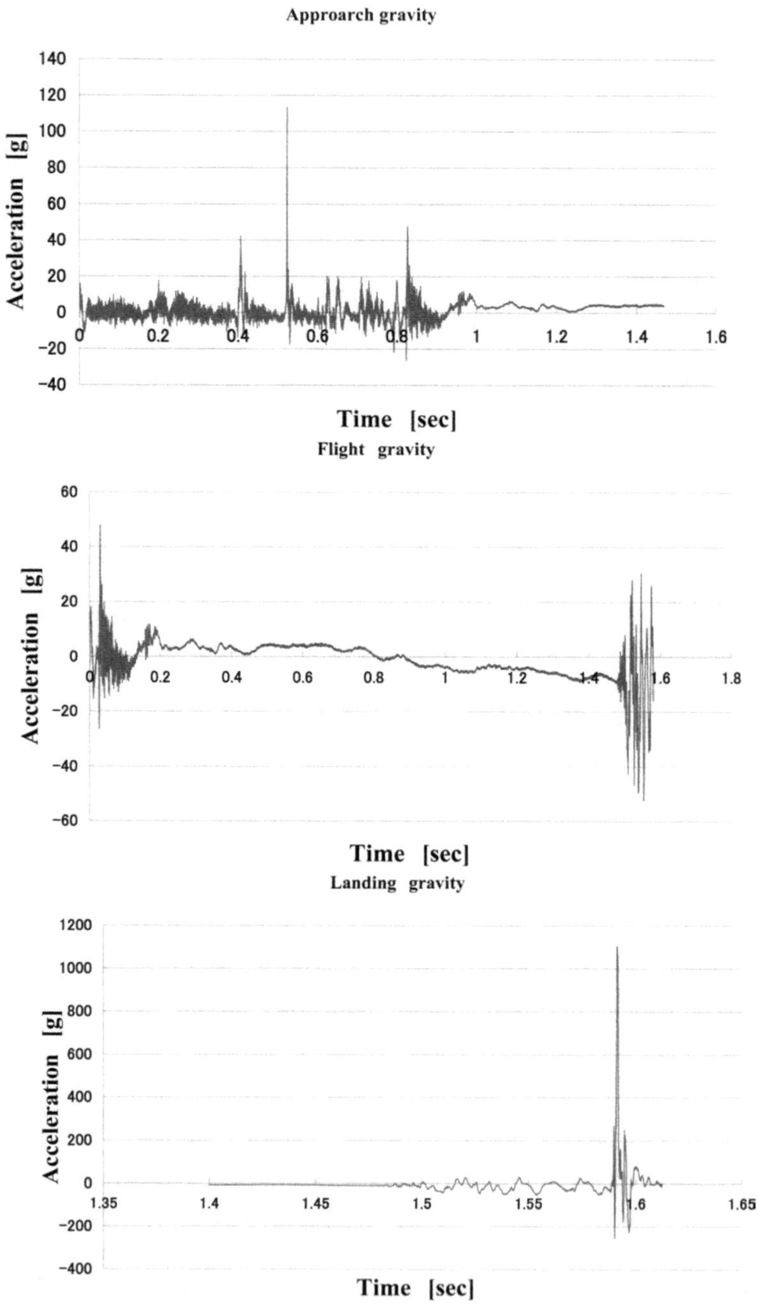

Fig.7. The accelerations of a jump ski in the upper figure to approach, the middle to flight and the bottom to landing.

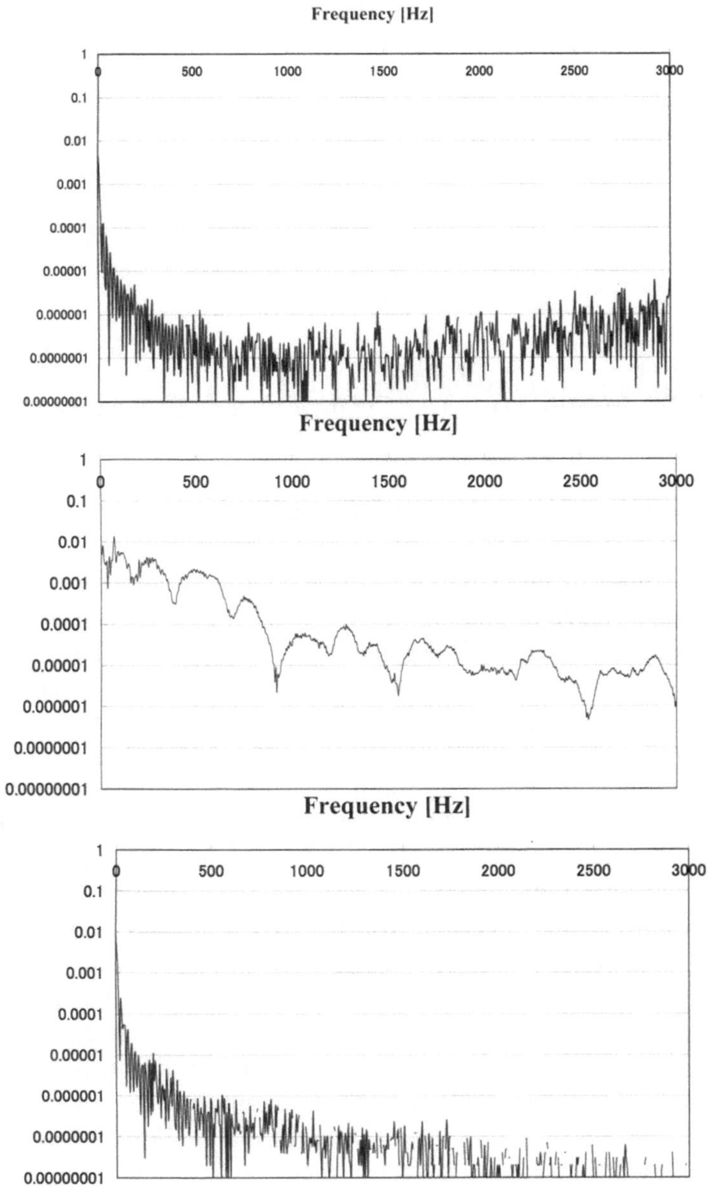

Fig.8. The vibration frequency of a ski in the upper figure to approach, the middle to flight and the bottom to landing. (Logarithmic axis)

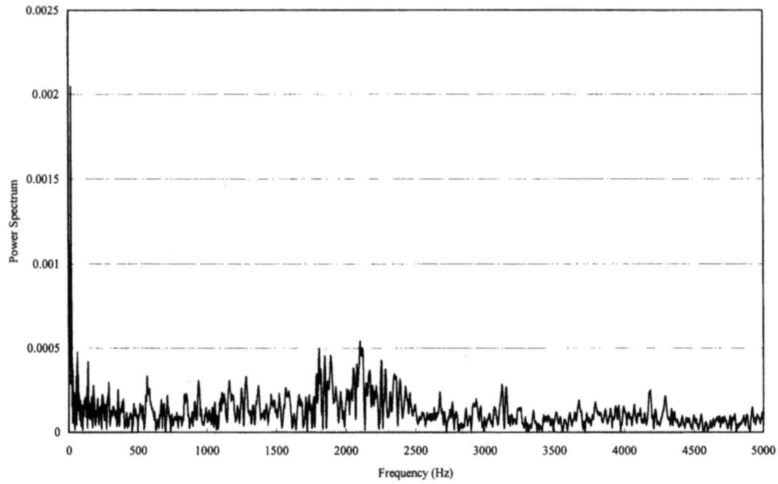

Fig.9. The vibration frequency of jump ski in flight

The first purpose of this study was to develop a system for measuring the flexural vibration of a jump ski in actual ski flight. The system developed in this study was within 2.0kg and could measure the flexural vibration of a ski plate without negative effects on the flight performance of a jumper in the medium-hill schanze. In an normal-hill or the large-hill schanze, however, the weight of this system may yield negative effects on the flight performance. Recently, jump skis which are unequipped with self-propulsion mechanisms such as a jet engine, have tended to be ultra-lightweight. The specific gravity of the jump ski in this study was 0.54 [g/cm^3] and was ultra-lightweight. Furthermore, the weight of a jumper has tended to be lightweight in comparison with 10 years ago in order to decrease the gravitational force. Therefore, a ski jumper has been required relatively high muscular strength per unit. These tendencies have based on the principle of the ideal flight object. According to these tendencies and in order to avoid negative effects on the performance in flight, the trend toward to lightweight system is likely to continue into the foreseeable future (The newest system in 2007 has a weight of 0.75kgw. Refereed in Shionoya(2007)).

The second purpose of this study was to measure the flexural vibration of a jump ski in flight using the developed system. As mentioned above, the system in this study could measure the flexural vibration of a ski plate without negative impact on the flight performance of a jumper on a medium-hill schanze. In alpine ski, the vibration of a ski plate is a very important factor in ski driving (=sliding) and controlling. As in the previous studies, the natural frequencies of an alpine ski plate were between 50 and

200[Hz] in the impulse response. Furthermore in actual skiing, a frequency range of between 100 and 200[Hz] was detected when an expert skier was skiing. On the other hand, this frequency range was not detected when a non-expert skier was skiing. It has been thought that these phenomena were related to the physical principle of a ski sliding. When the amplitude of the frequencies between 100 and 200[Hz] of a ski plate in skiing is low, the driving velocity in skiing is remarkably slow, because the contact times between the ski and snow surface increases, and then with it the amount of the snow surface destruction. Frequencies over 100[Hz] in skiing have been thought to be necessary for effective ski driving performance.

On the other hand, the natural frequencies of a jump ski were estimated between 10 and 200[Hz] in this study. However, these frequencies were not detected in flight. Furthermore, these were not detected from approach to landing except for a brief moment immediately after take off. The maximum frequency of a jump ski in approach and flight was almost 10[Hz]. The previous alpine ski studies mentioned above suggest that the driving performance of skis to the present time has been less than optimal, because the contact time between a ski and snow surface –and the amount of snow surface destruction--has increased in 10[Hz] of the frequency. A Japanese ski jumper, who was participated in the Winter Games Olympics, remarked that jump ski plates having a high driving performance frequently suffered from low flight performance. On the other hand, he noted the reverse was also true-i.e., that those with low driving performance often had high flight performance, too. These remarks are of similar interest as the results of the vibration frequency of a jump ski in flight.

It is not understood, however, what frequencies are necessary (or not) for a ski jump from this study alone. The natural frequencies may be an index for the most fitted physical characteristics of a jump ski, as is the case with an alpine ski. How to optimize the necessary frequency or the most fitted physical characteristics in a jump ski is a subject worthy of future consideration.

5 Conclusion

This study is summarized as follows;

1) The system for measuring the vibration of a jump ski in flight was developed and the flexural vibration of a jump ski in actual flight was measured on a medium-hill schanze using this system.

2) The acceleration amplitude on approach was higher than that in flight. Immediately after take off, however, the acceleration amplitude became lower. Natural vibration frequencies, as estimated in the impulse response, were not detected in approach, flight and landing.
3) In flight, a jumper lifted the top of his ski immediately after take off and then took a flight, falling along the landing hill, the base line of the acceleration curve fell in accordance with flight time. The natural vibration frequencies were detected for a brief moment of the immediately after take off. These frequencies, however, were not detected in flight.
4) At landing, the acceleration amplitude of over 9800[m/sec^2] was detected at the top of the ski. Furthermore, low frequencies between 40 and 70 [Hz] were detected at the top of a ski at landing, because the jumper landed on the tail of ski.
5) How to create an even light-weight system and how to optimize the necessary frequency or the most fitted physical characteristics in a jump ski will be subjects of future research.

References

Asai.T (1998) Computational fluid dynamics analysis of the V style ski jumping" DESCENTE SPORTS SCIENCE. Vol.19, pp246-253.
Kitamura.T (2006) My preparation for World Nordic Ski Championship. 30th anniversary lecture of Nagaoka University of Technology.
Physical Education Instruction Center,(1992) " EMG analysis of International elite ski jumper in Japan in actual ski jump" The bulletin of Physical education instruction center in Hokkaido University. Vol.4, pp49-62.
Physical Education Instruction Center,(1993) " EMG study on V style ski jumping" The bulletin of Physical education instruction center in Hokkaido University. Vol.5, pp17-21.
Sakata.T., et al(2000) "Improvement for damping properties of a ski" Proceedings of the Symposium on Sports Engineering. No.00-38, pp167-169.
Shionoya.A (2005) Flexural vibration analysis of ski plate in curving turn. The impact technology on sport. Vol.1, pp131-136,2005.
Shionoya.A (2007) Flexural vibration of a jump ski in actual ski flight. Proceedings of the International conference on winter sport (Commemorative conference on Sapporo World Nordic Ski Championship in 2007). In print.
Tanahashi.T (1992) "Mechanism of ski-sliding-Effects of materials and vibration on friction-" Mechanical Engineers. Vol.95(888), pp45-48.

Subject Index

C

D

G

H

I

J

K

T

U

MIX
Papier aus verantwortungsvollen Quellen
Paper from responsible sources
FSC® C105338

If you have any concerns about our products,
you can contact us on
ProductSafety@springernature.com

In case Publisher is established outside the EU,
the EU authorized representative is:
Springer Nature Customer Service Center GmbH
Europaplatz 3, 69115 Heidelberg, Germany

Printed by Libri Plureos GmbH
in Hamburg, Germany